IN THE NAME OF SCIENCE

IN THE NAME OF SCIENCE

BY

H. L. NIEBURG

CHICAGO 1966

QUADRANGLE BOOKS

 For information, address: Quadrangle Books, Inc., 180 North Wacker Drive, Chicago 60606. Manufactured in the United States of America.

Library of Congress Catalog Card Number: 66-11868

FIRST PRINTING

Typography and binding design by Vincent Torre

. . . to Meg, Bill, Pat, and Liz—
in spite of whose importunities
this book was written—and to Janet,
without whom . . .

PREFACE

In three hundred years of accelerated technological innovation, science acquired a revolutionary mystique. Crowned by liberating and transforming success, it became a form of secular religion, tending toward popular dogmatism and uncritical authority. Like all things, science and technology are essentially *amoral* and their uses ambivalent. Their miracle has increased equally the scale of both good and evil. The human situation has vastly benefited, but so too has "evil" attained monumental potential. Man himself remains a moral primitive.

This is a book about federal science policy and government contracting. But the inquiry leads inevitably to fundamental questions about the changing nature of American society. In the last two decades the public investment in science and technology (research and development—R&D) has swelled to massive proportions. The pace, direction, and resources of innovation have assumed an unprecedented role as an adjunct to government in defense, economic growth, public services, and general social management. Science in effect has been nationalized.

The fading of *laissez-faire* science is inseparable from other aspects of the major reordering of the conditions and organization of human life in recent decades. The emergence of the United States as the outstanding power in a restless world, the domestic trend toward economic giantism, the persistent problems of a slump-prone economy, the unfinished business of social justice, and the growing complexity and interdependence of all aspects of life—these and other factors have shaped the role of public science in new and sometimes obscure ways.

Government today is the economy's largest buyer and consumer of goods and services. The public investment in science and technology has grown largely by contracting with private organizations, mostly industrial profit corporations. Government contracts for goods and services have become increasingly important devices for intervention in public affairs, not only for procurement but also to achieve a variety of explicit or inadvertent policy ends—allocating national resources, organizing human efforts, stimulating economic activity, and distributing status and power. The federal contract has risen to prominence as a social management tool since World War II, achieving today a scope and magnitude that rivals or overshadows traditional instruments (subsidies, tariffs, taxes, direct regulation, and positive action programs) in its impact upon the nature and quality of American life.

This evolution occurred quietly through a series of improvised reactions to specific problems, achieving its present stage after little consideration of any far-reaching implications. Of the more than $130 billion a year that passes through Washington, research and development comprises a substantial share ($16 billion in 1965), second only to defense. But its place in the contract system exceeds its quantitative measure, because R&D contracts have become the wedge that opens the door to larger procurement awards and in effect controls the fate of companies, corporate managements, financial elites, geographical areas, and political leadership. Scientific research and development are open-ended undertakings.

This is also a book about politics. In the best sense, the political process is the heart and soul of national life and shapes the future. The truths of conflict and accommodation between the motives and ambitions of men are difficult to ascertain. Like sex, politics is divisive and full of explosive potential. When battle is done, the myths of consensus flow over the scene like the innocent sea. Too much current discussion of public affairs errs in the direction of the isolated lurid parable which indicts this or that man or group, or toward burnished generalities and abstractions. Public debate often fails to communicate the actual abrasive contact of men and groups, their motives, actions, and adjustments. Just as a psychiatrist must deliberately poke old hurts and graceless compromises, so this book seeks to describe the political encounters that anchor the abstractions in reality and arrange the lurid details in meaningful mosaics. We seek to discern broad and concrete social issues and policies in the rough interstices of politics.

We will delve into the pork barrel of favoritism, profit, and waste in federal contracts, in search not of sensation and scandal but of political relationships. Isolated instances of impropriety and illegality are at best peripheral matters. They transpire ubiquitously and shallowly in all human relations. Far more important are the human relations themselves, the on-going operation of a system which has acquired legitimacy and stability, whose participants not only retain respectability but inspire general admiration.

The making of a book is the work of many hands. For some congressmen and government officials my gratitude and respect will be apparent in the pages that follow. A few names deserve mention here: Dr. Jerome Wiesner, Dean of the Massachusetts Institute of Technology; Edwin Diamond, associate editor of *Newsweek;* Ruth Adams, managing editor of the *Bulletin of the Atomic Scientists;* Dr. Manuel Gottlieb, professor of economics, University of Wisconsin at Milwaukee; Dr. Melvin Kranzberg, editor of the journal *Technology and Culture;* Dr. David Cohen, professor of history, Case Institute, and countless colleagues who counseled me by argument and example; Karl Schultz, a graduate student who footnoted and indexed this book and spent many hours in library detective work; and my editor, Ivan Dee, whose contribution to this book was tasteful, sensitive, and indispensable, albeit invisible. Finally, gentle reader, I owe a heavy debt to you, which I hope the lines that follow may partially repay.

H. L. NIEBURG

Milwaukee, Wisconsin, 1966

CONTENTS

IN THE NAME OF SCIENCE

CHAPTER

I

THE SCIENCE-TECHNOLOGY RACE

The motives of nations are composites of fears, hopes, bargains, and the ambitions of many different men and groups. The world is now witnessing a science-technology race in which this maxim is apparent. The competition contains elements of astonishing achievement as well as those of a teen-age drag race. The underlying military dimension is unmistakable. Both sides, East and West, fear that the other will through new technology gain a diplomatic advantage and upset the present strategic stalemate; both devote substantial resources to weapons research and testing programs in order to gain for themselves or deny to others the advantage of lead time.

In effect, the science-technology race is a transfigured, transmuted, and theoretical substitute for an infinite strategic arms race; it is a continuation of the race by other means. Built into this equation and secondary to it is the need to maintain a healthy economy. Fear of stagnation, the habit of massive wartime spending, the vested interests embracing virtually all groups, pork-barrel politics—all are aspects of what has become deliberate government policy to invest in the "research and development" empire as an economic stimulant and a public works project.

Admittedly, a valid national interest is at stake. Industrial capacity (including scientific and technical skills) is clearly related to military strength and is therefore one of the main components of bargaining power in diplomatic maneuver.

Throughout their history, the Western nations have held sway over the destinies of the world through superior technology. Industrial development and naval technology were the keys to great-power status from the Napoleonic Wars through possibly World War II. During most of the nineteenth century, British pre-eminence in the world rested on unequaled naval strength and the ability to act as the power balancer among rival blocs in continental Europe. Until the Spanish-American War of 1898, American security depended upon a coincidence of interest with the British in denying the powers of Europe hegemony in the Western Hemisphere. At the turn of the twentieth century, the U.S. outgrew its dependence upon the British and undertook to raise its own Navy to police the hemisphere and maintain its interests elsewhere, notably in the Pacific where the U.S. exploited the waning power of European colonialism to consolidate its own influence. The Kaiser's naval ambitions were a contributing factor to the conditions that ignited World War I, during which German submarines proved themselves a match for British surface-vessel superiority; as a result of naval stalemate, the war shifted back to the trenches. Here technological parity brought an impasse until British introduction of the armored tank and the addition of American hardware and men forced a decision for the Allies.

Between the great wars, the leading powers relied upon their technological superiority to deter strong challenges to the status quo. But Nazi Germany was able in a decade to make itself master of the newest technology, combining the dimension of air attack with armor and mobility on the ground. While a young Charles de Gaulle fought in vain for the integration of new technology into French strategic plans, and Billy Mitchell was being court-martialed for his views on U.S. airpower, Hitler's war machine acquired an aura of invincibility. It retained its momentum until the Allies, once again dependent upon U.S. industrial capacity and manpower, reversed the technological advantage.

It is not surprising that at the close of World War II the United States relied upon its technological superiority as a major basis of strategy. It was clear that the Soviets intended to maintain tremendous conventional military forces which the United States and Western Europe recognized as a threat to the political status quo. The U.S. atomic monopoly was seen as an adequate counterforce which obviated the matching of Soviet manpower. The pattern of the post-war balance of power became Soviet massed manpower versus the Strategic Air Command and an industrial base for rapid mobilization of manpower if and when required. But Soviet achievement of atomic weapons in 1949 and thermonuclear weapons in 1953 undermined the rationale of Western strategy. The reverberations of Sputnik I in 1957 can only be understood as a fear of the apparent reversal of Western technological supremacy. What had really happened was less drastic and should have been recognized earlier: the emergence of technological parity between the two major centers of power. David E. Lilienthal: ". . . we must fully understand that not only is our nuclear monopoly gone, but with it the once widely used tranquilizer that there is *any* technical achievement an advanced industrial country, and notably the Soviet Union, is incapable of attaining. We should be realists enough to look at a world in which there is no such thing as a technical monopoly except for a brief period of time, among even the lesser industrial nations of the world." [1]

Subsequent attempts by the United States to regain technological pre-eminence have not changed the equation but have only driven the arms race to a higher level. Each side denies the other any diplomatic advantage from its investment, each side girds its economic resources for the next round of what has become a burdensome and self-defeating exercise.

Strategic stalemate is a dynamic and not a static balance. New weapons systems which supersede nuclear missilery could fundamentally modify the equation; but it appears unlikely that either side can long enjoy the advantages of a one-sided scientific breakthrough in new offensive or defensive capabilities. The technological environment of world diplomacy between the great powers appears to have reached a dynamic equilibrium, leading to a mutual need for stabilizing the arms race. The Test-Ban Treaty of 1963, the détente underway between the major Cold Warriors, the mutual cutback of plutonium production and the leveling of missile forces, the beginnings of an effort to call off the space race (the latest phase of strategic technological competition)—all represent increased understanding by political leaders of both nations of the implications of technological parity.

The break-up of Cold War blocs, the emergence of China as a problem common to both Russia and the United States, the divergence of economic and political interests among the NATO allies, and the melting away of bi-polar myths in the turmoil of nationalistic politics throughout the Middle East and Africa have made it more and more difficult for the great powers to manipulate the political situation and to find decisive advantages. As the diplomatic maneuverability of the great powers has declined, the myth of "prestige" has risen to an almost paranoid obsession. American determination to better the Soviet Union in space expresses a residual refusal to accept Soviet parity as a first-rate power. Somewhere among its enormous panoply of resources and industrial riches, the U.S. seeks to find a method for turning these to decisive diplomatic advantage; but the objective proves increasingly elusive.

In more than a figurative sense, the science-technology race represents an attempt to escape from the obscure and unpredictable daily challenges of world politics through a great effort to vindicate our technological manhood and pre-eminence. Both the Soviet Union and the United States have become muscle-bound by great military power and increasingly helpless to direct the outcome of crises throughout the world. Both are involved in a self-fulfilling prophecy of technological competition reaching beyond the rational limits of security and economic development. The motives of the science-technology race reach into a realm of almost pure fantasy which could, if carried far enough, bring social and political regression and historically unequaled waste of national resources. Meanwhile, the normal problems of domestic and international politics, the urgent population explosion, the unsatisfied needs of newly awakened masses, the widening gap between affluent and underdeveloped areas (within as well as between nations), the restlessness and revolt of young people and excluded minorities, the pollution of earthly elements, the perennial cycles of drought, flood, and earthquake—all the real problems are subordinated.

Technological power (which cannot confer decisive military advantages) and a narrow concept of prestige (tied to a single endeavor) cannot effectively meet the problems of our time. It is not enough to do something well; it is more important what one chooses to do. Attempts to impress the world with power are a preoccupation that cannot fail to raise the threshold of others' resistance against designs for influence. Maxwell Taylor put it succinctly: "Security is something like happiness. It is one of the by-products that comes from purposeful living as a nation." [2]

There is a valid incentive underlying government support for scientific R & D (research and development): the central and legitimate motive is military. Advancing knowledge and enriching the civilian economy are spin-offs which should be exploited as part of the total effort, but there is no imperative dictated by technology itself. Must we invest over 90 percent of our scientific resources in strategic defense and space? Mutual anxiety has accelerated the drag-race aspects of the science-technology race with the result that both sides are showing signs of resurgent rationality, searching for ways to control a politically mindless competition which threatens permanently to distort both societies—and which neither can win.

The Russian people are doubtless proud of their space pre-eminence maintained for eight years, but as the debates and political changes within the communist leadership indicate, the Russian people are also intensely interested in lower meat prices and indoor toilets. Public opinion polls show that the majority of Americans have never been convinced of the high priority granted to space in their own country. Congressman Chet Holifield, chairman of the Joint Committee on Atomic Energy: "I have seen what I think are very worthwhile programs in applied science, which have direct benefit to the people of America, put on the shelf for the pursuit of some fantastic objective . . . in the meantime the people are denied benefits which are just around the corner. . . ." [3]

The recent revival of rationality offers a hope of eventual recovery from Cold War-excited fear and wish fantasies, although the Vietnam conflict may well reverse the trend. In many areas the science-technology contest has been declared a draw, and continued efforts have been reduced and internationalized. The frenzied search for prestige in the Eisenhower Atoms-for-Peace program has ended in disappointment. The overstated proposition that nuclear reactors might soon transform the world and give the United States an infallible tool of international leadership has been thoroughly discredited. It is now generally recognized that the Atoms-for-Peace program may in the long run have contributed to the spread of the military atom by promoting interest in atomic research and power. In any case, it has done little to help the advancement of backward countries. It may in fact have retarded economic development by concentrating scarce technical manpower and materials in research of little immediate value, emphasizing political prestige at the expense of solidly based but less glamorous investment.

High-energy physics, controlled thermonuclear reactions, exploration of Antarctica, and meteorological and geodetic satellites have been removed from the wraps of national secrecy and made part of international agreements in which both sides (and other nations) now share the results of scientific discovery. The same trend manifested itself in President Kennedy's 1963 invitation to the Russians to join with us in the exploration of the stars and the planets, including the moon. This initiative proved premature but points the direction which the space race may take as its expense, difficulties, and doubtful diplomatic and social utility are more clearly established.

Myths Examined

The logic of this trend has a reverse effect upon those with vested interests in maintaining the science-technology race. It intensifies the passion with which they cling to increasingly challenged myths. As a result the myths are more and more formalized and reinforced as if they were self-evident truths capable of challenge only by the ignorant, the shortsighted, the troublemakers, and the treasonable. The National Aeronautics and Space Administration, the Air Force, and aerospace contractors have been busy pamphleteering Congress and the public in an effort to overcome growing skepticism about space budgets.

The break-up of Cold War blocs, the eruption of world political problems common to East and West, and growing recognition of the limits of both American and Russian influence—these facts are shunted aside. The Russians, it is argued, have discovered a new route: "Having declared that nuclear war isn't necessary to achieve world domination, they hope to accomplish the same end by dominating space." [4] NASA Administrator James E. Webb repeatedly asserts that Soviet diplomacy on earth is primarily supported by Russian space exploits: ". . . some people doubt very much if the Russians would have moved as aggressively with respect to Berlin if they had not had such successes in space." He calls for support of the U.S. space program as a means of forestalling "such difficult problems." [5] He claims that American success in the confrontation over Cuba was due to the U.S. space program and "the image of John Glenn's successful flight." [6]

In the NASA lexicon, space is the new decisive diplomatic arena: ". . . if we do not have a position of pre-eminence in space at the bargaining table of the nations, or in the minds of the world's leaders, the decisions that bring peace or war are not apt to be of our making." [7] Webb argues that the apparent dissipation of the Cold War in recent years is the result of "our rapid rate of advance" in space, which has "denied the USSR many of the benefits and many of the options which the Soviets expected their space program to provide as part of their forward thrust toward world domination." But, he continues, "it would be a mistake to feel that the Russians are no longer pursuing an active and vigorous program, or that they do not see many advantages of a vigorous space effort." [8]

According to space enthusiasts, it follows that the U.S. will never achieve security until it has achieved irreversible pre-eminence in science and technology. Such pre-eminence, they say, is within our grasp, if only we determine to seize it. Then all of the nagging problems of international politics will be solved. "Nations might become extroverted to the point where the urge to overcome the unknown would dwarf their historic desires for power, wealth and recognition —attributes which have so often led to war in the past." [9]

Such statements are reminiscent of the now discredited view that atomic bombs had created a "new world." During the period of the American A-bomb monopoly, the U.S. was frequently distressed by the Soviets' continued use of traditional power diplomacy. We were forced hastily to improvise non-atomic means of dealing with recurrent crises. When the Soviets detonated their own atomic weapon we ascribed their success to espionage and moved on to super-bombs to re-establish our unchallengeable position. When the Soviets came up with the super-bomb about when we did (or perhaps a little earlier), we were ready to ascribe it to the lack of enthusiasm of certain American science advisers who were muddle-headed enough to permit "the uninspired and plodding scientists of a backward Russia" to deny us pre-eminence. Even then, the U.S. held to the notion that there were advantages to be gained by a continuation of the strategic arms race; we pushed forward with "more-bang-for-the-buck" programs to multiply and diversify our stockpiles of warheads. "Massive retaliation" became the doctrine by which we hoped to win diplomatic power with the superior redundancy of our strategic systems. We managed to overlook the continuing success of Soviet non-nuclear diplomacy, the improvisation of our own diplomatic instruments, and the implausibility of a military doctrine that inhibited no one so much as ourselves. The advent of Soviet nuclear missiles and satellites gave us an alibi for the errors of this period and drove us to remarkable achievements in missilery before the end of the sixties, as we once again sought unchallengeable superiority in warheads and delivery systems.

The Kennedy administration, however, perceived the lessons of the fifties and moved from 1961 onward toward more realistic diplomatic capabilities. The Strategic Air Command and vulnerable surface-based missiles were downgraded in favor of a finite number of second-strike Polaris and Minuteman systems able to react after a Soviet first strike.

The massive-first-strike philosophy, which deterred all forms of diplomatic action and political change, was virtually eliminated. At the same time the new administration moved to create non-nuclear capabilities to deter other forms of Soviet diplomatic pressure. Both moves tended to stabilize the nuclear arms race. This was codified in 1963 in the Test-Ban Treaty and in other joint actions.

The Space Cult nevertheless elects to keep alive the arms race in space. It is true, they admit, that the full meaning of space cannot be forecast. But "would the Russians be going to this expense, putting this kind of money and effort into the space program, unless they thought there was the necessity for a man in space from a defense standpoint?" [10] We must, they argue, be pre-eminent in every kind of space activity that technology makes possible, because we need "insurance against surprise."

This kind of thinking leads to an anxious, item-by-item comparison of American with Soviet achievements and causes every special-interest group to formulate its claims in terms of the danger of the Russians doing something first. It leads to the false standard of "What are the Russians doing?" for evaluating proposed U.S. projects. It leads to respected scientists appearing before Congress and calling for larger appropriations in their particular specialty because, to quote one such scientist, "In this nuclear age we simply cannot afford that one fine day they might learn something fundamental that we simply don't understand." [11] It leads to such statements as that of Senator Stuart Symington: "If the USSR can use women as well as men, and we use only men, they of course would have a tremendous man/woman power advantage over us." [12] It leads an otherwise responsible engineer to call for federal support to test the "historical-cosmological" theories of Dr. Immanuel Velikovsky, author of *Worlds in Collision:* "I further believe that if American scientists do not soon engage in this work, Russian scientists will—thereby scooping the U.S. once again." [13]

In diplomacy, national prestige is a very real factor. It reflects the respect and influence a nation enjoys in its dealing with the world. But no one is fooled by "image-making" or by advertising gimmicks. "Prestige" is in no way comparable to the calculated exposure of a celebrity in the mass media.

Mr. Webb sees his role as public relations man for the American people, his main function to conduct NASA as a "tie-in" for selling America's image to the world. Congress is frequently puzzled when the Russians manage to get more propaganda mileage from space than does the U.S. (even though we chalk up a significant number of "firsts" and have already overtaken their lead in manned flight and heavy booster development). Dr. Simon Ramo says that public relations is "half the value" of the space program, and the U.S. should allocate "some of the total resources to a major extension in the methods of information dissemination to insure that we do capture all the benefits that such a pioneering program must carry with it." [14]

But regrettable news items, not necessarily accurate or significant in themselves, reflect other aspects of our society and dull even the brightest moments of our national program. The story of Captain Edward J. Dwight, Jr., Negro pilot, allegedly dropped from the astronaut training program because of racial discrimination, may have more powerfully affected the American image abroad than the Gemini flight that occurred in the same week; [15] similarly, the United States' agreement with the Union of South Africa to enforce a segregation policy in the manning of the three U.S. tracking stations located in that country; [16] the story told by Senator Stephen Young of a Negro constituent who proudly took his family to Cape Kennedy to pay homage at the shrine of American space effort and was allegedly refused service across the street in the motel dining room owned by the astronauts.[17] And in the summer of 1965, NASA officials tore their hair when a man in a 13½-foot sailboat (crossing the Atlantic in seventy-eight days) stole the front pages of the world's newspapers during most of the historic eight-day Gemini mission (around the world 121 times).

No public relations technique can conceal some of the hysterical attributes of the U.S. effort. President Eisenhower stated the problem in a press conference the day after his Farewell Address, noting that there is "an almost insidious penetration of our minds that the only thing this country is engaged in is weaponry and missiles—and I'll tell you we just can't afford to do that." In the same manner, overemphasis on prestige tends to have a reverse impact upon the intended world audience.

The advocates of escalating the science-technology race also assert as an article of faith that science can solve any problem, provided only that a sufficient number of scientists be mobilized and given enough money.

Congressmen are generally in favor of science and defense in the same way they are united in their opposition to cancer and the common cold. A special myth of Congress tends to reassure legislators that the power of the purse continues to overshadow the White House. Thus Congress often adds large sums to programs not in the budget and contrary to administration policy, or slashes other agencies with a flat percentage cut, to demonstrate dissatisfaction with executive programs. Webb has played the game remarkably well, protesting reductions of NASA budgets by stating the exact number of weeks each cut would delay the lunar landing (then refusing to break down NASA requests in terms of any reasonable cost factors).

There is also a myth that the signing of research and development contracts with private industry virtually ends government responsibility and automatically insures that goals will be achieved. There is little regard for available national resources and competing needs in the process of appropriating and contracting. As a result, the government is in the position of competing with itself, within and between agencies and programs, causing a skyrocketing of R&D costs—brought about by too much money seeking too few resources of scientific manpower, specialized equipment, and facilities.

High appropriations for NASA steal resources from other departments, weakening not only conventional weapons development but agriculture, medicine, and all the requirements of the civilian economy. The traditionally strong cadre of civil service engineers and scientists has been virtually depleted. The universities are holding their own with great difficulty, while research and development for purely civilian needs has been rendered prohibitively expensive for all but the great industrial empires. In the words of Dr. Vannevar Bush: "It is folly to thus proceed. Great scientific steps forward originate in minds of gifted scientists, not in the minds of promoters." [18] Dr. Lawrence Kavanau, special assistant to the Director of Research and Engineering, Department of Defense: "This situation cannot be remedied by applying more dollars indiscriminately to space research and development, nor by adding a variety of new major programs which tend to further dilute available resources and aggravate the management problem." [19]

This faulty accounting reflects a belief in the inexhaustible wealth of America and the unique "know-how" of private industry. By forcing the Soviet Union to ever-higher levels of the science-technology race, it is argued, we are proving the superiority of our system and leading theirs to bankruptcy. In the words of Dr. Lloyd Berkner: "America with its far more powerful economic apparatus can maintain the pace of the contest with much less strain on its system." [20] This is a common refrain in congressional committee rooms. NASA officials argue that our space effort will redound beneficially in every way to our overall economy, while it will steal butter and eggs from the average Russian. There is no doubt that to some extent this is true, but there may be advantages on their side. There are indications that in terms of gross national product the Russian defense and space effort imposes a heavy burden but is conducted much more economically than ours. According to Donald Hornig, President Johnson's Science Adviser, the U.S. investment has exceeded 2 percent of the gross national product, while the Soviets' has stabilized since 1963 at about 1 percent. It is also by no means certain that it will serve American interests or international stability to raise Soviet defense and space costs beyond their breaking point. The result may harden the Soviet regime or augment pressures for international adventures aimed at keeping us off balance in areas where they enjoy advantages.

Extravagant claims are made for a direct pay-off from space exploration, citing the analogy of Christopher Columbus' discovery of the New World. Obvious factors render these speculations absurd. We already know that the moon and other parts of the solar system are intrinsically more hostile to life than the most miserable spot on earth. Life on top of Mount Everest would be sheer luxury in comparison. The economic impotence of space arises from the simple fact that the energy required to lift something to the moon is more than twelve hundred times that required to move the same weight on the surface of the earth. On purely economic grounds, leveling the Rocky Mountains or colonizing the North Pole makes much better sense.

Finally, the advocates of an infinite science-technology race always return to the lyrical and mystical theme: man is driven by a purely spiritual urge which motivates him to do whatever is challenging and inspiring for its own sake—to climb mountains because "they are there." (This was the same answer given by a Don Juan when asked why, left alone with any woman, he tried to seduce her.) Dr. Harold Urey, a responsible scientist, tells of overcoming his reluctance to support the Apollo moon project with the thought that men would do it anyway through some ineffable compulsion. He did not wish to be found using his influence to prevent "the building of great cathedrals." Urey asks: "Why do any of these things? Why not just make use of our wealth to furnish luxuries to live comfortably, to have a purely material existence? The answer is that people are just not made that way. They do wish to do magnificent things. . . ."[21] Lloyd

Berkner cites the most fundamental reason for his support as: "Its influence in stimulating man's spirit and raising his sights as a big stride toward a freer and fuller rationality. . . . Man yearns for bold steps to press the conquest. To do less would shrivel his spirit, diminish his stature." [22] Poet Robert Frost exulted: "I don't know why, but it's glorious. . . . Perhaps it will improve communications between stock exchanges." [23] In the same vein, an Atomic Energy Commission official argues for a massive bomb shelter program which ". . . would be a true symbol of national vigor. . . ." [24]

It would be presumptuous to dispute any solemn declaration of faith. As a personal religious matter, the issue is removed from public policy, unless of course someone injects the First Amendment. But those who wish to practice this faith force a compulsory tithe upon the nation.

Cathedral-building is not necessarily a symbol of social vigor nor of highest human aspirations. Quite the contrary, it often is associated historically with decadence and incipient decline. As a kind of make-work program, it postpones the confrontation of the populous poor with the few rich without lessening the sum total of misery. Instead, it increases that sum through forced labor and heavy taxes. Societies afflicted by institutional rigor mortis may create brilliant monuments in their death throes, but these are tokens of folly as well as of beauty, and neither lift the spirit of the people nor save their rulers.

Writing in *Science,* Fred J. Gruenberger offers a number of standards as "a measure for crackpots." One of these he calls the "Fulton *non sequitur,*" a negative test by which, he says, "the true crackpot can frequently be spotted." The crackpot argument goes: "They laughed at Fulton. He was right. They are laughing at me. Therefore, I must be an equal genius." [25] It is remarkable how often the Fulton *non sequitur* turns up in congressional testimony as the scorn of the scientific community for Project Apollo has grown.

Space Weaponry—Real or Imagined

NASA's founding charter emphasized its peaceful mission in scientific research and space exploration. The Space Agency was something of a compromise between Air Force advocates who saw space as a new military arena, and those who wished to check Air Force control of the nation's space efforts. President Eisenhower and his scientific advisers sought to avoid extending the arms race into space.

The new agency, however, soon manifested subservience to the Air Force, and, with varying degrees of explicitness, the military dimension has largely dominated the U.S. effort.

Ambiguity of NASA's primary mission continues to be part of the official line, whether stated by NASA (Dr. George E. Mueller: "We are providing the base which permits the nation a freedom of choice to carry out whatever missions the national interest may require—be they for national prestige, military requirements, scientific knowledge, or other purposes." [26] Webb: Our policy is "to develop the space program as a civilian peaceful effort . . . but always pressing with the kind of technology that would permit us to move rapidly in the military field if we were required to do so. It is a little bit like 'keep your powder dry' with respect to the military side. . . ." [27]) or by the Air Force (General Bernard Schriever: "We must have necessary strength to insure that space is free to be used for peaceful purposes . . . both the military and civilian aspects are vital" [28]). As one scholar observes, "Obviously the goals of peace and the policy of pursuing only peaceful purposes are secondary rather than primary." A proper characterization of our space effort would be: "Since we have a space program, it is desirable to give reassurances concerning it and to utilize it to promote goals that we would want to be pursuing in any case." [29]

Is there an authentic military requirement in space? Of three categories of military systems (offensive, defensive, and passive), clearly defined military requirements have been identified only in such passive functions as communications, reconnaissance, meteorology, and geodesy. These functions duplicate capabilities already existing on the surface of the earth and are not of decisive importance; all could be perfected without massive investment in more powerful boosters, life support systems, and maneuverability in the interface of atmosphere and space.

NASA's Webb frequently states that the Apollo Project will cost only $2 billion, since without it we would have to invest at least $18 billion in the military space capability we get as a bonus on the $20 billion investment. This rhetoric collapses in view of the fact that the nation's total space investment is already well over $20 billion. In 1965 hearings of the Senate Space Committee, Dr. Harold Brown, former chief scientist of the Defense Department, said that if NASA were not doing it (at a cost over $5 billion a year) the Defense Department would have had to spend some "hundreds of millions a year," perhaps even "$1 billion a year to develop that technology."

There is a legitimate military requirement in science and technology, but it is different in kind from that advocated by the cult of an infinite arms race. Military men, struck by the fantastic accomplishments of the wartime A-bomb project and by rocketry, appear convinced that the fantastic itself is now within their reach. They would like a massive science-technology effort to realize every weapons system they can dream up. The Air Force has seriously favored super-missiles with giant hydrogen warheads able to maneuver in space; neutron bombs able to kill people without harming property; troop-carrying rockets capable of reaching trouble spots in minutes; nuclear planes able to stay aloft for weeks at a time; space bombers in orbit; satellite interceptors to patrol the vast reaches of near-space in order to deny it to others (using the analogy of air supremacy in World War II); military bases on the moon, or on large planetoids orbiting the earth, or assembled in orbit; death-ray weapons based upon the principle of the laser (coherent light); "super-cars" able to traverse all earthly media and maneuver in and out of orbit; and, most recently, anti-gravity bombs. The list is practically inexhaustible. Serious discussion and debate have surrounded such proposals as weather control, by which the generals may whip up a batch of fog at will, or deluge the enemy forces with hail and rain—even possibly small earthquakes or tidal waves; doomsday machines (a concept now made ridiculous by "Dr. Strangelove"); and anti-matter warheads with power thousands of times mightier than today's biggest H-bombs. As Congressman Craig Hosmer, a friend of the Air Force, says: "The Air Force has been for everything that ever came along at one stage or another." [30] Most of these fantasies have been rejected by the Department of Defense as "blue-sky" proposals of doubtful strategic need, prohibitive cost, and questionable scientific feasibility.

But the Air Force wants to move swiftly into the full range of all conceivable new weaponry: "We've got to move into new systems as soon as they are technically feasible." We should begin to fund deployment and production even before the weapon itself has been fully developed, in other words, on a *concurrency* basis, "because I think that in the long run we will save money by doing it that way." [31] Concretely, the Air Force has been pushing urgently for such "military space operations as inspection and negation of hostile systems and means of defending or reducing the vulnerability of our own space systems." [32] General Thomas D. White, former Air Force Chief of Staff: "I wish we would move faster on the satellite inspector and interception. We soon may need to verify what the Soviets have put into space; we may someday want to shoot it down." [33] Dr. Walter F. Dornberger, former head of the German V-2 program, now vice president and chief scientist of Bell Aero-Systems, has painted a vivid picture of the nightmare world: "We may not even learn that a Russian interception took place. They may deny everything and we have no means to prove it. . . . Out of this first military action in space a kind of hide-and-seek war will result. . . . The Russians will not rest until they have obtained control of the space effort of the free world. Our space systems, such as reconnaissance, weather, communication, navigation, detection, and interception satellites, which are so vital for our defense and the survival of our nation, will be wiped out of space." [34]

Dr. Alton Frye, Air Force scientist, says that a satellite interception capability for the United States will "open a new vista of political opportunities," enabling us to lay down conditions that the Soviet Union must meet if it is to have access to space. If our conditions are rejected, "the United States will be compelled to destroy every vehicle which the Soviets fire into orbit or in a trajectory toward our territory." [35] General Thomas Power predicts that in the event of a total nuclear war, the U.S. will require a "really survivable manned control structure," such as might be found in a maneuverable space station.[36] Dr. Edward Teller has called upon Congress to adopt a program for establishing a "large and independent colony on the moon, as a means of having a working base in space and control of near-space. . . . A nuclear reactor should be developed to operate on the moon, eventually to furnish the power to extract water from the moon's soil." [37]

Admiral William F. Raborn, former chief of the Polaris fleet, wrote in the U.S. Naval Institute *Proceedings* that weather control would be a more significant weapon than nuclear bombs: "The capability to change the direction of destructive storms and guide them toward enemy concentrations may exist in the future arsenal of the naval tactical commander." [38] The House Committee on Science and Astronautics, praising the weather photos of the Tiros satellite, slipped into hyperbole: it "could lead to an early weather control capability . . . which would provide the United States with a great deterrent to war. . . ." [39]

The idea of dropping a bomb from a satellite or shooting it from the low gravitational field of the moon is based on a naive analogy with the dropping of explosives from aircraft. A bomb dropped from a satellite or the moon will not fall to earth: "It will reach the earth only if it is pushed away from the satellite with the right amount of force and with controlled direction. The whole launching operation would be more difficult than the launching of an ICBM from the earth, and the flight would take longer." [40] The notion that a missile overhead could reach the earth in less than twenty minutes, as does a manned satellite after retrofire, entertains the misconception that such a low orbit could be maintained indefinitely. Actually, the orbits used for manned flight through 1966 were such that even without retrofire a launching platform's orbit would decay through friction and burn up in the atmosphere within a few months. As for using the moon as a kind of bargain-basement space platform, you must first get your stuff to the moon, and it takes two hundred pounds on an earth launching pad to deposit one pound of payload on the moon. A projectile fired from the moon would present the same problems as landing a projectile on the moon from the earth: the missile would fly in a curved path a distance of roughly 360,000 miles, giving the target some 60 to 120 hours' warning. The unreliability of control and guidance devices and their vulnerability to enemy countermeasures also present great obstacles.

In resisting Secretary of Defense Robert McNamara, the infinite arms-race advocates have made their position clear: "It is vital that we hold the lead at whatever cost"; [41] that "we must be protected against the unlikely as well as the likely"; [42] and that "we must maintain not only a qualitative, but also a quantitative advantage. . . ." [43]

The strategic doctrine of the 1960's recognized the futility of the search for infinite security. Although U.S. military power has steadily increased, "the national security has been rapidly and inexorably diminishing." This did not result from inaction on our part but rather "from the systematic exploitation of the products of modern science and technology by the USSR." [44]

The self-defeating nature of the arms race is vividly shown by Secretary McNamara's estimate that, against Soviet attack, we would find it virtually impossible to provide anything like complete protection no matter how large our nuclear forces. By spending up to $25 billion more, we might reduce our fatalities from 150 million to perhaps 80 million—assuming 1970 population and force levels. But, by increasing their offensive missile forces, the Soviets could prevent us from achieving even this level of protection. And they could do it at far less increased cost for offensive forces than our extra cost for defense.[45]

It is the legitimate business of military men constantly to look for new weapons; but it is also the business of civilian leadership to control this process in terms of other values, to prevent large-scale development or production of weapons systems that lack proven feasibility, cost analysis, or strategic utility. As Dr. James A. Van Allen comments: "Apart from military missile systems and the transit system of navigation for submarines . . . the field for direct military applications of space techniques appears . . . to be one of much loose talk and sweeping generalization. . . ." [46] Dr. Jerome Wiesner, President Kennedy's Science Adviser: "Many of us feel that there are no demonstrated uses for man in space and that many of the proposed space missions are things which might be better done in other ways. We are developing capabilities and large boosters . . . that are in excess of our ability to visualize in use. Uses may emerge, but at the moment, I would say that the capabilities exceed the known military needs." [47]

In 1965 Secretary McNamara reiterated that the Apollo program had "no direct military worth." Apart from the modest utility of near-earth satellites for communications, weather photography, and reconnaissance, the chief justification of manned space investigation is to eliminate the military unknowns in order to stabilize this environment as we have nuclear weapons. Harold Brown: "I think we are at least as likely as the Soviets to find any . . . breakthrough if it exists. I rather doubt that it exists. As I say, for seven or eight years we have been looking for something really new that space would show us that we could do that we can't do any other way. I don't think we have found it."

No nation can ignore the development of new technology which may give decisive advantages, but the argument easily corrupts to serve other motives—the need of a military service to maintain its role in spite of the fact that its existing missions are becoming technologically obsolete; the need of industry to maintain its R&D contracts; the need of various congressmen to maintain the health of defense industries in their districts or to acquire new industries as the result of new government contracts, and so on. So far, according to Dr. Brown, no decisive advantages have been found in the military uses of space. The argument that we must "avoid surprise" means abandoning judgment completely in a futile search for infinite security.

Under Robert McNamara, the Department of Defense has attempted to subordinate military ambitions to independent civilian judgment. Resentment by the brass, instant and fierce, has resulted in subtle and savage bureaucratic infighting conducted mostly out of public view in the White House, the Pentagon, and Congress. Both Presidents whom he has served have largely supported McNamara's efforts, though sometimes softening his decisions in compromises with Congress. His control has clearly become necessary, for "the technical input is only one and not necessarily the most important . . . the inputs of military experience, fiscal and international considerations all enter." [48] Military technology is limited by many factors. "First, we have the question of what we can accomplish within our available economic means and not just what one would predict would be possible with our present scientific understanding. Second, military technology is only justifiable insofar as it contributes to our national security, which immediately takes one into areas of national and international politics wherein the scientist or engineer has no special competence." [49]

The cost of development of a new weapons system may run to hundreds of millions and billions of dollars: for example, three experimental B-70 aircraft cost $1.3 billion. Well over a billion dollars has been invested in the Nike-Zeus anti-ICBM system, and the problems of development are far from solved. Each major scientific program in defense and space costs roughly $1 billion for the R&D stage alone. Obviously, as a number of alternative projects appear to become technically feasible, the nation must choose. Not only are there limited resources, but there is also the important question of allocating these resources for the positive ongoing values of our lives—of which defense is necessarily a very limited and negative part. A man may build a house and, out of concern for burglars, build into it every conceivable safeguard to counter every conceivable means of breaking and entering. Such a house would fail to achieve security against all contingencies; it would quickly cease to be a house at all, and no one could live in it.

The military pursuit of blind technology has in the past twenty years wasted billions of dollars (for example, the abortive six-hundred-foot radio telescope begun by the Navy at Sugar Grove, West Virginia, canceled before completion—cost, about $50 million; the Navaho missile project—cost, $750 million; the nuclear-powered aircraft—over $1 billion; the Advent communications satellite—over $150 million; the nuclear propulsion space booster—$2 billion, and so on). As for the concurrency method of R&D, Harold Brown declares: "I think concurrency is justified only in a very few vital programs . . . unless one can show that a project is vital, that is, that without it the United States stands in very serious danger of failure of deterrent, say, or of an enormous loss of military power—unless that is the case, I do not think concurrency is justified. I think that you should . . . show that it is feasible and that it will work before you go ahead and produce it." [50]

McNamara has forced the Department of Defense to distinguish between what is possible and what is needed, between what *could* and what *should* be done. By adopting the concept of budgeting in "program packages," he requires the services to establish a strategic need. He is able to spot whether the proposed allocation of funds is justified or whether wasteful redundancy and overlap exist. The old system of lumping all military appropriations together was abandoned in order to make visible the relation of R&D to approved requirements, and to provide a means for evaluating the worth, cost, and promise of various technical approaches to the solution of military requirements. Dr. Brown: "The question which really confronts us here in the United States is how can we within the framework of democratic processes make the wisest possible decisions in those matters in which scientific and technical questions hinge upon government activity and vice versa. . . . I believe that gradually there is emerging within the Executive Branch of our government a rational, reasonably safe approach to the very great problem." [51]

But it is hard to make this kind of rationality stick, especially when there are so many vested interests involved. For example, while the President's Science Advisory Committee could agree on canceling the nuclear aircraft project, this was hard advice for the congressional Joint Committee on Atomic Energy to swallow, especially after its years of promoting the project against a reluctant administration. In the collision between the scientists and the committee, Eisenhower stalled for two years and Kennedy an additional year until the cancellation became politically possible. Meanwhile, more than a billion dollars had been spent. The stories of McNamara's cancellations are each of them Herculean sagas—the Skybolt missile, the Dynasoar space glider, the B-70 bomber, the satellite interceptor program, and other classified space weapons, including a bomb-in-orbit proposal. Some of these stories we will relate in future chapters; in brief, each program proved to be an India rubber ball—no sooner had it received a budgetary *coup de grace* when it bounced back in another place or form.

As it had in developing nuclear weapons, with Sputnik I the Soviet Union again made the point that it was a first-class power able to remain so. Soviet science has always been strong and is as competent to explore the frontiers of technology as that of the richer U.S. Scientific-technological achievements that have important diplomatic consequences will not be ignored by either side, and initial advantages gained by one or the other will soon be overcome. Parity in space therefore is inevitable—but it is within the power of the U.S. to influence the level and cost of parity.

The quest for infinite security will force a higher level of the arms race, perhaps a level less stable than at present, and with possible secondary effects distorting the political and economic systems of both countries in unfortunate and unpredictable ways. This was the lesson President Kennedy expounded in the year of the Test-Ban Treaty. If the trend toward larger hydrogen warheads and more systems for their delivery had been permitted to continue, it would have meant deeper bomb shelters and larger retaliatory weapons—which in turn would mean deeper shelters, which in turn would mean still bigger bombs—and the music goes round and round. It took the world two decades to recognize and at last control this madness. The moon project has provided for both sides a continuation of the same hysterical behavior in a sublimated form.

The massive U.S. investment in science and technology as a means of diplomatic prestige or military power contains an implication that the United States cannot live happily in a world which contains equally great political powers. They may pursue contrary political interests, opposing their will to ours, forcing us (as we force them) to limit our objectives, to compromise our wishful thinking in the face of the real alternatives of action; forcing us to tolerate ugly situations which we may not have the power to resolve immediately as it suits our whim. If this implication dominates the American mentality, the future is bleak indeed.

Diplomacy tests the basic strength of a nation, not alone its arms and men but also its imagination, industry, resourcefulness, and the quality of its society and government. In all these areas we have advantages and built-in strengths. At the same time we have been learning in the hard school of frustration and occasional failure that good intentions do not automatically bring peace and justice in the world, that our concept of peace and justice may not be universally shared, that there is no quick, cheap road by which good may overcome evil through some transcendental process. The bright hope of the Kennedy years appeared to many to be a fulfillment of a new maturity in facing a complicated and dangerous world.

In spite of the agonies endured by the nation-state system during the last hundred years, it appears to be more full of life than ever. The strategic stalemate between East and West may have liberated the small nations from domination by the great as never before, placing a premium on non-alignment and causing a break-up of Cold War blocs. With a number of nations engaged competently in the advanced and endless frontiers of science, we cannot hope to be first in everything. The mature approach is to choose always what *we* want to do, to select what is best for *us,* and to do that competently and well, learning from what others in turn may do confidently and well.

To what extent will space help America to solve the problems of NATO, improve its relations with underdeveloped nations, and resolve domestic tensions in constructive and sustainable ways? We should continue a healthy and balanced support for basic science and for technological projects that evidently serve national needs. But there is no magic in science or in space, and the plain language of democracy must be reasserted against the worshipers and beneficiaries of an infinite science-technology race.

CHAPTER II

CELESTIAL ROULETTE

President Kennedy committed the U.S. to a moon journey in this decade; but since 1961 an appreciation of the obstacles and dangers of the trip has grown with our competence and knowledge. Soviet space plans have become increasingly vague, and there are indications that NASA anticipates a graceful waiver of the time limit due to slippage and delays beyond its control. Yet both sides move ahead, reciprocally goaded to persist even as both see problems and costs grow larger, and as hoped-for military values evaporate. Both feel cautiously for a space détente and hedge their bets by gradually shifting emphasis from far- to near-space, from interplanetary manned flight to unmanned probes and to careful manned experiments in low orbit.

The Unknowns of Man in Space

Dominated by political commitments, the U.S. lunar landing timetable drives ahead into a variety of unknown perils, any of which may upset the gamble. The unknowns include the moon's surface, the nature of deadly solar flares in space, the unproven techniques of space maneuver and control, the limits of mechanical reliability, and the still undetermined reaction of man to this hostile environment. The engineering design of the entire Apollo Project has already been completed, frozen, and placed into hardware production, including the vehicle that will be used by the astronauts to descend from a lunar parking orbit to the surface of the moon (known as LEM, for *L*unar *E*xcursion *M*odule). This concurrency was made necessary by the Kennedy commitment to accomplish the project during this decade.

Among the nightmares of NASA scientists are a number of dire possibilities concerning the nature of the moon's surface, a quality still unknown in spite of the magnificent Ranger photographs supposed to dispel such doubts, and in spite of the successful soft landing of the Soviet Luna 9. Delay of the American soft-landing probe will leave the enigma intact until it is too late to modify existing hardware. There was a recent decision for further preliminary surveys by a lunar orbiting vehicle before the final commitment of human lives; but the present timetable of Apollo will not permit re-design and new hardware procurement until 1972 at the earliest, and the lunar orbiter itself is slipping behind schedule.

Inspection of the Ranger photographs shows that the moon's surface is quite smooth even down to a scale of a yard or so. The irregularities that appear can be described as gentle undulations. Its weight-bearing strength remains unknown: given present knowledge of soil physics and the structure and composition of the moon's face, the sum of existing evidence allows inferences as wide apart as dust and tea biscuits.

Luna 9, whose payload was small and light, lurched after landing as though breaking through a thin crust. Its photos led NASA evaluators to conclude that the landing surface was rock froth ("vesicular lava") of undetermined cohesiveness. Luna 10, orbiting the moon since April 1966, carries a gamma ray spectrometer which may narrow but cannot eliminate speculation. Laboratory experiments with volcanic ash have proven inconclusive. In spite of thousands of close-up moon photographs (provided by Rangers VII, VIII, and IX), NASA's photo interpretation team is divided and uncertain. One team member, Dr. Harold C. Urey, suggested that the dimple craters shown by photographs were formed by finely divided material "slumping" into voids or subterranean crevices beneath the lunar surface. "I am scared of a manned lunar landing, because of the enormous number of slump features Ranger IX shows."[1] Scientists at the Jet Propulsion Laboratory suggest that the lunar surface "may collapse catastrophically when contacted by rocket exhausts or other contaminations."

The moon pictures are "like a mirror . . . each experimenter holding them up and seeing his own theories reflected in them." The NASA evaluation team is agreed that the moon's material is very porous near the surface, but there is sharp dispute concerning its weight-bearing characteristics and its roughness, two of the three men contending that the evidence does not contradict design assumptions.

Field studies are still going on at NASA's Ames Research Center and at the White Sands Missile Range in New Mexico. Unarmed missiles have been rained on sand and rock of varying degrees of cohesiveness and particle size. So far the evidence suggests that the predominantly circular, well-developed, and smooth craters characteristic of the moon are created only in fine-grained and loosely bonded stuff; craters in rock or strongly bonded material have irregular, sometimes elongated rims. A Soviet scientist told a 1965 scientific gathering of observations that indicated the moon's interior to be four or five times as radioactive as the earth's, with a dust layer some twenty feet deep on its surface.[2]

Other problems have appeared as our instruments have improved. There is considerable evidence that the moon contains many "hot spots" of volcanic activity and that it is the target of heavy showers of micrometeoroids. Dr. Homer E. Newell, NASA Associate Administrator: "Being a gravitating body, the moon serves to focus these particles. We know from laboratory experiments that such particles impacting at the surface at the high velocity they have, gouge out holes many times the volume of the particles, and the material of the hole then is ejected as a large number of particles. It may be that the surface of the moon will turn out to be an environment of fast moving particles that can penetrate space suits."[3] NASA's Director of Manned Space Flight admits that "we are still guessing," in spite of the fact that "the lunar excursion module design is in fact frozen at the present time. We are going to go ahead and build it."[4] A useful Ranger-type mission might attempt to photograph the impact craters of earlier Rangers, an objective suggested by the Jet Propulsion Lab in late 1964; but so imperfect are the mechanics of space-shooting that the notion was rejected as having little likelihood of success.[5]

The Surveyor soft-landing and the lunar-orbiting probe are expected to increase our knowledge of the moon's surface during the period 1966-1967. New findings may revise the guesswork built into present commitments: "There is always the possibility that when we really know the situation, it will be much more adverse than we expect. This is always a possibility, in which case we . . . would have designed the wrong thing."[6] In that event the "engineering program would have to be redirected"[7] and the moon landing put off for this decade.

Bursts of protons from solar flares are the most dangerous but not the only radiation hazard. Cosmic rays reach the earth-moon system at relatively low intensities, although with high energy. Van Allen radiation is also a relatively low hazard because of the spacecraft's rapid passage through this belt-like region. But such low-level radiation may still cause damage to transistors and control circuits. The most insidious are the protons associated with sunspots. The astronaut will be most exposed to these on the surface of the moon, where he has only his space suit as a shield. On the moon's surface, the astronaut will receive about ten times the dose he would experience in the command module, and a solar flare lasting but a few hours would put the astronaut beyond specified safe limits. If there be strong likelihood of a series of solar flares, NASA considers it may be necessary to abort the mission. A system of radio and optical telescopes is now at work around the world trying to monitor the scatter of space radiation, but we have not solved the problem of actual protection against radiation sickness, sterilization, or death.[8]

The year 1965 ended the two-year period of the quiet sun and began a new eleven-year cycle of agitation with solar proton flares raging through space, expected to climax about the time set for the U.S. expedition. The U.S. Naval School of Aviation Medicine reports that proton streams from the sun in 1969-1970 could range "from deadly downpour to light shower." Chance of a lethal dosage was calculated as three in ten thousand; chance of moderate but incapacitating "rain" as one in one thousand; and chance of light shower as one in one hundred. Super-flares may occur at any time and cannot be predicted. In 1964 Major General Don R. Ostrander, Air Force Commander of Aerospace Research, observed that great strides had been made in predicting the onset of solar flares, ". . . but there is still a large gray area in which we do not know whether conditions will be safe or unsafe."

"The solution," says Dr. Van Allen, "lies only in the direction of shielding, and that is a monstrous engineering enterprise." [9] A NASA advisory committee has concluded that "shielding of the crew for Project Apollo is not possible within the time and weight limitations of the project," and that the crew "will simply have to accept the relatively low probability of encountering a major solar flare during their relatively brief excursion to the moon." [10] A geneticist comments: "We have to let the people know there is a genetic risk. Again we have to learn to live with it because we cannot prevent it. . . . Possibly we should hope that the people who do go into space for any extended period of time . . . would not have any more children." [11] Senator Clinton P. Anderson, chairman of the Senate Space Committee, recently switched to a "go-slow" position on Project Apollo, declaring that it would be "sound" to delay the flight into the 1970's in order to improve safety: "We don't know all there is to know about these flares." In spite of bland assurances to Congress, there is a general consensus that the full dangers of radiation are simply unknown. It is another area in which the first astronauts will be the guinea pigs.

We have also yet to confront and master the problem of space maneuverability and navigation. No earthly analogy applies to space maneuver. An orbiting object cannot simply increase or decrease its speed on a fixed course; an increase in the direction of flight will result paradoxically in a slower orbiting speed in a higher orbit, while a decrease (brought about by exerting energy against the direction of the flight) brings about an increase in orbiting velocity in a lower orbit. Docking depends upon the ability to create a situation of two independent intersecting orbits which will bring the spacecraft and its target to the precise point at the precise moment. This depends upon the accuracy and speed of electronic calculators able to read positions at a given moment and program the necessary orbital change. Even an infinitesimal error will defeat this maneuver, not to mention the unsolved difficulty of establishing accurate base points from which to calculate relative position.[12] This has been accomplished in earth orbit; but the problems will be infinitely greater in the vicinity of the moon.

In practically every area of this infant technology, problems are still to be solved. The reliability of electronic or mechanical components is never perfect, and Apollo lacks any means of on-board repairs. Most assemblies are difficult to repair even in a well-equipped factory on earth, impossible to repair in space. Weight limitations prevent complete redundancy of mechanical systems, and it will be almost impossible to deal with a major mechanical failure or fire en route. There will be many phases of the Apollo journey in which a rapid abort will not be possible if something goes wrong. The public is already aware of the problems of reliability. Within one hour before John Glenn's historic flight five items went wrong,[13] and malfunctions have marked every manned flight, fortunately none disastrous at this writing.

The selected mode of ascent for Apollo involves a boost from Cape Kennedy directly into a "parking orbit" around the moon, accomplished by the mighty Saturn V rocket, 365 feet high, three thousand tons fully loaded, whose five engines will develop a thrust of 7.5 million pounds. From the parking orbit, two of the three spacemen will descend to the moon in the lunar excursion module, whose small rockets will make possible a soft landing in a selected spot. After a quick stroll over the area, the astronauts will re-enter the module, the upper portion of which will lift off for a direct rendezvous and docking maneuver with the mother ship. The module will then be abandoned in space while the mother ship propels itself back to the earth's gravitational field.

From a technical point of view, the maneuvers of the lunar excursion module present the greatest challenge. In its lunar parking orbit the spaceship will be traveling at a rate of 3,500 miles per hour with respect to the moon's surface. The module must reduce speed to 10 miles per hour or less in order to avoid being smashed against the moon. The braking maneuver must be accomplished entirely by rocket engines, because the moon has no atmosphere to help slow the rate of ascent; this will be an entirely different experience from the re-entry maneuvers practiced in earth orbit. During this time the module must maintain continuous radio contact with earth stations in order to determine its exact position from microsecond to microsecond. Blasting off from the moon and making a direct rendezvous and docking with the mother ship is even more complicated. Throughout all these maneuvers there will be no margin for error. The slightest deviation from precision timing and navigation will mean disaster for at least two of the men.

The decision to go to the moon by way of lunar-orbit rendezvous is still subject to bitter debate. It may have been a necessary and right decision, for all of the available routes contain equally great technical problems. But the notion of having men attempt the difficult maneuvers so far away from the earth and its tracking stations, so far away from the capability of mission-abort, as the first experiment of its kind in the moon's vicinity—this is especially chilling. Dr. Jerome Wiesner strongly objected to the lunar-orbit rendezvous decision as technically unsound, suggesting that direct-ascent or earth-orbit rendezvous might be better. In addition to its danger and uncertainty, Wiesner charged that such a route was technologically "a dead end," with limited value for other kinds of missions.[14]

The successful rendezvous of Gemini 6 and 7 in late 1965, and Gemini 8's successful docking in March 1966, were brilliant technical accomplishments. On subsequent Gemini flights (through 12) direct-ascent rendezvous and docking will be practiced in order to achieve in earth orbit as much applicable experience as possible to aid the Apollo mission. But the critical differences between earth and lunar orbit prevent real confrontation of the unknown until human lives have been irrevocably committed.

Unnatural haste makes waste. NASA's response to congressional budget cuts has been, for example, to reduce reliability testing programs for the Saturn boosters: "We are planning to fly only seven or ten Saturns before we call the Saturn operational. We hope to accomplish this by a very intensive ground test program." [15] In March 1966 Webb reiterated his faith in the timetable but pointed out there were no extra rockets for the usual amount of testing or for substitution in case one fails. The testing of earlier boosters required some hundred or more test launches. But NASA wants to keep Apollo on schedule, at least on paper. For the same reason, NASA has abandoned plans for sterilizing the space vehicle. Sterilization had been contemplated earlier in order to prevent molestation of the moon's organic materials.[16] There is also some danger (not considered significant) that organic materials which might prove harmful to terrestrial plant and animal life could be brought back to earth on the spacemen's return.

Finally, the adaptability of man to space is full of unknowns. In spite of the encouraging results of Russian and American space flights, scientists recognize that man's exposure to this environment has been very limited. It is impossible in an earth-bound laboratory to simulate subgravity conditions for long durations, or to scale human beings up or down. Soviet scientists doubt man's ability to survive in extended weightlessness for more than two weeks. Dr. Tau-Vi Toong, consulting Air Force scientist, writes: "The human heart might possibly degenerate to such an extent that its beat slows down to something corresponding to the normal heartbeat of an abnormal giant on the earth. The situation might become even worse should he again return to the earth's gravitational field with his weakened heart." [17] Space Administrator Webb admits that should it become necessary to insert artificial gravity in the Apollo design, we would for this reason alone fail to make the moon in this decade.[18]

The Gemini missions appear to have established man's ability to operate in a weightless environment for at least fourteen days, almost twice as long as required for the first Apollo mission. But during extended flight the astronauts' rate of heart action slows to less than half that on earth, which suggests that long-duration flights will make readjustment to normal conditions precarious.

Human disorientation, decalcification of bones, decreased cardiovascular tone, and other possible dangers will no doubt be clarified by future experiments (the most important being the Apollo shot itself), but the status of the American program is eloquent testimony to the patchwork of guesstimates upon which our enormous investment is founded. The major arguments for manned as opposed to instrumented space ventures arise from economic and political rather than strictly technical or military motives. Unmanned probes are considered inferior in prestige and scientific value. Moreover, since they would cut the size of the national investment by over 75 percent, they are resisted by aerospace contractors who see in Apollo a substitute for diminished defense contracts. These considerations have prevailed. The arbitrary decision to shoot man to the moon was made before the state of the art warranted it and before any convincing arguments had been cited.

The ultimate in "concurrency" was a proposal that the U.S. land an astronaut on the moon *prior to* development of the means of retrieving him. Later, when the means have been devised, the brave precursor may be rescued. This would be "a very hazardous mission, but it would be cheaper, faster, and perhaps the only way to beat Russia." [19] Similarly, Wernher von Braun has proposed the same method for beating the Russians to Mars. This kind of thinking pervades the actual planning of the Apollo mission.

Predictions are foolhardy, and so are large wagers upon intangible and unpredictable outcomes. Even a cursory glance at the remaining unknowns of Project Apollo, the slippages that have occurred and will continue to occur, raises questions about the realism of the present timetable. The obstacles will not quickly dissolve, and the amount of money spent will not by itself determine the rate of progress. Webb is already trying to create an alibi into which he can retreat should future events make that necessary, shifting the blame to congressional budget cuts, a tactic designed to pressure the legislators into maintaining NASA's funding level.

Is Russia in the Race?

There is much evidence that the United States has now slackened its timetable in the race to the moon, and that the Soviet Union has never been in the race. NASA and its advocates take pains to deny this in order to maintain the momentum of the American investment, but the evidence mounts.

After the shock and humiliation of Sputnik I, the U.S. took up the science-technology challenge as an area of ultimate Western advantage and superiority. The urgent emphasis placed upon the space race contains an element of wishful thinking: we want the science-technology race to be important because we believe we can win it; we are not sure how to deal with other kinds of problems in which we find ourselves competing for influence with the Soviet Union. Scientist Hans Bethe says we have been "running an arms race with ourselves. The idea that the Russians might get weapon X has had the same impact on us as if they actually had this weapon." [20] We would like to be able to choose the weapons we want to be decisive in world diplomacy, weapons in which we feel sure of ultimate superiority. We may convince ourselves of the diplomatic decisiveness of victory in racing to the moon, and we may in fact win the race—but find that we have not won very much else.

We are well on our way to reversing the Russians' initial advantage in heavy boosters and have demonstrated clear superiority and versatility in our space capabilities. We are already, Webb declares, past the "critical midpoint in our effort to achieve space pre-eminence. . . ." [21]

We put 37,700 pounds (including the final stage of the booster) into low-earth orbit early in 1964 with the Saturn I, and we now have several other heavy boosters in our stable (including Saturn IB and Titan IIIC). In 1967 we expect Saturn V to be test-fired, a giant designed to carry thirty tons on planetary missions. Until mid-1965 all Russian exploits employed the same basic rocket booster generating a thrust of between 800,000 and 1.2 million pounds, sufficient to accomplish a number of near-earth missions with payloads up to seven tons, but not enough to land men on the moon. This Soviet vehicle was brought into operation in the late 1950's as a result of an ICBM program that contemplated extremely heavy hydrogen-bomb warheads. Only in 1965 did the Russians exhibit a larger vehicle, the Proton I, which orbited 26,900 pounds. Adding the final-stage rocket casing, Proton I appears to exceed the lift power of Saturn I by a slight margin. The timing suggests that their heavy booster program was largely a response to the U.S. challenge and was not undertaken until after the West raised the ante after Sputnik I.

There is no doubt of our equaling and perhaps surpassing Soviet capabilities during the late 1960's, but Russian science and technology is competent to the challenge. If the Soviet Union makes the political decision to do so it can prevent us from running away with the show. Whether it has made this decision or not we don't know; there has been a deliberate ambiguity in Russian declarations. "What we do see very clearly," Webb reported to Congress in 1965, "is that they are getting in a position and it appears intend to maintain that position where they understand the environment, their engineers know how to apply the knowledge that comes from the understanding of the environment, and they will have the ability to select options most useful to them. . . ." Although there is no evidence that the Russians are building a Saturn V, they may learn to strap together available boosters in order to lift anything they want, he judged.[22] The Proton I suggests that if the U.S. has one, the Russians will have one too, sooner or later.

Soviet declarations altogether express ambivalence and uncertainty—as though they sought to parry with U.S. intentions, testing our willingness to join with them in ending the costly and meaningless competition, just as we have managed to do in nuclear weapons and in other areas. Their careful diplomatic feelers have been consistently met by a firm restatement of U.S. determination to press on to the moon regardless of what they do.

A great debate on Soviet intentions opened in July 1963 when Sir Bernard Lovell, director of Britain's Jodrell Bank Observatory, returned from a three-week tour of Soviet space facilities. He told the American press that "there is a great deal of discussion in the Soviet Academy as to whether it will ever be worthwhile getting a man on the moon. . . . I think at the moment, the Americans are racing themselves." [23] The Lovell statement came at a critical time in the Apollo Project, because many kinds of opposition were coming to a head in the scientific community and in Congress.

Sir Bernard quoted the president of the Soviet Academy as calling for internationalization of moon exploration.[24] Lovell expanded his report in an interview with *U.S. News:* the Soviet Academy, he said, saw "certain insuperable difficulties—first, they did not at the moment see how it was possible to protect any lunar voyages from the lethal effects of solar radiation; secondly, they did not think that it was economically feasible with present techniques to land sufficient material on the moon to give the necessary protection against solar radiation and at the same time to enable a scientific program to be carried out and also to give the chap a reasonable chance of getting back to earth." Most important, he continued, "they have decided that with a soft landing of instruments they can extract nearly all the scientific information they want long before there would be a chance of getting a man there to do it." [25] President Kennedy commented: "There is every evidence that they are carrying on a major campaign and diverting greatly needed resources to their space effort. With that in mind, I think we should continue." [26] Shortly thereafter, however, he seconded the Soviet proposal, formally inviting Khrushchev to join the U.S. in space cooperation.

Kennedy's response created confusion and consternation in Congress and placed Space Administrator Webb in the position of discounting Soviet overtures while seeming to support those of his boss. In the softening of congressional opinion, the scientific community spoke out against the Apollo Project as never before, and a few independent legislators challenged the tendency to identify the manned lunar landing program with motherhood and the flag. Senator J. William Fulbright attacked the program's timing and objectives on two grounds: "First, it is not at all clear that the Russians are trying to beat us to the moon; second—and more important—it is even less clear that it would be an irretrievable disaster if they did. . . . Most emphatically, it would not change the course of history . . . the competition between freedom and dictatorship is a great deal more than a competition in technological stunts." [27]

Khrushchev entered the debate by declaring categorically that the Soviet Union would not race the United States to the moon: "At the present time we do not plan flights of cosmonauts to the moon. I have read a report that the Americans wish to land a man on the moon by 1970. Well, let's wish them success. And we will see how they will fly there and how they will land there . . . and most important how they will get up and come back. We will take their experience into account." [28] Dr. Edward C. Welch, executive secretary of the National Space Council, responded with what soon became the official NASA line: "Khrushchev's announcement was a very wise pronouncement. Whether they plan to go to the moon or not they have seen some slackening of the moon program here and have made a strategic move to capitalize on this situation." [29] President Kennedy, three weeks before his death, told the press that "he would not make any bets on the Soviet intentions." [30]

NASA faced an annoying dilemma. Webb had repeatedly told Congress that there *was* a moon race and that the Soviets were doing their best—but we were going to win because the Soviets lagged in developing anything like our huge boosters. This argument was intended to support the contention that only continuous massive expenditures would assure our ability to move ahead of the Soviets. But now it appeared that the Soviets were not in the race. With

Congress in a mood to save money and Khrushchev wishing us well, Congress was altogether prepared to cut the NASA budget, which they did by half a billion dollars.

The successful launching in mid-1965 of Proton I led to NASA speculation that the large Soviet "space station" might be a forerunner of a springboard from which lunar flights would be launched. Mstislav V. Keldysh, president of the Soviet Academy, sought to still such notions, declaring it too early to fix a realistic date for any manned programs beyond earth orbit.[31]

NASA weathered the 1963 crisis, and the official line returned to an emphasis on the danger of surprise, dwelling on each new sign of Soviet activity as a threat. Recurrently, officials echo the view expressed by von Braun in 1964: ". . . I am convinced that they are about to come up with something quite formidable." [32]

It would be a mistake to base our estimate of Soviet intentions on pronouncements and claims by their high officials, space scientists, and cosmonauts; their statements comprise a variety of motives ranging from firm program commitments to nebulous high hopes. Soviet spokesmen have openly proclaimed their long-run goals for many years but always shy away from exact target dates—such expressions as "in the not distant future" and "relatively soon" are used instead. Interest in a manned moon flight and other space exploration has been frequently asserted, but with the disclaimer that there are no concrete plans for cosmonaut flights. Khrushchev in October 1963: "Soviet scientists are working on this problem. It is being studied as a scientific problem and the necessary research is being done."

Shortly after Sir Bernard's sensation, Khrushchev told a group of American businessmen that the Soviets were still moonward bound: "We never said we are giving up our lunar project. You are the ones who said that." In June 1964 *Pravda* quoted Khrushchev: "We plan to fly to the moon in the foreseeable future. Not to live there, of course, but to see what is going on there." Professor A. A. Blagonravov, chief Soviet delegate to the United Nations Committee on Peaceful Uses of Outer Space, clarified the Premier's remarks, telling a news conference in Geneva that the Soviet Union had no plan for landing a man on the moon "within the present decade." [33] After the change of leadership, new party secretary Leonid Brezhnev declared that his country did not "regard our space research as an end in itself, as some kind of race. In the great and serious cause of the exploration and development of outer space, the spirit of frantic gamblers is alien to us." [34]

A sounder method of evaluating Soviet intentions is to look at the actuality of their space and science programs. They have demonstrated a broad and diversified interest in both manned and instrumented flights—comparable in most respects to the U.S. program. They appear to have made a decision to carry on work in all areas of scientific, military, and technological interest. All in all, according to Charles S. Sheldon II, member of the National Aeronautics and Space Council, the Soviet Union "operates its program in a conservative manner" with a high degree of skill and originality.[35] In spite of the enormous strides taken in the American program during 1965, the Russians continue to show more advanced recovery techniques. Sheldon comments: "Overall, the Soviet program seems to have almost as much variety as our own. . . . We have seen enough . . . to be reasonably certain that they have the general scientific and engineering capability of mounting advanced manned lunar and planetary programs." [36]

The abortive 1963 attempt to cool the space race may or may not have been sincere on the parts of both Kennedy and Khrushchev. The former's untimely death prevented any constructive follow-up to determine real Soviet intentions. President Johnson has been a firm advocate of the necessity to proceed with the Apollo program regardless of Soviet plans. And so cooperation with the Soviets in space has been limited to the exchange of certain kinds of scientific data and meteorological photographs.

The Apollo program endured and maintained its thrust despite the necessity of cutting out some non-Apollo related projects in order to hold the annual budget at the congressionally ordained limit of $5.2 billion. Soviet ambiguity may reflect not only uncertainty of our intentions but also their adoption of the U.S. strategy in reverse—let Americans exhaust their energies and solve the expensive scientific problems, while the Russians move with deliberation along the developmental pathway cleared by their American colleagues. Rather than falling for the U.S. ploy of raising the ante of the space race, the Russians may be conserving their resources for their own needs or for a diplomatic initiative with a more concrete return. On the other hand, we may be forcing them to imitate us as we imitate them (for "insurance against surprise") in an escalating, sublimated arms race which is self-perpetuating and inescapable.

"Let Us Go to the Moon Together"

The drag race to the moon has not yet lost its nationalistic fervor, but the future seems to point in that direction. Since the Kennedy-Khrushchev initiative a number of baby steps have been taken to internationalize space science in the areas of medicine, weather, survey of the earth's magnetic field, and experiments with passive communications satellites. In the United Nations Committee on the Peaceful Uses of Outer Space, both nations have concurred in allocating frequencies for space communications, undertaking preliminary studies of a world-wide civil navigation satellite system, registering orbiting satellites, banning nuclear weapons in orbit, and scheduling a major international conference to be held in 1967.

There has been a selective and cautious reduction of secrecy on the part of both, but both have enforced secrecy for all activities faintly pertinent to military work and "counter-intelligence." Both nations have been accurate in reporting successful flights to the United Nations, with the date of launch and orbital calculations. (The U.S. names the launch vehicle and reports the fact of failure; but we do not report military payloads nor state their purpose. The Russians do not name their launch vehicles, and their descriptions of spacecraft have become increasingly sketchy.) In areas of collaboration, American scientists have not faulted their Russian colleagues in the completeness and accuracy of the technical data which, after a slow start, began to flow in 1964. This has been especially useful to both nations in the area of space biology and medicine in which the U.S. Space Agency and the Soviet Academy of Sciences recently agreed to joint compilation and publication of findings.

The initiative toward internationalizing space has been interpreted by some as a betrayal of the national commitment to establish American pre-eminence in space and to plant the American flag on the moon before the end of this decade. For example, Congressman Thomas M. Pelly charged that President Kennedy had completely reversed U.S. space policy. In this he was joined by the majority of both houses, which from 1963 onward began to write into NASA appropriation bills a ban on joint Soviet-U.S. space cooperation. Space Administrator Webb and Air Force spacemen were also lacking in enthusiasm for the administration's position which they viewed as a plot to cut back their budgets. As the President's spokesman to Congress, Webb was in a ticklish position, asking Congress to remove the ban on the grounds that the President was not entirely serious in his proposal. In 1963 he told the House that there had been a "misunderstanding" about the President's position and that "no one has in mind a sudden decision leading to a joint exploration." [37] A softened version of the ban has become standard in the Independent Offices Appropriation Bill: "No part of any appropriation made available to the National Aeronautics and Space Administration by this act shall be used for expenses of participating in a manned lunar landing to be carried out jointly by the United States and any other country without the consent of Congress." [38]

Apart from peripheral areas of collaboration, the frequent exhortations for cooperation in space by the leaders of both countries have been largely ceremonial, accompanied by recriminations in which each blames the lack of real progress upon the other.

The science-technology race has been internationalized in practically all areas where it has been clearly demonstrated that no strategic advantages were to be gained by going it alone. This is true of Antarctic exploration, high-energy physics, peaceful uses of atomic energy, research into controlled thermonuclear reactions, and so on. Although space is tending in the same direction, its strategic nullity is far from being established. First, all space exploration is dependent upon the same technical capabilities that underlie strategic missilery. As stated by Donald Hornig, "Central problems in cooperation haven't stemmed from our own will, but from the fact that space activities are also related to rockets, which in turn are related to military problems." [39] Second, space is closely tied to such passive military functions as targeting, reconnaissance, and early warning. Third, the technology of manned space flight is still in its infancy. Neither side can make a confident judgment about the indispensable strategic utility of manned space systems. The reciprocal anxiety of the unknown forces both sides to continue high-level investment in space experiments aimed primarily at denying the other any strategic breakthrough. This is the real motivation of both nations and serves to justify a military requirement for research and development of a manned orbiting laboratory. It does not, however, justify the moon project, which mingles many diverse motives unrelated to strategic requirements.

A space truce and substantial international cooperation will not be possible until it has been clearly demonstrated that space confers no decisive strategic advantages and/or such military uses as may be found cannot be denied to either side. Only this evidence will spark a common interest in reducing an expensive and essentially useless competition. That time is bound to come for space as it has for missiles and warheads. The softening of U.S. determination in the moon race, the administration's budgetary lid on large-scale funding for extended moon exploration and interplanetary travel beyond Apollo, and the beginnings of space cooperation all point in that direction.

The administration has shown a growing skepticism toward claims of strategic decisiveness. Dr. Hornig recently told the Senate Space Committee that "at this stage the moon looks to me more like Antarctica or the Sahara Desert than it does a gold mine. So that until we know more . . . I would personally think that a maximum of international cooperation would be wise and simple, and in that respect . . . our activities in Antarctica might be a model." [40] Harold Brown, now Secretary of the Air Force, interprets Russian motivation as essentially the same as ours: "It is the same mixture that we ourselves have . . . I would put military goals reasonably high among them. . . . I don't myself conclude that they see an overriding military capability, strategic capability coming out of space." [41]

In late 1965 a White House conference met to consider new international programs which the U.S. might implement in fulfilling its commitment for the International Cooperation Year (ICY). A variety of proposals emerged which, if acted upon by government, will accelerate the trend toward the internationalization of space. Among these was a proposal to create an international launch site on the equator which would enable all nations to take advantage of the earth's rotation for peaceful space purposes; an international multipurpose navigation satellite system; a world-wide meteorological system integrating satellites with ground stations, ocean buoys, and balloons; cooperative study of the earth's atmosphere and of the solar system; and a broadening of satellite communications.

Sir Bernard Lovell has strongly advocated the internationalization of all space experiments: "The human race . . . faces a critical situation in the next few decades, because there is at the moment little evidence of the moral or legal controls which must be enforced if man's continued life on earth is not to be jeopardized by the accidental or intentional results of space research." [42]

CHAPTER III

THE POLITICS OF SPACE

As space has commanded a growing share of American resources, so has it also been infused with the immemorial dynamics of politics. The pulling and tugging which somehow resolves (for good or ill) the substantive issues of public policy must be understood in order to pierce the fog of clichés. The politics of space provides tales that run from farce to tragedy, from ugly, calculated press leaks and forced resignations to quiet acts of courage by obscure or eminent men—and the inevitable travesties of both. These "interface problems" are meaningful parts of larger social questions; in an area of such great uncertainty and speculation, there is no substitute for the political process. Here the maneuvers and alliances of pressure groups play a determining role in the trial-and-error probing of the national future.

There is historical irony in the fact that the Air Force won its independence from the Army on the eve of the decline of air power and the rise of missile power. During the period 1954-1958 all three services developed missiles on a crash basis, each coveting a major role in the national space commitment and bitterly resisting other claims. By 1958 the Air Force emerged as dominant and anxious to extend its empire. By a variety of means, it succeeded in penetrating NASA during its infancy. But faced in 1961 by the unknown quantity of a new administration, the Air Force launched a campaign to maintain its role. The Apollo decision and Defense Secretary McNamara's efforts to subject the military services to administrative control led to a period of fierce infighting which climaxed in mid-1963.

With the Test-Ban Treaty and new overtures to cool the space race, the Air Force rallied its allies for a major assault on what it termed "unilateral disarmament," calling for massive investment in new strategic systems of all kinds. But the presidential candidate who adopted their slogans was resoundingly defeated. As the Apollo Project slipped into reduced urgency, McNamara held out support for "the study phase" of a manned orbiting laboratory (MOL); in the same breath he killed the Dynasoar project. During all of 1964 and until mid-1965, the Defense Department conducted low-intensity studies of MOL, seeking by a series of compacts between McNamara and Webb to disentangle Air Force influence in the Space Agency.

In mid-1965 President Johnson ordered immediate steps toward creating a thirty-day, two-man experimental MOL for launching within three years. This move symptomized both a de-emphasis of Apollo and a tighter grip on military ambitions by civilian leadership. By hedging its bets against the possibility of technical delay or failure in Apollo, the nation implied that the original decision to leapfrog the Russians might have been a mistake. What was needed instead was a painstaking steppingstone approach to resolve the unknowns of space in order to eliminate hidden military incentives and stabilize this new arena of the arms race.

Jupiter vs. Thor-Atlas

The Army Air Corps became the independent Air Force in 1948. In World War II rocketry had developed importance primarily as a form of lightweight artillery, but the desperate Nazi V-2 gamble in the final stages of the war suggested future advantages for heavy offensive rockets. Later technological advances (miniaturization of nuclear warheads and electronic means of control) perfected ground-to-air weapons, destroyed the future of manned bombers, and made inevitable the shift to strategic missiles. German rocket experts captured by the U.S. Army formed the embryo of our postwar program. The Air Force organized Project Rand (later to develop into the RAND Corporation) in order to develop a competitive claim to these new weapons developments.[1] Faced by modest R&D budgets, the military concerted efforts to persuade the War Department to support a major commitment, but they were blocked by the return to "civilian normalcy." In 1947 the Air Corps suspended active interest in rocketry, placing primary emphasis on strategic bombers and air-breathing ram or pulse-jet guided missiles. The Army and Navy continued exploratory in-house work (i.e., in their own laboratories) in rocket artillery and V-2 technology.

The first Soviet atomic detonation in 1949 forced a high-level reactivation of interest in space,[2] but the program foundered against the problem of size and weight of atomic warheads, the cost of very heavy booster development, and the lack of signs of a similar Soviet undertaking. Three circumstances combined during 1953 to force a change: the feasibility of lightweight thermonuclear warheads was confirmed; "we had information on what the Russians were doing, and it was pretty clear that if the program were to take beyond five to seven years a serious threat to the United States would exist";[3] and an intense competition was unleashed as the Air Force sought to overcome the Army's initial advantage (embodied in the improved V-2, the Redstone missile) in what appeared as a struggle for mission survival. The Army and Navy were already overshadowed by the young and aggressive Air Force which had been assigned the primary role in delivering atomic bombs, the nation's chief strategic reliance. As early as 1953 Defense Secretary Charles E. Wilson appointed Trevor Gardner to head a committee "to eliminate interservice competition in development of guided missiles. . . ."[4] The space race had already become as much an interservice as an international competition.

The Missile Evaluation Committee, headed by scientist John von Neumann, became in 1954 the forum of conflict. Two separate long-range rockets were underway: the Army's Redstone-Jupiter held the lead, while the Air Force had reoriented the Atlas to reduce weight-lifting requirements. The administration favored Air Force claims to missile jurisdiction, leading the Army and Navy to join forces behind a satellite launching program—Project Orbiter—which could demonstrate that a shorter route to a missile capability could be based on the existing Redstone.[5] The issue blew into the open in the fall of 1955, when President Eisenhower assigned a "DX priority" (the highest) to ICBM programs, "both land and sea-basing to be considered," leaving ultimate jurisdiction open and permitting parallel programs.[6]

The Army-Navy defensive alliance against the Air Force, based on developing an IRBM for both ground and ship launching, broke down as the Navy recognized the technical infeasibility of using a liquid-fuel missile (such as the Jupiter) at sea. By the end of 1956 the Navy was given permission to investigate solid fuels.

Proposals for U.S. participation in the planned 1957 International Geophysical Year (IGY) provided a new focus of battle. (This was neither the first nor the last instance in which the military invoked the name of "science" as window dressing for what was basically a bureaucratic power struggle.) Three separate proposals for a satellite launcher now stirred government: the Army Redstone, the Air Force Atlas, and a new Navy proposal for a rocket to be developed from the shorter-range Viking. Another committee was established to review the proposals. Unable or unwilling to resolve the deadlock between the Army and the Air Force, the committee compromised on the Navy proposal. This was to lead to the dismal failures of the Vanguard in 1958-1959, as the U.S. sought in vain to counter the sensation of Sputnik I.[7] According to some, the Air Force deliberately attempted to sabotage the Navy's Vanguard mission by hiring away the Viking engineering team for a project of its own. In the words of John P. Hagen, Vanguard director: "After our letter of intent had been given to Martin, the unfortunate fact came to light that the original Viking engineering team in the Martin Company had been broken up. Unknown to the Navy, Martin had received a prime contract from the Air Force to develop the second generation ICBM Titan. Some of the leading Viking engineers were placed on the much larger Titan program." [8]

Following the IGY decision, the Army-Air Force space race intensified: each withheld information from the other and treated the other's representatives as spies; each circulated and/or leaked memoranda and confidential information on the virtues of its own systems and the faults of its rival among members of Congress and to the press; each eagerly publicized examples of the other's mismanagement, waste, and stupidity. The Army continued to cast a covetous eye in the ICBM direction, working up from the intermediate range Jupiter. The Air Force, which had started with the intercontinental range Atlas, worked downward in range with the Thor, guaranteeing a continued clash of interest at both IRBM and ICBM ranges.

The Thor succeeded in challenging Army IRBM responsibility, and a decision was made to fund both Thor and Jupiter on the cooperative basis of "full exchange of technical information" between the services. The cooperation that ensued resembled that of a married couple preparing a divorce action. The contestants sabotaged each other's programs, barred each other from contractors' plants, deliberately withheld from each other critical technical information (such as that concerning fuel-pump and nose-cone design). In later investigations, Army rocket boss General John D. Medaris cited a Convair report (on radiant heating from engine exhaust gases) which the Army considered important for the Jupiter, but which the Air Force withheld as an internal document. The Army had to wait sixty to seventy-five days for reports which were not even contested by the Air Force. "Sometimes the release of a report was delayed as much as ten months," and after a personal request to Air Force General Bernard A. Schriever, "after it was a dead issue, we would get it." [9] The services, especially the Air Force, used their industrial contractors to promote their own causes in Congress and in public forums. In addition, Air Force contractors successfully undertook to pirate technical manpower from the more extensive in-house establishments of the Army and Navy.

In 1956 Trevor Gardner resigned as a protest against Defense Department unwillingness to place all its bets upon the Air Force. The Army went through the charade of court-martialing Colonel John C. Nickerson for releasing secret information which revealed Air Force and contractor irregularities. An alliance was now in the process of forming between the Air Force and the Democratic Congress, both of which began to parrot slogans of "missile gap," charging the administration with encouraging "wasteful rivalry" by not giving the Air Force a clear mandate and adequate funds. All attempts to clarify the situation ended in a worse tangle than before. When it was decided that the Army would continue to produce Jupiter for the Air Force, the latter sabotaged the rocket in order to destroy its technical credibility.[10] Army R&D was restricted to missiles having a range up to two hundred miles, but this was ignored and the Army continued IRBM development. The Air Force was granted operational control of *all* offensive missiles, the Army limited to defensive and tactical systems, but no line was drawn as to which was which.

This unseemly disarray continued to agitate the bureaucratic jungle, forcing the resignation of a frustrated but serene Secretary Wilson, who was succeeded by a frustrated and disturbed Neil H. McElroy, who presided over the mess at the dawn of Sputnik I. Finally despairing of the administration's ability to resolve the problem, he threw it into the lap of Congress and resigned, leaving Thomas S. Gates, Jr., as the post-Sputnik caretaker of what became a *de facto* Air Force victory over its sister services.

Air Force Ascendancy, 1958-1961

The Soviet Sputnik and stalemated U.S. service rivalry created the opportunity for intervention by President Eisenhower and his scientific advisers, supported by those elements in Congress and the nation opposed to Air Force doctrines of massive retaliation, infinite deterrence, and escalation of the arms race.

This intervention led to the creation of the civilian Space Agency as a means of checking Air Force ambitions and avoiding the extension of the arms race into space. Unfortunately, domestic and international circumstances defeated the executive goals. Eisenhower in his last two years retreated behind slogans of fiscal responsibility; U.S. international prestige declined in the face of new Soviet initiatives which we lacked means to contain; and a "missile gap" panic was inspired by leading Democrats, becoming a symbol of the broad spectrum of dissatisfaction—latent since the end of World War II—that was rising throughout the land to demand change. In the midst of this surge, the Air Force maneuvered with all the dexterity acquired in its postwar political combat to capture NASA and subvert it to its own ends.

The swift orbiting of the Soviet satellite generated even swifter political ellipses in Washington. Secretary of Defense McElroy happened to be at the Army ballistic missile agency when Sputnik I was announced. The arsenal team took advantage of this situation: promising a satellite launching in sixty to ninety days, General Medaris and von Braun won McElroy's approval to fire a Jupiter bearing the Explorer satellite, a proposal which had been consistently rebuffed for over a year. In January 1958, some eighty-four days after the go-ahead, the Army team successfully launched the first U.S. satellite, raising Army hopes for another more promising round in the struggle with the Air Force. Capitalizing on the acclaim for Explorer, the Army proposed a multi-stage rocket with a large, clustered first stage as the "national integrated missile and space vehicle" for future satellite programs.[11] The competition now became a fight for ascendancy in *non-missile* space ventures, and the Army hoped to recoup its losses in this area. The Air Force immediately took up the cudgels to acquire this domain for itself, arguing that missiles and space boosters were inseparable.

The old (since 1915) and respected National Advisory Committee on Aeronautics (NACA) adopted a resolution seeking to pull together this divisive eruption. It recommended that the national space program be made a joint effort of the Department of Defense, NACA, the National Academy of Sciences, and the National Science Foundation, with NACA responsible for research and scientific operations while the Defense Department concentrated on military aspects. The congressional Joint Committee on Atomic Energy, the most clearly science-related legislative group, called for "the peaceful conquest of space" under the jurisdiction of the Atomic Energy Commission. Unwilling to permit the already powerful joint committee to usurp this new dimension, majority leader Lyndon B. Johnson created a Special Committee on Space and Aeronautics in the Senate, and another was formed in the House. Johnson made plain the congressional intention to assume authority for the key decisions concerning space and missiles which now confronted the nation.

In the course of extensive hearings, debate, and cloakroom caucuses, it became plain that some kind of new institution with great authority was required for a massive new effort to overtake the Russians. Army and Navy spokesmen supported the Air Force position that the new agency be within the Department of Defense, hoping to use it as a means to undermine the growing Air Force space monopoly and win an enhanced role for themselves. Strong opposition to military control of space developed in some segments of the Congress and among scientists and other groups which contended that space should be held in civilian control, as had been the atom under the AEC. The American Rocket Society, the National Society of Professional Engineers, the IGY Earth Satellite Panel, the Federation of American Scientists, the American Association for the Advancement of Science —all put their weight behind this view. The anti-military sentiment came to a focus in the newly created President's Science Advisory Committee (under the leadership of James R. Killian, Jr.), which proposed that NACA be made the nucleus of the new agency. The view was adopted by the President and recommended to Congress in April 1958.

The services vigorously opposed the proposal but, faced by the growing consensus in Congress, soon devoted their energies to limiting the new agency to laboratory research and scientific testing, thus leaving actual space operations to the military. Failing in this, they concentrated on blurring the division between peaceful and military uses of space so that they might later resolve such ambiguities in favor of their own independent programs.[12] In this gambit they were successful; as finally enacted, the National Aeronautics and Space Act of 1958 left the issue obscure.

The Advanced Research Projects Agency (ARPA) was established at the same time by the Department of Defense to conduct research and development independent of the separate branches in order to end interservice bickering and impose topside control. The office of "Director of Defense Research and Engineering" was created at the assistant secretary level, endowed with broad powers of coordination over all branches and direct authority over ARPA.

From this point onward, a mobilized cadre of contractors swept over Congress, intent on converting both NASA and ARPA into Air Force subsidiaries. They shared Air Force fears of Eisenhower "fiscal responsibility" and were never enamored with Army-Navy strict contracting controls. Their success was notable: Dr. T. Keith Glennan, president of Case Institute of Technology, was named administrator of NASA at the urging of the missile industry. His connection with the Air Force can be traced through the chairman of the Case Institute Trustees, Frederick C. Crawford, who was also chairman of the board of Thompson-Ramo-Wooldridge Corporation, which was serving as systems-engineering and technical director of all Air Force missile work, and had been created at the instigation of the Air Force and on the basis of anticipated Air Force contracts, a story to be detailed later.

The Air Force placed its own aides throughout the NASA hierarchy. For example, Richard E. Horner, Air Force Assistant Secretary for R&D, became associate administrator of NASA.[13] Meanwhile, the Army and Navy saw their own facilities transferred to Air Force management under nominal NASA authority. Project Vanguard was transferred in toto to NASA, as was the important and long-term Army contractor, the Jet Propulsion Laboratory of the California Institute of Technology. NASA and the Air Force pushed aggressively in late 1958 for the richest Army plum, the Redstone Arsenal. However, bitter Army opposition forced a compromise under which it was agreed that Redstone would remain in the Army but act as a NASA contractor. This proved to be merely temporary, since in the waning months of the Eisenhower administration, NASA swallowed the arsenal, renamed it the George C. Marshall Space Flight Center, and forced the resignation of Army Missile Chief Medaris.

According to General Medaris, the scientific and engineering group working at Redstone was forced to choose between a transfer to the

Air Force and a transfer to NASA. It was deterred from the former alternative by "some very dangerous possibilities in that course of action. . . . We recognized the dependence of the Air Force on the aircraft industry, and we have seen those interests continually attack the concept of government-operated in-house major activities in the missile and space field. With, I believe, some justice we were afraid that if the Air Force did take the von Braun organization it would be under continual pressure to get it out of the hardware business and restrict it to engineering." [14] As General Medaris resigned to write his book of anguished protest, the Air Force coup was completed with the assignment of an Air Force officer (Major General B. R. Ostrander) to superintend the installation.[15] It will be seen in another chapter that the fears of the von Braun group were warranted, as NASA set out to limit—if not eliminate—the in-house capacity of both the Jet Propulsion Laboratory and Redstone.

The Air Force not only continued to manage its own programs but also implanted itself in managing new programs for NASA. This was done under the so-called single-manager plan: NASA contracts flowed to companies deeply engaged in Air Force contracts, and the Air Force assumed plant-resident control of all such programs. Further, the Air Force missile manager, Thompson-Ramo-Wooldridge's Space Technology Lab, assumed an important part of NASA systems engineering, as did its non-profit successor, the Aerospace Corporation. In the words of Ivan A. Getting, Aerospace president: "We had responsibility through the Air Force for the Mercury launch. Our responsibility included the integration of the booster, modifications of the booster, the safety of the pilot, the astronaut, the computation of trajectories, and actually approving the flight on the morning of the launch event. So you can see we did work for NASA in this case." [16]

ARPA also tended to become an Air Force rather than a Defense Department instrument. According to General Schriever, 80 percent of ARPA's dollars were being spent through Air Force contracts within a year of its founding.[17] But Schriever remained unhappy with ARPA's existence and asked that it be abolished altogether in order to "bring the operator and developer together under the same tent." [18] In late 1959 the Defense Department announced the transfer of space projects from ARPA back to the separate departments as "a step in the direction of greater clarity." All future development, production, and launching of military space vehicles would be assigned to the Air Force (with payloads developed by the several services according to their requirements).

The ascendancy of the Air Force was evident. Flushed with success, the Air Force sought the appointment of Charles L. Critchfield to the post of ARPA director. A key official of the Convair Division of General Dynamics (a heavy Air Force contractor), he agreed to accept the position only if he could retain his much higher Convair salary. The congressional outcry was too great, and Critchfield withdrew his acceptance. ARPA might have become a useful instrument of departmental control if the political forces had been available to make it such. Without effective presidential backing, such as McNamara was to have under two Presidents, the agency in no sense realized its potential for controlling research and development. Today ARPA still exists but is largely an empty and vestigial shell.

The Air Force coups of 1958-1959 led the Navy to solicit Army support for a plan to safeguard their remaining autonomy. On the Joint Chiefs of Staff they had a majority voice; therefore they proposed that a unified command for military space and missile development be organized directly under the Joint Chiefs. Although adopted, it proved ineffective. The rush of Air Force influence bypassed this institution on its way to grapple with higher centers of power in the administration and in Congress.[19]

New Broom, New Battles, 1961-1963

During the Kennedy-Johnson years the Air Force was caught in a bind which threatened to grow tighter than anything it had known and mastered during the Eisenhower period. President Kennedy and Secretary McNamara strove to re-define U.S. strategic doctrines, and the Air Force found itself increasingly blocked by the deliberate leveling of the U.S. investment in strategic systems, with an accompanying increase in conventional and tactical capabilities (largely to benefit the Army and Navy). The strategic revision provided the anvil upon which Air Force ambitions were struck relentless blows as McNamara sought to subdue the unruly services, to force the invisible politics of Pentagon corridors into the clear air of executive policy-making. In addition, the President's 1961 lunar decision charged NASA with a mission in some ways antagonistic to Air Force purposes. The tenfold growth of NASA appropriations and contracting created an independent constituency in Congress and industry, fostering in the Space Agency some impatience with Air Force meddling. In the face of McNamara's hard-nosed management team, the Air Force badly needed NASA as a means of sidestepping Defense Department authority. The device was to involve NASA in Air Force projects which had been denied formal "requirement" status by McNamara—such as large solid-fueled boosters.

In the last months of President Kennedy's life and increasingly under President Johnson, McNamara struggled to deal with this new and slippery form of insubordination. He set out to identify Air Force projects hidden in the vast maze of NASA contracts and to bring them under control by two methods: formal agreements with Webb by which each would have the right to approve new projects of military potential undertaken by either agency (this meant a Defense Department veto over NASA collaboration with the Air Force), and, if the project had merit, assigning it back to the Air Force, thus making it possible to subject it to direct management and budgeting controls.

The embattled secretary won some temporary measure of success, but the battles were never done. The Air Force demonstrated great political finesse, responding to the hardest McNamara punches like a feather pillow, puffing out in other places but fundamentally unhurt. The relation between NASA and the Air Force during the Kennedy-Johnson years amounted to a continuation of collaboration while covert wars were fought in limited areas of conflicting interest. Whatever incursions the Air Force made upon NASA's domain could not be too flagrantly in conflict with executive or congressional directives, yet a process of constant probing was necessary to determine the limits of mischief.

The new chapter opened a few days after the 1960 elections, when the Air Force began to organize its forces in order to withstand possible sweeps by the New Broom. The Office of the Secretary of the Air Force sent a confidential memorandum to all Air Force commanders and contractors pointing out the probability of a battle with NASA for the dominant role in space after Inauguration Day.[20] Among its first acts, the new administration ordered a review of the space program. The Air Force promoted rumors on Capitol Hill that the President was contemplating extreme measures which would subvert and destroy U.S. security. The Air Force viewed askance the agreement signed by Webb and McNamara in February 1961, under which each agreed not to undertake new launch-vehicle programs without the other's written concurrence. President Kennedy was forced to respond: "It is not now nor has it ever been my intention to subordinate the activities of NASA to those of the Department of Defense. . . . There are legitimate missions in space for which the military services should assume responsibility. . . ." [21] Hearings were conducted by the House Committee on Science and Astronautics into charges that the Defense Department was contemplating a shotgun divorce of NASA from the Air Force. The Air Force disavowed imperialist designs and pledged full cooperation with NASA, and the Defense Department gave assurances regarding its intentions. The committee was "happy to have these assurances from the proper officials in the Department of Defense. However, the Committee has a large bulk of printed material which derogates NASA . . . this would seem to throw the responsibility for slurring remarks about the importance or efficacy of NASA on non-governmental sources; but whatever the source, the Committee regrets such attacks as unwise." [22]

Decisions soon began emerging from the White House which tended to reassure both agencies and to postpone some of the harder issues. In March 1961 McNamara announced the assignment of all military space development programs and projects to the Air Force "except under unusual circumstances," although all departments might continue to conduct preliminary research and might be assigned operational authority of systems most closely related to their missions. The Air Force was reassured that it could continue to work on solid propellants for large space boosters while NASA would work on large liquid-fueled vehicles. Decisions regarding their future use were to be reserved until more was known about the basic technology.[23]

Close upon the heels of this period of careful deliberation, less than six months after inauguration, closely linked to the twin humiliations of the Bay of Pigs and the Gagarin space flight, his first hundred days empty of legislative accomplishment and facing an imminent economic recession, President Kennedy on May 25, 1961, announced his decision that the nation send men to the moon in this decade. This decision began a monumental budget climb for NASA; the Air Force sullenly explored the depths of despair. Within weeks, however, it summoned its shaken allies, their loyalty now divided by prospects of huge NASA contracts, and mounted a voracious counterassault. General Schriever informed the American Rocket Society: "I have been, am being, and, if the situation is not changed, will continue to be inhibited if our space efforts continue to be carried out under an unnecessary self-imposed national restriction; namely, the artificial division between space for peaceful purposes and space for military purposes." [24] The battle raged in Congress where both Republican and Democratic Air Force advocates decried the Apollo priority and charged the administration with ignoring the "purely aggressive and military" space capabilities demonstrated by the Russians. The real nature of the debate over military versus civilian space programs was diagnosed by Congressman George P. Miller, chairman of the House Committee on Science and Astronautics:

> It seems fairly well established that the real cause of all this squabble . . . stems from an in-house difference of opinion within the military establishment. The problem is not that our civil space program is retarding the military. On the contrary, it is enhancing it and will continue to do so in the future. The problem is that the military space enthusiasts have not been able to obtain all the green lights they want from their bosses.

Miller characterized the position of the Air Force as "not asking that we do twice as much—but that we do everything twice. I do not think that the economy will take that, but even if it would—it just does not make sense." [25]

The battle raged among scientists, too, most of whom denied the military potentials of space but now found themselves allied with the Air Force generals in attacking the pace and form of Apollo. They called for a cautious, long-range, scientific approach to space, based for the foreseeable future upon unmanned, instrumented probes of both near and interplanetary reaches of space. The battle raged within the administration itself, where Dr. Jerome Wiesner, Presidential Science Adviser, argued the risks of a too-specific commitment, foreseeing the difficulties and the multitude of dangerous gaps in the knowledge and technology required for the journey. The battle apparently raged also in the mind of President Kennedy himself, who a few months before his assassination sought a way out by inviting the Russians to join in calling off the space race and "going to the moon together."

In June 1963 the nuclear Test-Ban Treaty was signed, giving new impetus to the leveling and stabilization of the strategic arms race. McNamara's vigorous managerial controls now came to mean not only the integration of the services and the elimination of waste but, more important, curtailment of new R&D investments in strategic weaponry. Physicist Eugene G. Fubini, Defense Deputy Director (Research and Engineering), has described the change: the assumption that "we are still fighting a war of strategy" is mistaken and American security "can no longer be gauged by the continuous development of new strategic weapons systems." Progress in weapons development, he said, must now come in other areas.[26] Under these conditions, the downgrading of the Air Force appeared inevitable and with it the loss of control of billions of contract dollars. The best hope for the Air Force future lay in the chance that new "decisive" strategic systems might emerge from space exploration to force a renewal of the arms race. The flurry of "blue-sky" Air Force weapon proposals which now hammered the public and Congress can be understood within this context of events.

The nation was obliged to maneuver its weapons policy in a very delicate and limited range between reopening the arms race and losing potential military advantages in space. Seizing upon this legitimate area of uncertainty, the Air Force made increasingly extravagant claims for the decisiveness of space supremacy. With these claims came glowing descriptions of highly hypothetical new weapons systems and charges that, in the face of demonstrated Soviet superiority, the government was pursuing a calculated course of unilateral disarmament and surrender.

McNamara Makes Headway

The critical year was 1963. The vicissitudes of politics began to reflect changing conditions, and important decisions and realignments began to emerge from the tangle of political maneuvers and compromises. The key symbol of the change was the formal cancellation in December of the Air Force Dynasoar program. This courageous step, and the fact that it was done without major reprisals by Congress, revealed the changing tide of McNamara's fortunes. All the trends set in motion in 1961 began to come to a head, promising that the integrity of rational decision-making at the highest levels might still prevail. Almost three years of effort had gone into denying Air Force claims to being a fourth branch of government. At the time of his 1963 retirement as Presidential Science Adviser, Dr. Wiesner was asked about this problem. "We are getting into areas where I feel," he told the House Science Committee, "I really should be quite discreet. I really think we finally have made a good start in the direction of coordination in the space area. I think there is a good deal left to be done, but I am hopeful now that we are going in the right direction." [27]

A personnel change in NASA also indicated the direction of events: D. Brainerd Holmes, driving force of the manned space program, resigned in the face of increasing dissonance with Webb. This proved a fertile field for press speculation which Holmes did nothing to dispel. It appeared that he had indeed been forced out, and that his recurring conflicts with Webb were related to subtle shifts of emphasis in the Apollo program.[28] Most apparent was an increase in funding for the Gemini program which, in a year of congressional budget cuts, was made at the sacrifice of anticipated Apollo increases, suggesting Webb's reconciliation to possible slippage of the moon flight. This subtle reorientation coincided with the Defense Department's decision to cancel Dynasoar and to mount an extensive collaborative effort on Gemini.

The shift in NASA priorities was closely related to McNamara's fight to take command of the Department of Defense in fact as well as in theory. In January 1963 Webb and McNamara signed a joint agreement for establishing a "Gemini Program Planning Board" which would delineate the scientific and technological requirements of both agencies and monitor the actual program to insure that requirements were met.[29]

A *Washington Post* correspondent was quick to charge that this was merely a ploy to mollify congressmen "and prepare them for the cancellation of Dynasoar." [30] The Department of Defense responded with what amounted to confirmation: "When Dynasoar was begun we did not have the Gemini program. In fact until recently we did not have any mechanism for including the Defense Department . . . as an active part of management and participation in the Gemini program." [31] Webb told Congress:

> . . . there is the closest liaison between NASA and the Department of Defense. A structure has been developed which is designed to avoid duplication, to maximize the use of knowledge gained in research by both agencies. In the Gemini program NASA is working in close cooperation with the Air Force. In Apollo, although the moon is the ultimate goal, vast experience will be gained in near earth orbit before the first team of astronauts begins that exploratory journey.[32]

The slogan was no longer "the moon or bust"; rather, the moon became only "the ultimate goal" which now appeared somewhat more remote and less urgent.

It appears plausible that all of the converging forces that placed NASA and the Air Force in a state of crisis during 1963 provided McNamara and the White House opportunity to impose a higher degree of rationality upon the U.S. space program. NASA was faced with a pattern of priorities and a timetable which if rigidly adhered to would have meant spending more and more money on an increasingly unbalanced approach to space exploration which fewer and fewer people wanted. The 1963 shift meant a greater emphasis on those aspects of the overall effort that more and more people approved. Holmes appears to have taken the approach: Let's at least put everything we've got into the moon try, even if it means cutting down on other aspects of space exploration and limiting Gemini! Webb took a different tack: Yes, by all means, let's continue full tilt on Apollo, and let's appear confident of success in this decade. At least until the mounting slippages and inevitable engineering problems permit us a graceful line of retreat. But let us also hedge against the possibility of an Apollo stretch-out by continuing those programs that have strong support, especially by assisting the Defense Department in its evaluation of the military uses of space through an extended Gemini program.

As McNamara asserted his right to manage the Defense Department, a number of curious instances of the *modus operandi* of the Air Force in NASA and in relation to its sister services were laid bare, causing some Air Force embarrassment and blame-shifting. For example, Air Force interest in heavy boosters for orbiting military space stations was unable to win approval as a formal Defense Department requirement. Consequently, under the semblance of a NASA requirement, the Air Force supervised and managed a large-scale R&D program for 260-inch solid-propellant motors. The NASA lunar orbit decision in the summer of 1962 eliminated this alibi, exposing the fact that the program had been primarily motivated by the Air Force. Thereupon the Department of Defense canceled further development and the Air Force ran to Congress for aid in overriding McNamara or in forcing NASA to resume a fictional interest in advanced technology for its own sake.[33] Congress kept the program going under NASA until 1965, when the President sought unsuccessfully to squash it.[34]

The cases of Centaur and Advent were even more clear-cut. In 1958 the Defense Department approved the development of a military communications satellite to be placed in a synchronous orbit (that is, 22,500 miles from the earth). The Air Force was assigned responsibility for developing a space vehicle (Centaur) and the satellite (Advent). The vehicle was to utilize liquid hydrogen, a high-intensity fuel which theoretically would be capable (as the upper stage) of lifting a 40 percent greater payload. Liquid hydrogen is a highly unstable and volatile substance, but once developed it might fill the performance gap between the existing Atlas-Agena and the Saturn heavy rockets already under NASA development. Centaur was seen as a basic workhorse for a variety of missions. Air Force interest in the communications satellite was in part a means of providing justification for control of liquid hydrogen development, a significant element in enlarging the Air Force share of space responsibility.[35]

After formal approval of the project in 1958, the Advent satellite portion, which was proving technically deficient and overweight, was reassigned to the Army, while the Centaur was officially transferred to "NASA management and funding." But the Air Force was not prepared to relinquish either program and in fact did not do so. An Air Force project manager remained in charge of the program, "the original Air Force contracts were retained and NASA control over the program was exercised through the Air Force Centaur project manager and the Air Force contract administrator." [36] In 1960, when the von Braun group joined NASA, complete responsibility for Centaur was assigned to them. Nothing changed, however, since it was a paper transfer only; the Air Force project manager remained at the Air Force management center in Los Angeles but "reported to the von Braun group." Dr. von Braun was extremely unhappy with this and urged NASA headquarters at least to relocate the Air Force people nearer to Huntsville: "We were told . . . by NASA headquarters that the disruptive effect of pulling the technical people away from the contractual people may hurt the program more than whatever little input we could provide from Huntsville would help the program." [37] It was not until 1962 that NASA was in a position to challenge these arrangements and von Braun was able to acquire actual direction of the program. With good reason, a congressman asked von Braun: "Doctor, I am just wondering why the Air Force should have veto power over your recommendation in this instance when this group had been transferred to NASA, is under the direction of NASA, and NASA is picking up the tab. Now, you try to explain that to me for the record, will you?" [38]

The Army fared better in trying to take over Advent management. Until the end of 1961 the Air Force continued to control the contract, freezing out the Army entirely. But a McNamara team quickly changed this.[39] The Air Force had refused to provide information or to allow Army representatives to inspect the General Electric plant where the contract was being performed, yet on paper it was an Army responsibility. Army Assistant Secretary Finn J. Larsen: "I overrode the objections in December 1961. Under my orders Army representatives did go into the General Electric plant."[40] "This strikes the subcommittee as a curious restriction for a participant in such a venture to place upon the agency which has been assigned overall management responsibility for the program."[41] In early 1962, after the Army technical committee had finally studied the program, it was determined not only that the Centaur stage would not be available when expected but that the Advent satellite had been obsoleted by new developments during the lengthy period in which the Air Force had continued to spend Army funds. As a result the Army recommended terminating further development and substituting a lightweight satellite development program which could use existing boosters.

The Air Force not only endured such exposures but won a number of its objectives from such elaborate conspiracies: it attached itself to new technology and picked the contractor, who thereby obtained unique capabilities for holding future contracts, thus enhancing Air Force power for future maneuvers. The Air Force managed to keep many projects under its wing well into the McNamara reign. But one by one these were plucked from beneath the shrouds of military secrecy and reassigned to other agencies or to a more independent NASA—or left with a somewhat humbled Air Force.

The MOL Battleground

The Air Force is an indefatigable, resourceful, and agile boxer. Never accepting as final any ruling adverse to its position, it continues to dance, duck, and feint, always moving swiftly, whether in tactical retreat or forward with deft jabs and calculated roundhouses, sparring with the Defense Secretary like Cassius Clay tantalizing the Bear.

Dynasoar was out, but the Air Force think-factories were busy manufacturing all kinds of new proposals. As Senator Clinton Anderson put it: "It looks as if whatever the horse race gets into, a new horse gets on the track."[42] Most important and most energetically sought was the manned orbiting laboratory (MOL), which had been promoted as a large-scale project of great urgency for almost five years. During 1963-1965 McNamara forced the Air Force to shelve plans to move into early hardware development in order to await results of Defense Department participation in the Gemini experiments. But this did not deter the Air Force from "quiet" arrangements with McDonnell Aircraft, which continued to work on the MOL capsule throughout this time. This was done by the use of "company funds," paid for by the Air Force as overhead on other contracts or redeemed in future contracts. With Dynasoar successfully canceled and a joint Gemini program with NASA arranged, McNamara appeared to hold the means to resist the push for uncritical large-scale funding of new programs. The story of MOL through 1964-1965 was essentially the story of McNamara's successful diplomacy in containing Air Force ambitions without arousing the ire of congressional and industrial constituencies.

In the fall of 1965 MOL was finally given a green light. At an estimated total price tag of $1.5 billion, the Defense Department approved plans for five launchings of two-man Gemini capsules with forty-two-foot laboratory canisters attached, the first mission set for 1968. The MOL would maintain the astronauts in a shirt-sleeve working environment for up to thirty days in order to evaluate man's usefulness for military reconnaissance, weather-watching, missile targeting and control, and so on.

The MOL decision was far from being a conversion of McNamara to Air Force pretensions; in fact, it was quite the contrary. It represented an attempt to keep the military portion of space R&D visible and subject to evaluation and control, to hold the nation to careful scientific exploration of manned space flight as a steppingstone and proving ground for space technology. The procedure of the MOL decision took more than two years. It began with "design experiments on paper to answer these questions . . . : What can a man do that a machine can't do, then send a man up with those experiments. Having successfully answered the question and shown what he can do that equipment can't do by itself, then you design equipment that takes advantage of what the man can do and turn it into something operational." [43] Harold Brown declared that the program was not a race, "it is an experimental program and we need to take the time to insure that it is well planned before we commit ourselves to a lot of hardware." [44] The Defense Department, he said, "will require and will continue to require strong evidence of utility . . . that we have confidence of performance before we approve major investments on expensive space systems . . . proceeding to develop and build new systems only when there is reasonably clear justification." [45]

MOL emerged from pre-program definition as merely a slight augmentation of the Gemini effort, in which a laboratory module would be attached in order to provide space for experiment monitors and other equipment. McNamara made clear in mid-1965 that he would not release funds requested for fiscal year 1966 for hardware development until a full review was made of Gemini flights to determine the value of the operational results, and to assess the extent to which flight equipment proposals "can meet program objectives from efficiency, cost, and timeliness considerations." [46] The Air Force had asked for $175 million in fiscal 1966. This figure was cut to $150 million on the urging of Science Adviser Hornig, who supported McNamara's and Brown's urging to keep MOL under control.

Far from satisfying the Air Force, the decision limited MOL to purely experimental status and in effect was a brake upon ill-considered ambitions to rush into space with operational military systems of all shapes and sizes. Together with the presidential decision to eliminate from NASA budgets any hardware funding for missions beyond Apollo, the impact upon international relations overall may be ultimately to slacken the extension of the arms race in space. The unknowns of the new environment prevent a full truce between the rival space explorers, but the fever has to some extent been controlled and the frenzy slightly diminished. The MOL experiments arise from indications that Soviet space efforts are also going in this direction (rather than to the moon). The arms race in space may be temporarily intensified, as the Soviets charge, but if conducted and evaluated responsibly the MOL will clarify once and for all the strategic value of space. If any be found, both sides are capable of employing it, thus denying either a decisive advantage and leading toward eventual stability in this new environment. If none be found, the destabilizing effects of present space imponderables will be eliminated. In either case, the ultimate result will be to eliminate military panic and bring about a space truce which will make possible the internationalizing of future explorations. It is doubtful that hidden military motives in the Apollo program can be normalized in any other way.

A flurry of activity followed the MOL decision as the Air Force and its advocates sought ways of slipping out from under McNamara's constant vigilance to obtain immediate funding for such advanced space missions as large operational space stations, maneuverable air/spacecraft, and so on. Curiously, after fighting for years for MOL authorization, the military space enthusiasts now called for assignment of MOL to NASA where Congress could fund it uninhibited by McNamara's scrutiny. The space editor of the *Army–Air Force–Navy Journal and Register,* of all people, suddenly saw the assignment of MOL to the Defense Department "as an ominous harbinger . . . an indication that the military services may play a more prominent role in future space exploration. . . ." [47] Senator Clinton P. Anderson, chairman of the Senate Space Committee, wrote a letter to the President urging that MOL be combined with the extended moon program in order to save $1 billion. Skeptical of McNamara's sincerity in supporting MOL, the Senate wrote mandatory language into the 1966 Defense Appropriation Bill to insure that the $150 million not be used for any purpose other than MOL.

Under the urging of Hornig and the Science Advisory Committee, the President determined that decisions on programs beyond Apollo and MOL be reserved at least until 1967, when the progress of the program would provide more information upon which to base future undertakings. The issue was a keen one to the aerospace industry in view of the fact that Apollo funding requirements had peaked in 1965 and were expected to fall off precipitously thereafter. Webb sought an immediate $2 to $3 billion annual funding increase for space activities in order that "we could accomplish everything that is going to be required in our national interest short of manned flights up to the planets." [48] NASA was interested in manned flight to Mars and space stations in both near and far earth orbit. The Mars expedition alone was estimated by Hornig as requiring a $100 billion investment. He indicated skepticism of all these plans: "Although planning for an Apollo extension is needed, it would not seem wise at this time to commit ourselves to manned orbital systems of great size . . . we need more direct experience . . . so we can predict better what they will cost to develop and to operate. Second, we will have to define more clearly than we can today the scientific, technological, and other advantages which might accrue through their use." So far, he declared, "I have not been made aware of potential uses for large manned space stations which would justify their costs." [49]

The Air Force still has powerful resources for covert and overt rebellion against administration policy controls. But the conditions that fostered its supremacy no longer exist, and its industrial and congressional allies are bound sooner or later to go where the action is in protecting and advancing their own interests. The House Space Committee is demanding a quasi-executive planning and decision-making role in NASA, in order to block White House belt-tightening. As power moves increasingly into the White House and the Secretary of Defense's office, so the loyalties of the military constituencies will be divided and weakened, and the ability of the executive to work toward a broader concept of the public interest preserved.

CHAPTER
IV
INNOVATION AND ECONOMIC GROWTH

Scientific and technological innovation have brought revolutionary strides in human productivity, comfort, and longevity, providing new sources of power, processes, tools, products—all ultimately judged socially useful because they have satisfied needs more cheaply and effectively. Rapid scientific and technological innovation is responsible for the unprecedented wealth and well-being of modern civilization. It has made the exploitation of nature more profitable than the exploitation of man, making pluralistic mass societies possible and giving thrust to the revolution of expectations throughout the underdeveloped world,[1] providing the classical patterns of an emerging era as fresh and unfamiliar as any major historical epoch in its infancy.

For the advanced nations, however, the process of acceleration is now in the throes of major qualitative change. Scientific and technological innovation have reached a point of social ambivalence and diminishing returns. In mature, highly capitalized and concentrated economies, investment in innovation has been institutionalized at ever-higher financial levels; it is more indispensable but at the same time less productive. The wonders of a hundred years of unparalleled progress have begun to exact their tolls: the gap between the economically powerful and the powerless (both within and between nations); incipient depletion of the earth's most readily accessible minerals and raw materials; the population explosion; the poisonous by-products of industrial production which have been freely spewed into atmosphere, rivers, and streams; and the increasing inefficacy of individual initiative and private profit as a method for unleashing innovation for the benefit of the whole community—these are some aspects of the anticlimax of progress.

In all the advanced nations, the negative results of innovation threaten to overtake and overwhelm the miracle of modern science and technology. They require social control, national and international coordination, increased decision-making by public authorities for the allocation of resources and energy, and increasing restrictions upon the rights of private action and property. Whether by direct or subtle state planning, all modern nations are moving toward integrated control and a primacy of politics over economics.

The old formula (science-and-technology-equals-increased-wealth-and-happiness) is no longer useful in the advanced nations. An ad hoc, uncoordinated, improvised method of innovation can be positively dangerous—both because of its negative spin-off and because it may become a myth of powerful special interests to legitimate their status as private governments and planners, unaccountable to any representative political constituency and able to place their own short-run aggrandizement above that of the public.

The halcyon ways of the nineteenth century are gone and with them the period of enterprise, creativity, and energy which began in the Renaissance, destroyed old state systems based upon agriculture and landowning, and liberated human spirit and imagination as never before. The values bred by four hundred years of history face a pervasive challenge in this century. The unfinished task of the present generation is somehow to preserve the irreducible human values of this heritage despite the inevitable necessity of enlarged public authority, the overriding demands of security and national interests, the encompassment of individual lives and fortunes by impersonal and relentless mass organizations.

Science and technology have become more indispensable as they have become ever more ambivalent. Society can no longer afford the blind rush of progress; magnificent achievements in the past do not provide a model for the present. The mounting tolls can no longer be ignored, and they will not be solved by the mechanisms of *laissez faire* which spawn them.

Old Myths and New Realities

The requirements for economic growth differ for economies at various stages of development. The traditional analysis attributes growth to the following factors:

Increases in the size and technical efficiency of the labor force, that is, more persons are gainfully employed and more of them have specialized skills which augment productivity;

An increase in the quantity and quality of capital resources, that is, the wealth of productive machinery and power is raised, making possible greater productivity per man-hour and bringing new consumer products as well as further capital improvements into the productive process;

Advances in managerial knowledge, allowing more optimum combinations of all factors of production (land, labor, capital, raw materials, and management) in order to realize the advantages of large-scale mass production and to facilitate competitive distribution and marketing; and

Constant feedback and improvement in the actual techniques by which goods are produced and marketed.

Essential to each of these elements is the continuing process of technological and social innovation, whether based on systematic research and development or growing naturally out of the improvisations of a competitive system in which many enterprisers are constantly seeking opportunities for profit.

This analysis generally applies to the early stages of industrial development, when the most significant new products and productivity/cost advancements derive from the transformation of primitive, labor-intensive production (that is, based upon manpower, simple hand tools and home crafts, and a predominance of agriculture) to the introduction of new power sources and labor-saving machinery. A wide disparity of income between the technological innovators and consuming laborers permits the accumulation of new capital for further investment. This drives the process onward toward a general rise in the standard of living and an increasingly productive allocation of resources.

To be economically worthwhile and to have social value, the cost of research and development must in the long run be less than savings in other areas of production, so that prices may be reduced and consumption increased. Not all such R&D investment returns a profit or does so in the short run; but over a longer term, when capital accumulation permits, many great business organizations recognize the value of reinvesting a portion of current profits in research as a means of seizing new markets or remaining competitive with rival producers and products. In theory, competition forces the benefits of increased productivity to be shared in the form of lower prices and/or higher wages. This enhances purchasing power and effective demand, thus radiating the benefits of economic growth throughout the economy, raising general living standards, and softening if not erasing wide disparities of income.

In practice, this analysis no longer describes what actually happens. The very process of increasing efficiency by means of large-scale production and mass-marketing creates a drive toward concentration and quasi-monopolistic markets. The high level of capitalization, which should produce more goods at lower prices for more people, also erects a wall against new producers who might compete for established markets. This gives the manufacturer greater control over both the costs of production and the market price of his product, leading inevitably to fewer goods at higher prices.

The process of concentration is self-perpetuating. With the trend toward larger and fewer producers, raw-material suppliers and labor organize defensively in order to maintain their bargaining power. On the other hand, the producer seeks to expand his empire to include his suppliers and his distributors in order to maximize his own control. The result of this circular process is continued concentration of economic power. Business and labor both escalate into bigness, and big government is forced increasingly to mediate between them and to protect to some extent the bargaining position of the weaker, more decentralized, and geographically scattered segments of the economy, including the consumer.

The price structure ceases to reflect the laws of marginal utility and instead expresses the relative bargaining power of corporate organizations *vis-à-vis* each other, the consumer, and the government. The price structure becomes in effect a political fact, determined by political relationships. Each manufacturer seeks to maximize his position by manipulating those aspects of the process which he controls, namely the levels of production and prices. Suppliers who are economically divided and weak (like the farmer) can be exploited by more concentrated buyers who can drive down the cost of raw materials by refusing to buy at higher prices. The consumer is weakest of all, and the administered prices of the goods he buys are set so as to shift to him the cost of all the bargains which the manufacturer is forced to make with bankers, suppliers, and labor unions.

This is very unlike the theoretical market of Adam Smith, where the savings of technological innovation and increased productivity are passed on to the consumer, in turn increasing effective demand and production. In the quasi-monopolistic markets of a mature economy, production is set at a level which will maintain prices and acceptable profit margins in terms of costs, invested capital, and the attractiveness of corporate paper to investors. The result is artificial restriction of production in order to stabilize prices at a high level. Technological innovation and other kinds of cost-saving may not be passed along to the ultimate consumer, but instead may cause a cutback of production to maintain the same or a higher price level.

Of course, to the degree that an industry is threatened by competition from domestic and foreign producers, it is forced to lower prices and increase production, in spite of increased risks. This is traditional *laissez-faire* practice and means real economic growth. To a substantial degree, however, highly concentrated industries will absorb competition, rely on covert collusion with rivals, seek tariffs and quotas on foreign goods, and follow the opposite route. In this way, technological innovation may cause economic contraction and higher prices. In spite of enormous productive capacity, the American economy suffers from deliberate relative scarcity in many sectors of production, leading to underemployment of resources (including labor) and underconsumption of goods.

Stated briefly, the impact of contemporary science and technology upon economic growth is no longer simple, raw, and linear. In terms of social utility, research and development costs more and produces less. The very fact of diminishing returns makes increased investment necessary and inevitable; but it suggests that a wiser allocation of social resources is also necessary for fuller development of existing technological capabilities, to realize a greater equity in the distribution of well-being and the fruits of art and science to all citizens. It suggests that all resources, including those of innovation, must be conserved and directed at tasks that contain no incentives for private investment but only for public need, such as softening the gaps between overcapitalized and technologically deprived sectors of the economy and sections of the country; redressing the ecological balance of man in nature by restoring the wholesomeness of air, earth, and water; and expanding a balanced investment in all kinds of public facilities, including transportation, education, recreation, and health. The immediate prospect is that science and technology in many ways may make things more expensive, but the alternative is that they cease to be available at all or only to certain privileged groups.

Corporate giantism, a deepening inequality of income, and the growing alienation between the affluent majority and the thirty-four million citizens who are chronically underemployed, who are denied the means of social mobility, who lack the kind of motivation and hope that enabled earlier generations of the poor to claim a place and share of American wealth and respectability—this is contemporary reality. (In 1966 this fact was obscured by Vietnam war spending.) The structure of effective demand (income) allocates resources in the civilian economy; a narrow class of concentrated economic power is enforcing an allocation that defeats both economic growth and social justice.

Cartelization and quasi-monopolistic administered prices lead to the paradoxical situation in which liquid capital accumulates in the accounts of those least able to find new ways of consumption and investment. Most people have great unsatisfied needs but limited effective demand, leading to a vicious circle by which the centers of corporate power maintain profits by artificially restricting production and raising prices, further reducing such employment and income that might sustain higher resource utilization. To evade this circular effect, government is driven to increasing levels of pump-priming and subsidy in order to augment purchasing power or to maintain profits at a higher level of production. But the power to administer prices eventually soaks up additional income among consumers by creeping price inflation, leading once more to production at reduced levels and to the danger of more drastic contractions of economic activity.

Creeping inflation becomes a way of life, softened and maintained by periodic injections of public funds through tax reductions, welfare programs, subsidies to education, and the maintenance of employment and income levels through non-growth government expenditures for R&D, space, defense, and intellectual make-work projects. This dynamic distorts the public R&D investment—however justified in terms of valid government missions—into a species of pump-priming and gold-plating. This dynamic also endows the centers of concentrated economic power with a vested interest in government spending, including welfare programs that do not threaten the status quo. Thus the most powerful elements of American business have converted to Keynesianism—but this new Keynesianism produces growth only as a by-product of other processes which enhance the distortions of power and income, and postpone but intensify underlying problems, forcing the government to maintain an ever larger subsidy system to prevent recurrent crises.

Government contracts are an escape route from this stagnating civilian economy. The mounting cash in corporate coffers must be expended; but surplus capacity and, at administered price levels, a saturated consumer market induce anti-growth incentives. First and most important, large sums must be used for the politics of corporate survival in the jungles of the stock market, banking, and finance. Here the corporate managers engage in corporate empire-building which inflates the price of stock and (through mergers and acquisitions) hastens the process of concentration. The officers and directors of major companies invest huge quantities of cash reserves in the stock of their own company, protecting themselves against corporate raids and incidentally granting themselves and their political allies attractive options for instant wealth. Dividends paid to the general investor are more closely related to stock market politics than to actual earnings. They are disbursed only to maintain the attractiveness of corporate paper to a wide variety of buyers, thus diluting any threat to officers' and directors' control and maintaining a high cost threshold against raiding parties. At the same time, corporation leadership buys the stock of its suppliers, distributors, and competitors to safeguard its own position and bargaining power and, if possible, to raid or weaken the corporate base of rival dynasties.

The cash that remains beyond the needs of "corporate growth and diversification" is expended on such things as: astronomical salaries and perquisites for directors and officers; investment in R&D aimed largely at the promotion of new government contracts; and capital export to Canada and Europe in search of higher interest and profits than are available in the stagnating American economy. This process is recognized by William McChesney Martin, chairman of the Federal Reserve Board, as the key element in the U.S. balance-of-payments problem:

In general, a country suffers from a payments deficit when its imports of goods and services exceed its exports. But the United States does not spend more abroad on goods and services than it earns. On the contrary, it has record export surpluses in recent years, even after deducting all government expenditures abroad from its export receipts. But U.S. investors have lent and invested funds overseas that were larger than the export surplus. Hence, although the international wealth of the United States has increased, its international liquidity has declined; its gold reserves have dropped, and its short-term liabilities have increased faster than its liquid claims on foreigners.[2]

This is the fastest growing (20 percent annually), most virulent ($6.5 billion in 1964), and least essential component of dollar exports, evoking cries of imperialism among our foreign friends and leading to the imposition of compulsory restrictions by them and efforts by President Johnson, so far unavailing, to impose voluntary restraints.

Administered pricing and generous tax treatment ($7 billion of the $13 billion tax reduction of 1962-1964 went to investors and corporations) have created a capital reserve pool of extraordinary dimensions within corporations. The congressional Joint Economic Committee reported in 1965 that "managers have been able to retain earnings, often in excess of the needs of their existing businesses, even if the bulk of shareholders would have preferred to have such earnings paid out as dividends." The earnings are used for acquisition programs, extending the parent company's operations into new areas not related to its original operations. Joint Committee: "Managements can greatly extend their control over the productive resources of our country, including the labor force, without directly increasing the real wealth of the nation. In the process they reduce the number of independent enterprises and weaken an important foundation of our democracy." [3]

The flow of investment abroad is a search not so much for higher interest rates as for profits and a stronger position in an increasingly cartelized international market. De Gaulle boggles at the sizable chunk of French industry in American hands (examples: automobiles, 13.5 percent; telephone industry, 42 percent; refrigerators, 25 percent; sewing machines, 70 percent). In most cases U.S. investment aims not at a partnership with Frenchmen but at control. France is not alone in NATO in seeking ways to stanch this invasion. In Japan the welcome mat that used to be out is wearing thin, and deals involving U.S. management control are being actively discouraged. A blue-chip roster of American firms (including IBM, Texas Instruments, RCA, Du Pont, Scott Paper, General Motors, and General Electric) has come under attack in the Japanese press and parliament.

R&D as Gold-plating

Surplus corporate cash flows into forms of scientific and technological innovation which do not significantly contribute to economic growth. This is true of much of the trend toward industrial automation (also facilitated by tax benefits designed to encourage American industry to modernize its productive plant). In theory, automation reduces costs and increases production; in spite of the temporary displacement of workers with older skills, the end result should be beneficial. This theory operates to some extent in areas where the competition of rival products and an elastic consumer demand force a price reduction which passes on savings to the consumer.

But much of automation has little or no price-reduction effect. Instead, the investment in automation may reduce labor costs but augment capital and management costs. Where there is no threat of foreign or substitute competition, there is no need to evaluate automation plans in terms of cost advantages, and in many instances the development and application of these new techniques may have little effect or may in fact *raise* prices. Where cost savings are realized, they may not be passed on in lower prices but instead used to increase earnings. Some portion of automation is merely gold-plating of the productive process, although it goes into the gross national product as net growth.

Automation is growing fastest in areas where administered prices predominate. The result is higher rather than lower costs. In effect, the public is forced to pay a premium for unproductive automation which is unrelated to the relative values of the marketplace and represents an effort by management to reduce taxes and increase company equities. It represents overcapitalization similar to other frills that characterize corporate operations—such as private aircraft for management, public relations advertising, junkets, elaborate "executive training" programs of doubtful practical utility, and so on. In terms of genuine growth, the impact of automation is often the reverse: there is no proportionate increase in production and consumption; in fact, the accompanying reduction of payrolls means a net loss in buying power and increased imbalance in the vicious circle of high-level stagnation. The labor response to automation by demands for shorter work hours and higher pay is equally irrational so long as the nation contains a vast quantity of unsatisfied needs, unsatisfied consumer demand, and unutilized productive capacity. The result is higher income for fewer workers and chronic underemployment for others. These demands represent a defense mechanism for strong unions trapped by conditions which they have not the power to change.

The continued emphasis on "the search for new products" underlines the need for industry to renew an existing market by technological obsolescence, rather than expanding it by lowering prices and thus enhancing effective demand. It may be said ". . . that the creation of new end products, or of new industries, does not of and by itself contribute to increased productivity or to measured economic growth per se." [4] Improvements in technology may cause some consumers to shift their buying patterns, and a slight expansion of the real market may result from the filter-down process of their trade-ins. But the overall market may be narrowed as higher prices accompany the convenience and versatility of the more advanced products. By restriction of the potential market, the higher price will tend to be perpetuated even as large-scale production permits write-off of the R&D investment. This is because higher prices are necessary to compensate the producer for the loss occasioned by the drop-out of old customers. In addition, he must anticipate saturation of this limited market and must therefore maintain his R&D expenditures in order to introduce new improvements which will once again renew his market.

The effect of gold-plating is hostile to growth, for it tends to distort production more and more in the direction of the better-heeled income groups, forcing into obsolescence technologies whose lower prices might have tapped a larger but now excluded market. Gold-plating tends to cost more, serve the same needs, and render obsolete cheaper products. Things become more expensive and more convenient for those who can afford them and command the market. The resulting obsolescence of older forms of technology denies low-income groups the logical culmination of the process of increasing productive efficiency, a process aborted by the narrowing of effective demand.

Gold-plating in the consumer market also widens the disparities of consumption and production among geographical regions. A vicious circle of poverty, decay, and underdevelopment comes to afflict large segments of the American nation, while other parts are rising to ever-higher levels of prosperity. The same pattern recurs in the relation of the United States to the underdeveloped regions of the world. The gap that separates the highly developed economies from those of the overwhelming proportion of mankind is increasing, sharpening the political and diplomatic dilemmas of global politics.

Virtually all U.S. capital exported abroad goes to highly developed nations where risks are moderate and profits high, where it may have the bonus effect of protecting the U.S. market against foreign competition by enabling U.S. corporations to merchandize foreign goods under their own brand names or keep them out altogether. The rudiments have been laid for what could develop into a system of international cartels largely under American control. Far from aiding underdeveloped nations, this capital export is widening the gap between advanced and underdeveloped nations. In addition, the tendency toward gold-plating defeats the efforts of U.S. foreign aid and economic collaboration with the non-aligned world. Relatively primitive consumer and capital goods which are suited to their stage of development (for example, cheap, easily maintained horse-, man-, or engine-powered agricultural implements and simple water pumps) have disappeared from American production. Convenience items, elaborate packaging, and complicated machinery are of little value to these countries and serve only to pamper the parasitical whims of authoritarian elites. It is not the high cost of American labor that disables American goods in competing for these markets. American technology and efficiency could maintain a sound competitive position for many kinds of consumer goods in the world market (as evidenced, for example, by the success of the automobile industry). The chief disabling factors are gold-plated scarcity and the cushion of administered prices. Large segments of American industry need not confront the grindstone of cost competition; they do not have to mass-produce low-price manufactures for lower-income groups either in the United States or abroad. Encumbered with gold-plate, industry seeks to exploit U.S. foreign aid programs to inject sophisticated American products into backward nations. There these products quickly fall into disrepair and are pushed aside, or a permanent dependence upon American companies is established, sometimes making possible American-owned and/or -operated assembly plants and maintenance facilities, their profits assured by continuing aid and U.S. support for governing cliques whose position the communist agitators (with the aid of the resident "American imperialists") increasingly challenge with popular nationalist cries.

The Gross National Product

The overall ability of the American economy to produce goods and services (as measured by the gross national product—GNP) continues to increase at a yearly rate of around 3.4 percent (in 1965). Over the last two decades virtually all the industrial nations have enjoyed a growth rate roughly twice that of the United States.

The total American GNP has almost doubled from 1953 ($365.4 billion) to 1965 ($677 billion), with the greatest strides ($30 billion annually) since 1962. With continued government prodding, it is forecast that a rate between 3 and 4 percent is likely to continue throughout the 1960's. This is impressive but less than the average global growth rate which, according to a United Nations survey, is now 5 percent a year, paced by Europe, Japan, and the Soviet Union (8.5 percent in 1965), all of whom climbed above the average in the mid-1960's.[5]

While useful, the GNP can also be misleading, concealing more than it reveals. As the sum of the values of all transactions for goods and services, it reflects non-growth as well as growth factors. Most obvious is inflation, which enables more dollars to represent fewer goods and services: adjusted to constant-value dollars, the U.S. increase since 1953 is cut by a third. Even the adjusted figure is misleading because it is insensitive to increases in non-productive and non-consumable spending (such as space, armaments, jails, etc.) and does not reflect significant differentials of economic activity (such as waste, gold-plating, and other negative constituents) nor the composition and quality of effective demand (unemployment, unused capacity, population increases, etc.).

Federal expenditures on goods and services in defense, space, and atomic energy have accounted for 9 to 10 percent of GNP (and 85 to 90 percent of federal procurements) in the last decade. The annual rate at the close of the 1965 was $67.3 billion. According to a 1965 report of the Council of Economic Advisers:

> The real cost to our society of allocating productive resources to defense is that these resources are unavailable for non-defense purposes. Thus . . . the nation must forgo $50 billion of non-defense goods and services. . . . There are, to be sure, tangible and occasionally significant incidental benefits . . . to the civilian sector. However, there can be little doubt that the nation could have obtained these . . . at substantially lower costs and with more certainty if comparable research and training resources had been devoted to civilian purposes.[6]

Our present statistical tools are inadequate to measure waste, built-in obsolescence, gold-plating, and the non-productive but substantial inflation of stocks and bonds and real property. All add little to future productivity, yet substantially to the raw GNP.

The GNP does tell something about the turnover of the money supply, at best a side issue. It reveals the volume but not the kind or direction of economic activity—which may be going backward or sideways, and which may contain little true growth. The key issue is: who commands and enjoys the allocation of resources and for what purposes? In terms of the public interest and social values, does this allocation promise maximum equitable utilization of resources to achieve a balance of public and individual values which accords with some kind of politically determined consensus? The quality and extent of growth in meeting human needs is the only meaningful standard. Given the ambiguity of competing social values and the inadequacy of present statistics, the GNP must serve as a rough index of total output.

A useful indicator would be output per man-hour, which provides an index of growth in terms of productivity, eliminating population increases and variations in the employed labor force. Here the postwar growth has averaged 3 percent annually.[7] But sound growth must be measured by the amount of consumption it generates and the manner in which consumption is allocated throughout the population. During the same period, compensation per man-hour showed an average increase only half as large as productivity gains, while reported corporate profits doubled (to $74 billion in 1965—not including hidden profits, increased company equities, management stock options and perquisites, etc.). Between 1964 and 1965, personal income increased 7.6 percent while corporation profits (after taxes) soared by 19.3 percent.

The pathology of the U.S. economy lies in the fact that prior to the 1965 Vietnam surge and after five years of unprecedented rising economic activity, a full 12 percent of the productive capacity of American manufacturing was still idle while millions of jobless contended with increasing frustration, humiliation, and rage. By mid-1965, some 4.7 percent of the civilian labor force was officially counted as unable to find work (about four million breadwinners). In 1966 this slipped to 4 percent, largely due to the infusion of federal dollars made necessary by the escalation of the war in Vietnam. And it is recognized that actual unemployment was and is larger, including those who have exhausted unemployment compensation, the millions who are working short hours or part-time, and those, mostly young people, who have never been employed regularly. Real unemployment is at least several points higher than the official rate. Submerged in the statistical averages is the fact that in the mid-1960's unemployment had reached depression levels for certain groups. The official rate for non-whites has been more than twice as high as for whites, teen-age unemployment almost three times as high. The double losers are the Negro teen-agers whose lives are impaled on the switchblade statistics of a socially mindless growth rate which leaves them completely behind. These hapless victims augmented the reported GNP in the form of increased public expenditures for jails and police. A more accurate index can be found in the mounting numbers on relief and welfare rolls—a rate twice as high as population growth.

The distortion of the income curve has increased rather than diminished. In spite of the unprecedented high level of economic activity during the 1960's, unbroken by such cycles of recession as marked the Eisenhower years, the performance of the richest economy in the world has been dismal and discouraging. Qualitatively there has been little or no economic growth.

Viewed against the hard statistics of unemployment, population increase, and idle productive capacity, the lame performance revealed by the GNP before the Vietnam step-up took on the character of a cripple crying pathetically for help. Merely to keep abreast of the increasing labor force, the economy must grow roughly 1.7 percent annually. If it is assumed the GNP growth rate contains about 1 percent non-growth transactions (waste, jails, etc.), and if this and the increase of the labor force be deducted, the residual growth rate ("getting-ahead" gains) appears to be on the order of 1 percent (most of which serves the upper middle-income groups and their growing demands for recreational and status consumption). Vietnam spending is now taking up the slack. Without this infusion of jobs and purchasing power, the economy faced a continuation of high unemployment of both labor and industrial capacity. The feverish new spending puts money in the hands of people who will spend it. But it does not heighten civilian production proportionately; thus it quickens the pace of inflation and aggravates the fundamental problem which will return with renewed ferocity should peace break out in Asia. This fact almost guarantees increased government spending—for one thing or another—to defer the implacable underlying crisis.

The Decline of Private R&D

Civilian research and development has been an important growth factor during vigorous days of industrial development, bringing a fourfold increase in real income per capita during the period 1870-

1950.[8] Until World War I science and technology evolved through the ad hoc actions of private individuals, business organizations, foundations, and universities. At the turn of the century the first great giant fortunes accumulated in the American industrial revolution became available for such non-utilitarian enterprises as the advancement of learning, and for support of large industrial laboratories. Before this time modern, highly organized scientific labs, especially those devoted to physical and chemical discovery, existed only as isolated embryos. Edison had a small laboratory which, by the late 1880's, grew along with its industrial profits and patents. The Du Pont brothers established a chemical lab in 1889. These pointed the way to the future, but innovation came mostly through the individual ingenuity and informal labors of mechanics, technical men, businessmen, a handful of professional inventors, and engineers: "They turned out wonders for their age and helped bring about a vast economic growth. But however precious, their inventing and patenting was almost always a simple, amateurish, tossed-off thing, compared to the thorough, elaborate, perfected, scientific product of thousands, of even millions of man-hours of labor of highly trained scientists, engineers, and their assistants that constitutes the great and valuable bulk" of the innovation industry today.[9]

"We move toward a new era," President Kennedy declared in 1963, "in which science can fulfill its creative promise and help bring into existence the happiest society the world has ever known. . . ." [10] This hyperbole has become a persisting myth of our age; the record of the postwar period belies this faith. While R&D expenditures soared over 1,000 percent, the rate of economic growth has averaged below the previous fifty years when total R&D investment was low and government expenditures in this area nonexistent. "After 1953, while expenditures for R&D skyrocketed, the rate of increase in output per man-hour slumps," reports an economist who has been unable to identify a positive relationship between R&D and the rate of economic growth.[11]

This would suggest that the emergence of big science is part of the process of giantism in the American economy; it occurred at a time when the trend of rapid productivity increases was coming to an end. In fact, since World War II, while expenditures for R&D have increased exponentially, the increase of productivity has drastically slowed. "No positive correlation is evident between the national rate of economic growth and the national level of R&D expenditures." [12] In a mature economy, science and technology become more expensive, and also both less productive and more necessary. This is quite a different concept than the myth of science as a bountiful provider and a cornucopia of boundless wealth. Today, in fact, the dominance of government contracts and grants threatens to convert science and technology into a pump-priming grab bag which defers real structural economic reforms.

In his 1963 Economic Report, President Kennedy made a more sober appraisal: "The defense, space, and atomic energy activities of the country absorb about two-thirds of the trained people available for exploring our scientific and technical frontiers. . . . In the course of meeting specific challenges so brilliantly," he noted, "we have paid a price by sharply limiting the scarce scientific and engineering resources available to the civilian sectors of the economy." [13] Probably no other major segment of the federal budget is so little devoted to the production of tangible products and to the performance of direct utilitarian services than the $16 billion allocated to R&D. While it is true that this represents a small part (less than 2 percent) of the GNP, it represents the predominant part of the nation's investment in scientific and technical innovation. Apart from the military portion, the direct products of this effort "have been almost entirely intangibles, viz., new scientific and technical knowledge, human achievement, national pride and prestige." [14] According to Assistant Secretary of Commerce J. Herbert Hollomon, less than $30 million of federal funds is directed at R&D in the civilian economy, a sharp contrast to our international competitors in Europe and Japan. This fact helps explain why in recent years so much innovation in the civilian economy has come from abroad or been forced by foreign competition.

As a matter of fact, the federal government now supports nearly 70 percent of all research and development being done in the United States today, while only a little more than twenty years ago it supported practically no research and development outside a few of its own laboratories. John Rubel, former Assistant Secretary of Defense (for R&D), observes:

> . . . curious and even puzzling as it is—we are faced with major areas of economic stagnation throughout our nation precisely when national income, research and development activity, and federal support of science and technology are at an all time high. Research and development supported by and for industry seems to be receiving less support and to be attracting and retaining fewer top-notch scientists and engineers as federally supported R&D competes with R&D in the private sector. Many believe that efforts in the private sector are retarded by those in the public sector, and that spillover from military and space programs, however useful it may be in

some specific cases, is increasingly inadequate to compensate for the shift away from private support of R&D aimed at private goals.[15]

Government contracts and publicly funded R&D do increase productivity in specialized non-growth sectors, financing large-scale introduction of automated machinery, new machinery design, and so-called scientific management in firms able to win contracts. New products do develop (such as nuclear submarines and rocket engines), but they are highly specialized and have few civilian applications. New technologies and devices appear, and the expensive developmental work, which private industry might not find it economical to support, is done at public expense. Masers, lasers, infra-red devices, radar, integrated circuitry, transistors—all owe their existence to government funding. Some of this finds its way into civilian products and may, as was the case with transistors, result in higher production rates and lower costs, thus contributing to growth. (In the case of transistors, however, it was Japanese competition that forced American manufacturers to follow suit after a two-year lag.) New scientific and managerial techniques and new industries have emerged (data processing, solid propellants, systems engineering, etc.), but these have their fullest development in the government contract market where cost-plus financing insulates them against the grindstone of cost competition; in the civilian market they are tending to become a species of gold-plating, raising rather than lowering prices. Given the chronic weakness of the civilian market, most of these new techniques and industries have failed to demonstrate growth outside the military-space sector upon which they depend for their continued existence. The channeling of scarce R&D resources into non-growth government work tends to swell unrealistically, beyond legitimate and necessary governmental functions. It induces severe and debilitating effects on the economy, distorting the mechanisms of trade and commerce, creating symptoms of Orwell's *1984* syndrome.

R&D expenditures by private sources are small and getting smaller. In recent years these have totaled about $4 billion annually ($5.8 billion in 1965), but only $1.5 billion has aimed at increased civilian productivity.[16] The balance has been devoted to winning larger government contracts. Over half of this private R&D is performed by a handful of the largest contractors who receive a proportional share of the total federal R&D budget (in 1961, 50 percent was done by the nine largest contractors who received 62 percent of NASA–Defense Department contracts—over 80 percent of the $2.5 billion is expended by the top hundred government contractors). The big industrial R&D spenders are aircraft and missile manufacturers, electrical equipment and communication companies, and the chemical and allied products industry, together accounting for well over two-thirds of the total. Most private R&D is thus paid for indirectly by the taxpayer as part of overhead charges on federal contracts or as tax benefits.

Some economists hold that growth-oriented R&D of both government and industry has not only remained small but has actually fallen below pre-1953 levels.[17] The definition of "growth-oriented R&D" itself remains highly controversial; space enthusiasts and the military argue that all R&D, however remote from the needs of the civilian economy, ultimately pays off through its "spin-off" effects. Former Secretary of Commerce Luther Hodges has answered these claims: "By wrongly assuming . . . that research and development for any purpose—space, military, or whatever—automatically fosters economic growth, we have completely missed the point that this is a highly concentrated industry, restricted by purpose, by geography, by company. . . ." [18] In any case, 90 percent of federal R&D funds are devoted to defense, space, and atomic energy; and out of every dollar of industrial R&D, over sixty cents is public contract money (compared to forty cents a decade ago), while two-thirds of the rest is ultimately charged as overhead on these same contracts.

It is probably true that spin-off, however delayed and speculative, has some impact on the civilian economy in providing new materials, industrial processes, instruments, and scientific knowledge. But in the words of John Rubel: "The direct and efficient way to stimulate economic and industrial growth is to direct your efforts at new processes, new methods, new products, and new technologies that are needed in the private sector. . . ." Just as we cannot rely on the civilian economy to provide military and space technology as spin-off, it should be equally clear that "unless we devote ample resources to the pursuit of development in the non-defense sector for its own sake, we shall fall behind there also." [19] One should not attempt to defend the non-productive but necessary needs of government by overemphasizing its by-products. "If you want the by-product, you should develop the by-product," says Dr. W. E. H. Panofsky, director of the Stanford Linear Accelerator. "I think you could do it more economically and more effectively. If you want to develop high-powered radio tubes, then the best way to do so is to push the development of high-powered radio tubes and not to build accelerators which require high-powered radio tubes." [20]

Spin-off is real but cannot be relied upon to solve the underlying problems of the national economy. Something more is required. Advances in space, military, and other governmental technologies have many negative spin-offs and, far from accelerating economic growth, may well deter and reduce it. Assistant Secretary of Commerce Hollomon: "If we have some incidental benefits from a military and space effort, let's work on them, but let's not depend on them to grow the economy, either locally or nationally." [21] The government has defined the problem as the result of a failure of communications between government contractors and civilian producers. Some millions are now being spent to disseminate information on space and defense technology. So far these efforts have been little more than public relations gimmicks designed to disguise the fact that all the extravagant claims about spin-off are already discredited. NASA's Industrial Applications Program was characterized by *Western Aerospace Journal:* "Years of work and millions of dollars have resulted to date in a handful of dreary pamphlets and a few sheafs of mimeographed paper. . . ." [22]

The Republican minority of the Joint Economic Committee is concerned about the negative effects of the nation's R&D, which it charges "has raised serious questions about the proper allocation of scarce scientific personnel and resources and the ultimate effects upon economic growth." [23] Luther Hodges warns that the demand for scientists and engineers to perform government R&D has since 1961 increased more than 300 percent—ten times the increase in the civilian economy. It is estimated that by 1970 one of every four technically trained persons in the United States will be engaged in some phase of the space program,[24] while five of every eight scientists and engineers, and eight of every ten new graduates will end up working in non-growth sectors of contract R&D. NASA attempts to minimize this drain on the civilian economy by pointing out that the projected supply of technical manpower will be entirely adequate for both sectors. What it tends to overlook, as the House Select Committee on Government Research points out, is that "there may be shortages in specialized areas," and there is genuine need for concern about the "inordinate push and pull among the sectors that employ scientists and engineers." [25]

Alarmist projections of scientist-engineer shortages have not materialized; in fact, we now face a surplus. But there is a growing shortage of competitive job offers from civilian-oriented activities undernourished by both public funds and risk capital.

The countries of Western Europe, our competitors in the world market, are spending twice as much (proportionately) of their gross national product for direct civilian research and development.[26] "The nation must face up," says the 1965 congressional Joint Economic Report, "to the question of whether a disproportionately large share of scientific and technical personnel and resources is being devoted to the government's objectives rather than to the promotion of economic growth and the creation of new civilian jobs." [27]

There is a wide range of research-starved industry in the American economy where investment in innovation holds high growth potential, where foreign innovation is providing powerful competition. This country has the oldest stock of metal-working machine tools of any major industrial nation. Steel-making, shipbuilding, electrical machinery, printing, textiles, elevators, scientific instruments, large electrical motors, materials-handling and sewing machines, to mention a few, are beginning to fall behind—with inevitable doldrums facing individual firms through the loss of domestic and foreign markets. In the typewriter industry, for example, German and Italian products have captured 40 percent of the domestic market and a much larger share abroad.[28] Hydrofoil water transportation came to the U.S. from Italy and Britain; the introduction of miniature transistorized TV sets came from Japan; low-cost home videotape recorders are beginning to arrive from England. In the area of precision equipment, such as cameras and optical devices, U.S. producers have lagged behind German and Japanese equipment both in quality and price—many old and respected American brand names now occasion a sneer from the American hobbyist. A recent report of the National Engineers Joint Council urged attention to the nation's civilian research needs:

> Present system of allocating resources to U.S. R&D is producing imbalance . . . non-defense agencies in government do not have adequate research programs relating to problems in civilian sector of the economy . . . R&D by private industry is influenced heavily by government allocations . . . R&D efforts applied to creating new materials and products have been highly successful, but have not been matched by development of systems to utilize these products and materials efficiently.[29]

Basic research, the most fundamental component of the expanding technological frontier, is also being shortchanged. The congressional Select Committee on Government Research added its voice to that of the scientific community seeking to protect pure science from repeated raids in favor of the moon project and mission-directed research. Something slightly under 10 percent of all R&D goes for inquiries motivated by scientific curiosity; this investment, while often difficult to justify to Congress, provides the fount of future understanding, opening the new and unpredictable potentials that lie still dormant in the seedbed of existing technology. The preponderance of R&D goes into applied research and hardware development. Basic science is little understood by the public and frequently looked upon by Congress as silliness. Legitimate attempts by Congress to control runaway R&D budgets are generally cuts in basic science, where new breakthroughs that do not occur are not missed. The President's Science Advisory Committee has argued repeatedly, and mostly in vain, that pure science is not silliness and, since industry cannot be expected to invest stockholders' money in the search for knowledge, only government can insure its health.

In absolute terms, the number of dollars going into basic science has more than doubled during the past decade, maintaining a fairly stable part of total government R&D, from a low of 8 percent in 1956 to 10 percent in the 1960's.[30] But the increases largely reflect very expensive work in such narrow fields as high-energy physics, atmospheric sciences, and oceanography. There is relative starvation of support across the broad spectrum of pure science which is submerged by huge expenditures for accelerators, research airplanes and ships, and expensive laboratory equipment. In terms of funds available for the typical pure scientist (who operates on a comparatively low budget) and for performance of actual research (as distinguished from the costs of administration and equipment), we may be spending less today than we did before the massive advent of government funding: Gerard Piel claims "we are spending about one dollar on pure science out of every twenty available for applied science, as against the ratio of 1 to 6 which prevailed before the war." [31]

The space program, which commonly invokes the hallowed name of "pure science" to justify undertakings for which other arguments are lacking, contains little basic research. The bulk of the program, over $4 billion per year, is for engineering and hardware development, some for mission-related applied science. Except as a promotion gimmick for NASA to sell its programs to Congress, pure science is an afterthought, plugged in here and there to provide an improvised payload for testing booster or guidance systems. The space program "happens at the moment to be the most extensive illustration of what has happened to science through its absorption into a fast-building and gargantuan establishment," charges David E. Lilienthal. "Here the goals of the program are not scientific goals; they are political." [32]

Research conducted by private industry (traditionally the strongest growth factor of our economy) has been rapidly tapering off during the present decade. The march of qualified scientists and engineers into the insatiable maw of government contracts, the inflation of R&D costs brought about by "to-hell-with-the-costs" contracting practices, the spreading attitude of industrial managers of "letting the government do it for them," and the persistence of excess production capacity in existing operations—all have been key factors.

Only the largest firms can afford to maintain extensive R&D activities. In many important civilian industries the business firms are small and profits slim, making it next to impossible to support the technical staff of a non-essential frill like R&D. Assistant Secretary of Commerce Hollomon, whose efforts to get the civilian economy moving have met unaccountable congressional resistance, points out that such industries as textiles, lumber, leather, wood and clay products, machine tools, foundries and casting, and the railroads "have not supported nor performed much R&D and consequently [are not] well situated to participate in the advances of technology generated by other R&D efforts, or to maintain their relative economic strength internationally." [33] The building industry, he adds (which accounts for some $80 billion of the GNP), is so highly fragmented that it cannot finance research on a scale suitable for so large and important a segment of the economy.

For those large and increasingly concentrated industries which command the economy and conduct the bulk of private research, the attitude is: why spend your own money when this torrent of government money allows you to do the same thing and lets you collect a substantial fee for doing it? Any organization that fails to take advantage of this opportunity without having "a patentable and exploitable product immediately in view can hardly avoid feeling like something of a sucker." People in industry are very conscious of "being under constant pressure to develop each line of work into the basis of a research contract. This is known as making the research department pay its way." [34]

For the giant corporation this procedure results in an important bonus: experience, technical skills, and scientific breakthroughs that result from performance of government contracts become "proprietary." There is a deliberate tendency to avoid the formality of reporting to the government patentable processes or devices that emerge from the work, because of the greater advantages in asserting a proprietary claim. Several effects follow. Chief of these is that the corporation may claim "sole-source" status (as the only company fully qualified) for both follow-on R&D and hardware production. In addition, the proprietary claim enables the company to contract for systems engineering work in programs in which other companies may have development or production business from government, or to charge back to such companies or to the government fees for the use of the data. Further, the company may freely exploit the benefits of such R&D in the commercial portion of its business, or use its technical advantage to acquire new subsidiaries engaged in commercial production. As a result, the pattern of government R&D contracting not only defeats the purposes of the civilian economy but artificially parlays the costs of government procurement and tends to accelerate the giantism of already great industrial empires. This process contributes to the restrictive production-pricing policies of a quasi-monopolistic market, intensifying economic contraction, defeating to some degree whatever contribution government R&D might otherwise make toward economic growth.

Traditional R&D funded by private industry has paid for itself in the long run by new products, cost advantages, or improvement in competitive positions. A hard reality principle informed this system and forced scientific innovation in the long run to converge with the social interest in increased production, lower prices, and economic growth. The turn from this grindstone to the softer parameters of government contracting signals an insidious transformation. The airframe manufacturers cry out to government to fund virtually all the R&D of a supersonic commercial transport which, they say, "will pay for itself further down the road." The managers of the quasi-public Communications Satellite Corporation call upon the taxpayers to develop an economically feasible system so that stockholders will not have to wait too long to receive dividends. The electronics industry, which performs almost two-thirds of the total government-funded R&D, considers government contract work more important (and more profitable) than production for the civilian market. The notion that private industry can no longer afford to pay for its own innovation, that special tax privileges and direct and indirect subsidies are essential to its health, may mean that the creeping stagnation that has long been symbolized by the railroad industry is spreading through the whole economy. Or it may represent the corruption of a fundamentally sound system by the practices of a government and a Congress acting impulsively in response to the permanent crisis of the Cold War. In any case, the captains of industry have somehow reached the point, to state it bluntly, of decrying private enterprise and clamoring for its burial.

CHAPTER V

ENTROPY AND PUMP-PRIMING

Science and technology have turned a critical corner. We have been careless of their ultimate costs and now must pay the toll. Belief in a limitless future wanes as progress becomes less easy and the necessity for choice more compelling.

The pace, scope, and majesty of modern technology have wrought a revolution in man's ability to subdue and exploit this planet and its resources, bringing him at last into collision with nature's limits. In the thousands of years of his existence, he was forced to play a docile part in the balance of nature. As he mastered the arts of contrivance, he plunged ahead with unrestrained joy in discovery, invention, and ever-greater power to impose his own image upon the wild elements. He did not need to count the ecological consequences. He was moist and fresh in an ample universe preserved in a balance yet undefiled and full of opportunity. But as his powers increased and his numbers multiplied, the universe became a narrow and constricted place, littered with the debris of his fond hours. Man awakened one morning to recognize himself as a destroying biotype.

The Ambivalence of Science

The principle of entropy exacts its inexorable toll. Just as a slight change in the temperature of a pond or in the salinity of the ocean may shift the ecological balance and cause the death of great numbers of plants and animals, so the rise of science and technology has initiated ecological changes that have slowly corroded the physical aspects of the earthly environment. Human contrivance and artifice impose a mounting level of destruction—polluting the air, accelerating the aging of lakes and waterways, intensifying living space problems for a rampant population, increasing the sum total of emotional distress and psychological disorder as scientific and technological change jigger the world. If not yet, the turning point will soon be reached: the need for survival in a depleted and corrupted environment will compel the inescapable tasks of redress and conservation.

If homo sapiens is not to be a brief episode of evolution, his science and technology must soon be addressed to repairing the ravages of early exuberance. This may appear on the chart of the GNP as a net increase in goods and services, but in reality it will be an employment of resources and labor to undo, remake, and sterilize the contaminated fruits of progress.

Entropy, the second law of thermodynamics, expresses the principle that "something cannot be had for nothing"; every time man organizes the inchoate materials of his environment to achieve some end, he reduces the opportunities and options of his future. Entropy may be thought of as a measure of chaos or of order that is not man's. To overcome it, and to create an order which reflects his values and imaginings, he must exert energy. Although the sum total of energy available in the world remains constant (first law of thermodynamics), it tends to decay into forms less fit for conversion by human action. The creative processes of life operate against this trend, fixing meaningful order at some points while inevitably creating less order elsewhere.

Human life shows enormous propensity for adaptation, survival, and increase. It clings like Spanish moss to every niche and cranny of the earth's surface where organic matter can support the chain of life. In every setting the mortality rate maintains the human population in some kind of balance with the renewal of sustenance. Thomas Malthus noted three mechanisms which promote this balance: moral restraint, vice, and misery. Moral restraint, he admitted, was not likely to be practiced on a large scale. Vice (that is, birth control) he looked upon with distaste. Misery, therefore, was necessary and inevitable. Famine, war, and pestilence were its great instruments. Malthus was a reactionary; he proclaimed these truisms as ideological slogans against the purblind humanistic aspirations of the reformers and visionaries of his day. From the comforts of his parsonage, his pantry adequately stocked with beef and pudding, he pitied them for their illusions and with charitable equanimity contemplated the misery of the great majority of mankind. His dismal theorems were rejected by the onrushing triumph of science and technology which gave substance to humanistic visions and credence to the view that industrialization had forever repealed the ecological law.

But the twentieth century is enacting the Malthusian doctrine in a new and even more dismal form: if the only mechanisms that can check the growth of population are war, famine, and pestilence, then the ultimate result of scientific and technological achievement is to raise the limits. The improvement of the human condition is a transitory stage and, as the population resumes its geometric increase, the ultimate result is to enable a larger number to live in misery than before. In short, the long-term effect of science and technology is to increase the sum total of misery.

One of the first influences of modern technology is to augment the food supply, freeing energy for further parlays of innovation, including advances in sanitation and medical knowledge. There is a dramatic reduction of the death rate and an increase in longevity, causing a spectacular population rise. The result is unsettling when these advances (pioneered in the northern latitudes) are exported throughout the world without the accompanying introduction of social institutions, skills, and technology for a simultaneous improvement in the means of life. In the underdeveloped regions of the world, food production is being outrun by population growth at an increasing rate. The annual food deficit in the Far East and in Latin America has increased each year in the 1960's by 1 percent. In 1964 the food deficits were: Far East, —2.8 percent; Africa, —3 percent; Latin America, —3.9 percent.[1] Invisible in the statistics are the lingering deaths of those too weak to care or to count; but others survive who become walking political bombs.

The continuing rate of population increase added to food deficits provides a negative sum. It completely obliterates the slight economic growth occurring in most of these countries, deepening misery and intensifying political turbulence. Even politically stable countries launched upon great undertakings of industrialization (such as China and India) suffer from the erasure of progress; their stability is increasingly jeopardized by the contradiction between freshly awakened great expectations and deepening despair. Air Force Chief of Staff General John Paul McConnell sees in the worsening cleavage a strong argument for an infinite extension of American military power. "Beyond Vietnam," he told the 1965 Air Force Academy graduating class, "virtually all of Southeast Asia is seething with crisis and conflicts. . . . There are potential trouble spots in Africa with its many new nations, some fiercely nationalistic and antagonistic to the West . . . and wherever you look—in Europe, in the Middle East, and in our own hemisphere." These nations, short of food and land, will be "encouraged by the communists" to encroach on their neighbors, he declared; therefore the United States military establishment must continue to grow in quality and scope, with ever greater advances in military technology and professional competence.[2]

The world's population now stands at about 3.2 billion. In fifteen years it will be four billion, and by the year 2000 more than double what it is today. Over 90 percent of the increase will occur in the underdeveloped areas of the world. This expansion is unprecedented and cannot continue at the present rate much longer. It may become technically feasible to feed a vastly augmented world population; but the political requirements are not now at hand, and the real question may be not who will survive but whether survival will be worth it.

It used to be thought that industrialization would solve the problem both by an increased sustenance and a declining birthrate. The lower number of births during the Great Depression seemed to indicate a permanent trend toward population stability as a result of stable and improving family life. This now appears to have been temporary, and the families of Europe and the United States have resumed their geometric march, intensifying all entropic processes.

In the time it took to read the last paragraph, a hundred new babies were born and fifty-nine people died, increasing the world's net population by forty-one. In the course of twenty-four hours the increase is about 100,000. To provide all of them with one glass of milk a day requires the full output of eight thousand additional producing cows. The term "population explosion" has become a tedious cliché: the American people in their comfortable parsonages, their pantries full of beef and pudding, solemnly reiterate their concern and do nothing, thereby giving validity to General McConnell's approach to the problem. We are in effect endorsing the Reverend Malthus' invocation of misery as the only solution, preparing the military means by which the U.S. and other privileged nations can eventually fall like a pestilence upon the desperate peoples of the world to scourge them of the anti-Western and anti-white passions of their torment.

There has been no dramatic announcement, but in most of the

underdeveloped world famine has already begun. "Malnutrition" is chronic and becomes "famine" whenever headline writers wish. Such countermeasures as the nations have undertaken have been pennies thrown to beggar boys. Hunger may well be the single most important political fact in the second half of this century. Swedish economist Gunnar Myrdal foresees a world calamity not too far away: "It makes me afraid." The World Food and Agriculture Organization estimates that ten thousand people now die each day of hunger. The children of India, Africa, and Latin America are stunted and hollow-eyed, and only a slight touch of illness is needed to finish the work. There are others to take their place at once.

"Very few grasp the magnitude of the danger," declared Thomas M. Ware, chairman of the Freedom from Hunger Foundation. Egypt's new Aswan Dam will add two million acres of arable soil on either side of the Nile; but while the dam is being built, more Egyptian children will be born than can be fed by these acres. India, in spite of impressive progress and the miracle of democratic institutions, faced a 20 percent wheat shortage in 1966 and the imminent reality of millions of deaths from starvation and malnutrition. Ware said: "The catastrophe is not something that may happen; . . . it is a mathematical certainty. . . ." [3] The Committee on Population of the U.S. National Academy of Sciences issued a report in 1965 whose measured language indicated the urgency of the pending crisis. "Too many Americans regard the so-called 'population explosion' as a mild concern usually reserved for vague crises in foreign lands. . . . The crisis is more immediate and impressive to the two-thirds of the human race living in the less developed countries. . . . However, population growth does pose a crucial problem for the United States as well. . . ." [4]

For both the United States and the underdeveloped world, the only alternative to famine, pestilence, and war is real economic growth, reflected in more equitable distribution and composition of well-being, which might break the vicious circle of poverty-demoralization and motivate family planning as well as the knowledge of planned parenthood techniques. Even without religious and political opposition, which in 1965 knocked birth control out of much of the nation's poverty program, contraception will still be defeated by the culture of poverty—people who see no advantage or possibility of planning their own lives are understandably indifferent to the consequences of sexual intercourse.

Hesitantly but surely, the U.S. government is recognizing the necessity for birth control aid to its own citizens, as it has come to aid government and private groups working to control the birth rate in India and elsewhere. By the end of 1965 the anti-poverty agency had approved thirteen non-urban projects for distributing free contraceptives and advice on the rhythm method. In major American cities, however, applications for federal aid by the Planned Parenthood Association continued in political abeyance. Abroad, the United States announced itself prepared to supply technical, financial, and commodity assistance to support family planning programs, but not to supply contraceptive devices or equipment for their manufacture. By the end of 1965 only India, Pakistan, and Turkey had expressed an interest in this support.

The persistence of uncertainty in U.S. policy and the experience of such nations as India so far offer little hope for the future. Education, pills, and devices represent a fragmentary approach to a problem which embraces the most complex kind of social engineering and involves conditions of life which are slow to change and which are unresponsive to official programs and policies. In addition, it is doubtful that governments are really prepared to address the full range of underlying economic, political, and social issues.

The population problem complicates all the other aspects of entropy which in recent decades have transformed the gifts of science into costly dilemmas. The spoilage and depletion of environmental resources are rapidly reaching critical proportions. Advanced technology burns at an accelerated rate the limited stores of energy and material on the earth. Although these are far from exhausted, we are nearing depletion of the richest and most accessible forms. The search for new supplies, their acquisition and processing, are becoming more costly, and the trend will continue upward in future decades. Even the new technologies (such as atomic power and plastics) will not provide an escape; the early myths of inexhaustible and inexpensive energy and materials have not been fulfilled. Atomic energy is still expensive, and the depletion of high-grade ores will make it more so within a few decades. The science and technology of new energy sources and materials promises at best a reprieve from entropic processes, but at a higher, not a lower, level of cost.

The "affluent society" is in danger of being transformed into the "effluent society." While the U.S. has undertaken national programs to prevent the nuclear pollution of the atmosphere and the biological contamination of Mars, we are doing perilously little about the more obvious and imminent perils that face our own natural environment, a resource which is not expendable. The condition of America's water is a national disgrace. Hardly a single river system is not already polluted, many of them dangerous both to wild life and to humans. The great valleys of America are becoming septic tanks at the end of enormous sewage systems. A honeymoon at Niagara Falls no longer evokes the lyricism of love but, with its unsightly sewage and noxious odors, the realities of married life.

The magnificent fresh waters of the Great Lakes, whose size alone would seem to defy poisoning, are going rapidly. Lake Erie, the worst, has experienced radical changes in just two decades. It is as if a person suddenly started to deteriorate from overeating, over-drinking, and fatigue; a cancer-like growth of plant life and bacteria is rapidly springing in a thickening mass from the accumulation of untreated raw sewage that has flowed into the lake for a generation. With a bacteria count four hundred times the safe limit, swimming has been banned for many years and fishing is unpleasant and unrewarding, as are other water recreations. The noxious vegetable growth steals oxygen from the water and kills most water life, drifts ashore and rots on beaches, clogs water-supply intakes and taints drinking water even after treatment. In 1953 there was a sudden disappearance of Mayflies (whose larvae used to provide food for pickerel and cisco), ruining half of the fresh water fishing business in the country. In areas of the lake without oxygen, there have appeared such sinister creatures as fingernail clams, sledge worms, and blood worms. "What should be taking place over eons of time," a Public Health Service report said, "is now vastly speeded up." [5]

The other Great Lakes are similarly afflicted, although the decay has been slower because of their greater depths. In all of them swimming at public beaches has been banned from time to time, and their ecology is drastically upset. Wild rice disappeared along the lake shores thirty years ago; more recently the crawfish began to go, then will pass the muskrat, and finally lakeside property owners. The Public Health Service estimates that over eighteen million fish died in 1964 from water pollution, more than a 100 percent increase over the previous year. The destruction is ascribed not only to municipal and industrial wastes but to the run-off of pesticides uncritically applied to agricultural land for almost a generation.

Meteorologist Morris Neiburger, former president of the American Meteorological Society, sees all civilization passing away, "not from a sudden cataclysm like a nuclear war, but from gradual suffocation in its own waste." The multiplying waste poured into the air by the burning of fossil fuels is greater than the earth's atmosphere can contain, and "a century from now, it will be too poisonous to allow human life to survive. . . ." Nothing will be done, he said, until it is too late. "Mankind will sink to its smoggy doom through inertia and irresponsibility." [6] The carbon dioxide produced by combustion of fossil fuels is expected to increase in the earth's atmosphere by 25 percent by the year 2000, enough to produce measurable changes in the earth's balance of heat absorption and loss, and significant changes of climate.

New York City leads the nation in sulfur dioxide released in the air (part of the average sixty tons of dust per square mile which fall on the city every thirty days). New York air also contains a high proportion of cancer-producing hydrocarbons, the breathing of which is equivalent to smoking two packs of cigarettes a day. In the winter of 1963 air pollution was thought to have caused about seven hundred fatalities from Asian flu. Annoyance, irritation, and inconvenience are the least of the effects we are experiencing. Experts agree that although the precise physiological effects are not understood, great sections of the country are already experiencing a general increase of illness and deaths traceable to environmental corruption.

In the mid-1960's the nation at long last girded itself to confront these problems. But a careful examination of the new "clean air, clean water" laws and the monies appropriated to implement them reveals a major preoccupation with additional scientific study of these problems. The anti-pollution bill almost died until a compromise was reached deleting the definition of a water quality standard, thereby removing the heart of the law and promising years of continued research and controversy over this issue. Likewise, no federal standards were set for air pollution, and the administration retreated at the behest of automobile manufacturers from effective measures to control car exhaust fumes—pending further research. The new Environmental Science Services Administration (including the Weather Bureau and the Coast and Geodetic Survey) is so far largely charged with organizing and coordinating large-scale government research and finding ways in which space satellites may be applied for such work.

There is already sufficient knowledge for action programs. But such programs would interfere with traditional concepts of private property and traditional divisions of political jurisdiction. For the most part, such programs will be prohibitively expensive for the local community and the private businessman. The salvage of Lake Erie alone could cost at least $20 billion. Each such undertaking involves a "national interest" and demands an allocation of resources on a national basis; but the mythology of the vested interests obstructs any bold national decisions or the creation of a consensus to facilitate action. The result is that the administration and Congress appropriate moderate funds for continued research, as if thereby to find some magic solution which will eliminate the problem without massive expenditures, without enhancing federal authority, without interfering with local sovereignties and vested interests, and without flawing the "rights" of private profits and property. Here is a new motive in the contemporary dedication to science: science as an alibi, science as a political delaying tactic, science as an evasion of responsibility, science as an escape from reality, science as a tool of ignorance, shortsightedness, and procrastination.

R&D and the Business Cycle

On Capitol Hill, research has become a magic password for getting billions of dollars. Just as (in Jerome Wiesner's words) "the armaments industry has provided a sort of automatic stabilizer for the whole economy,"[7] so, as the arms race has slackened, have space and science programs become a new instrument by which the government seeks to maintain a high level of economic activity. There is implicit faith that technological innovation of any kind will redound to the growth of the civilian economy; in any case, Congress will vote money for R&D which it will not vote for direct attacks on the problems of economic stagnation; and the infusion of federal dollars will at least maintain existing employment in the vicinity of aerospace industries and thereby avoid adding another depressant to the general economy.

In spite of its enormous panoply of productive resources, the general economy is infected with administered prices and relative scarcity, leading to underemployment of resources and underconsumption of goods. The vicious circle which results generates cyclical movements in the whole economy, with periodic recessions and booms, and with longer-term movement toward imbalance in income and economic power. The wages of labor are not only a cost factor but also the key factor of purchasing power and effective demand. Effective demand lags behind rising prices and leads to restricting production—in order to preserve acceptable profit levels. This increases unemployment and thereby further cuts demand, an effect which may be diluted by random opposing movements in other industries at any given time, but tends toward simultaneous oscillations in many industries. Higher prices and restricted production in many relatively concentrated industries come to be locked in cadence with basic industries like steel and auto, resulting in a recurrent tendency toward recession, deferred only by international emergencies, defense and space spending, tax cuts and government-sponsored creeping inflation, and so on. Such pump-priming permits profits to inch ahead before costs—thus maintaining the level of production—and purchasing power grows through the payrolls of the government and its contractors—allowing at least those who are employed to maintain their purchasing power and effective demand. Through expanded welfare programs, government reduces the distress of the unemployed and of those whose fixed incomes are swallowed up by creeping inflation.

Until quite recently, federal pump-priming largely subsidized the accumulation of capital in order to maintain employment and induce expansion by fostering new investment. Tax benefits and traditional forms of subsidy coexisted with the more formidable tool of federal procurement and contracting. Since 1963 a fundamental policy shift has begun which hopes to expand the consumer market through tax relief at lower and moderate income levels, and through fiscal and program tools to inject greater equity of effective demand into the chronically depressed segments and sections of the civilian market.

There is growing agreement in government that the rate of capital formation is already more than adequate. In the period 1960-1964 labor income increased 23.1 percent, while interest income increased 48.6 percent and dividends 38.9 percent. Undistributed corporate earnings must be added to reveal the real extent of structural disparity: after payment of taxes and dividends, the amount which flowed into corporate treasuries increased by 35.5 percent.[8] Investment opportunities for individuals have been diminished by the fact that corporate cash has been in excess of investment needs. Despite a rise (27.3 percent) in corporate investment in plant and equipment since 1960, even larger cash reserves were retained in corporate treasuries during this period, with a cumulative total in excess of $8 billion. This means that far too much income is being allocated non-productively, finding its way neither to productive investment nor to consumer demand, serving only to pyramid paper values of equities and real properties. At the low end of the income scale, twenty-six million consumer units shared the same aggregate income as did two million at the top; and the share of those at the bottom has been decreasing.[9]

The relative contraction of consumer purchasing power is a serious drag on economic stability (not to mention positive growth). In addition, the inflation of paper equities further increases the future share of income derived from interest and dividends, thus steadily worsening the drag-factor. Some economists believe this mechanism to be primarily responsible for the modern pattern of cyclical contraction and general stagnation of the American economy. This has become the official doctrine of the administration. Gardner Ackley, chief economic adviser: "It is the task of fiscal policy continuously to concern itself with the strength of markets in the economy, trying always to promote the generation of sufficient total demand for goods and services to allow the economy to produce at its full potential." [10] The Council of Economic Advisers believes that at the present time "there seem to be sufficient incentives and resources for the enlargement of capital facilities. . . . If anything, capital facilities may be tending to grow more rapidly than the expansion of the ultimate consumer markets for the output of these facilities." [11]

There is no longer a question of federal responsibility to make effective use of all policy instruments in order to manage the economy; the new controversies concern how, when, how much, for what ends, and in whose interest.

A recent Presidential Commission on the Economic Impact of Disarmament concluded that the "release of defense resources provides us with the opportunity to step up our long-run rate of economic growth." The commission pointed out lines of action: investment in physical capital, both private and public; investment in human capital (education, health, and crime prevention); and technological innovation in the civilian economy.[12] The administration has adopted the concept of "full employment surplus" (FES) as a means of monitoring the magnitude of federal actions. The FES represents the total tax revenues which would be yielded if the economy operated at income levels consistent with the target of 4 percent unemployment. The government is obliged to adjust spending and taxes to maintain maximum employment levels.

At present the Johnson administration anticipates that the GNP will continue to grow at the recent rate only if the government determines federal spending boldly on the basis of an anticipated increase in revenue of $7-10 billion annually. If present trends continue, the GNP in 1970 will be about $870 billion: on the basis of current tax tables this would yield some $165 billion in revenues. If federal spending or tax relief fails to act in anticipation of this growth, it will probably depress the growth trend and precipitate a recession.

Thus fiscal policy must stride ahead with programs to increase federal and private spending by $7-10 billion annually in order to prevent the drag of overall demand. Simply defined, the FES is the additional margin of combined federal spending and/or tax relief which must be built into national fiscal policy each year in advance of the nation's economic performance.[13] The government is determined to generate this new demand by some formula divided between tax relief (especially for the low and moderate end of the income scale) and new programs, ranging from defense and science through welfare.

The hard choices in allocating the FES will substantially influence the quality of American life and the success of its economic system. Already the contractor community is striking out for a major share of the FES, spreading its tentacles into every possible field of large-scale scientific exploration and insisting that only its "unique" systems-engineering capability, created for space and defense, promises success in new programs of water conservation, waste management, urban redevelopment, and so on. The war in Vietnam threatens to soak up most of the 1966 FES, granting the contractor community a reprieve from the trend toward a leveling of space and defense spending, and threatening to defer most of the positive alternatives that are beckoning fiscal policy.

Without fundamental structural reform, the FES program could result merely in a continuation of high-level stagnation whose underlying pathology may deepen. What could result is creeping inflation, geographical imbalance, increasing social rigidity, underemployment of human resources, and—to stave off incipient collapse—growing allocation of FES to maintain a patchwork of federal subsidies at all points of the income curve. Whether they be called science, education, health, or whatever, they will constitute in fact a form of permanent dole, not so much a cure as a substitute for economic health, covering with poultices and band-aids the eruptions of a sick society.

In the first half of the 1960's, federal intervention has succeeded in avoiding dramatic cyclical movements, but it has failed to drive the economy onto the broader ground of fuller employment and has been forced to improvise one-shot fire-brigade methods of temporary impact. All these methods pay tribute to corporate wealth and power and deepen underlying causes of imbalance and stagnation.

The obscure and persistent ailment afflicting American society arises from the basic fact of the concentration of economic power in the private hands of a few who are unaccountable to democratic controls. They intrude the perpetuation of their influence and the priority of their own shortsighted special interest into all aspects of public policy. This is the fact that makes creeping inflation an inevitable part of government's efforts to grapple with the nation's problems. Corporate power absorbs each year's crop of constructive federal initiatives, making it necessary for the nation to run faster and faster in order to stay where it is. The myths of private enterprise, the separation of political and economic decision-making, the refusal of most public leaders to confront the full implications of the government's responsibility to integrate and direct national life—these lead to the denial of social planning, while government employs its ascendancy on behalf of narrow special-interest groups which operate as private governments. While the government plans fragmentedly, it upholds the deliberate and covert planning of private corporation government. Public values and interests, as well as public control, are subordinated to the demand for profits and the vicissitudes of politics in the exclusive corporate and financial community. It is income distribution in the last analysis that allocates resources, and it is this allocation which determines the distribution of power and values among Americans, and the real quality of national life and institutions.

In a recent exchange with Gardner Ackley, Senator Paul H. Douglas touched this underlying fact: "Is it not true . . . that the attempt to cure unemployment by fiscal means is apt to push into the background any emphasis upon anti-trust policies as a means of increasing competition and getting prices reduced?" He cited the example of England, where cartelization and monopoly were the underlying causes of chronic stagnation, a fact which Keynes and other British economists deliberately ignored: ". . . You could never get them to admit that monopoly or quasi-monopoly was a cause of continued unemployment, and they turned continuously to fiscal policy and monetary policy as a means of offsetting this weakness in the society about them." Senator Douglas suggested that the "new economics" of the 1960's has contributed to a "comparative indifference to competition and has led to anesthetizing people" to the need for far-reaching structural reforms. Ackley agreed, adding that the British economy "today undoubtedly suffers from stagnation of innovation, from rigidities, and from a non-progressive structure which at least in part is attributable to the high degree of concentration in British industry." He refrained from making a similar blunt statement about the U.S. economy, expressing the hope that "preoccupation with fiscal and monetary policy . . . would not divert attention from the structural problems of our economy, one of which is the problem of competition."[14]

The new economics accepts "planned inflation" at a rate fluctuating around 1 percent a year as a necessary and desirable effect of fiscal policy. It is primarily a political device and a symptom of economic concentration, seeking to stimulate economic activity by maintaining arbitrary and privately determined profit margins. Whether by direct subsidy of profits or by support of consumer spending, creeping price inflation serves to keep business profits slightly ahead of costs without production cutbacks. Under conditions of administered prices, business responds to declining sales not by dropping prices (which might maintain sales and production levels) but by raising them and cutting back production, thereby reducing employment and unleashing the process of downward escalation. Government-subsidized creeping inflation seeks to avoid this syndrome, but in so doing it fails to stimulate substantive economic expansion and inadvertently expedites the process of concentration.

Government's collusion with the process of concentration may be seen in such instances as the President's economic guidelines for wage increases linked to prices rather than to profit. Prices do not represent income, and any universal guideline which excludes profits —which are income and which do allocate resources—is seriously unbalanced and inadequate. Efforts to reverse price increases in basic industries (by threatening to dump strategic stockpiles) represent an indirect and ad hoc form of economic planning. While preventing additional profit grabbing in the face of strong markets (mostly dependent on government spending), the device fails to touch the existing extravagant profit margins and the fundamental imbalance between profits and wages. Only a tax policy or other means of restraint to maintain a balance between profit and wages has a bearing upon the relationship between investment and consumption. In this instance, the government's self-denial or misconception serves to undermine its avowed objectives.

Another instance may be seen in the government's large reductions of income and excise taxes. Both showed an expansionary impact. The income tax cut not only bypassed structural reform but aggravated structural inequities, giving disproportionate relief in the upper income brackets. The excise cut was immediately undone by an even more massive and regressive increase in social security taxes. In both cases gradual inflation of prices soaked up increased purchasing power and is tending to dampen consumption. In fact, the excise tax savings in many broad areas were never passed on to the consumer at all but retained as added profits while production and sales remained static. Only Vietnam war spending has partially deferred these effects.

The process of creeping inflation demonstrates its circular effect in the form of the swelling burden of private and public debt. The tremendous market of unsatisfied demand which could be tapped by increased purchasing power is evidenced by the expansion of credit institutions, a means of adaptation to the chronic imbalance of productive capacity and effective demand. Pyramiding of consumer debt postpones even more severe shrinkage of the civilian economy and creates a broader interest in continuous inflation. In the long run this tribute attached to future income accrues to the same centers of economic power, further augmenting their share of future income at the expense of the consumer. The growth in the national debt and the tendency of local units of government to borrow for current needs through bond issues both have the same impact—building into urgent public expenditures an automatic income-concentrating device which reinforces narrow control over national resources.

Unsatisfied with the existing rate of income concentration, banking groups have for years been pushing for higher interest rates and tighter credit, arguing that the key issue of American economic health is the balance-of-payments problem which they ascribe to higher earning power abroad. The action of the Federal Reserve System (in December 1965), boosting the discount rate in spite of administration opposition, realized this long-term demand and was bound to aggravate the pathology of economic concentration. America's national debt, public and private, is now more than $1.3 trillion, and a .5 percent rise in the interest rate will eventually add over $6 billion a year to creditor earnings. In time it may sentence millions of workers to unemployment and businessmen to bankruptcy. By banker fiat, this action alone meant an early diversion of a portion of the FES into interest on the national debt (now about $10 billion a year) and additional cost for federal procurement from contractors who pass along to government much of their capital costs.

Some aspects of government procurement and contracting reflect another instance of collusion with the processes of concentration. Spending for defense, space, and science, although springing from legitimate national needs in some cases, tends to swell into a form of special-interest pump-priming. Little of the product flows back to the consumer market, while most of the wages and salaries do. Purchasing power is thereby invigorated by a means that does not simultaneously increase civilian production. Artificially high administered price levels can thereby be maintained or even further inflated by pumping money and jobs into the economy without commensurately increasing supply and production. Government contracting assumes a significant role as another subsidy to creeping inflation and capital formation. Harold Urey's comparison of the space program with the building of the Parthenon or Saint Peter's is applicable: pyramid building is a form of expending accumulated wealth (deferring the hunger and discontent of the people) without modifying the fundamental distribution of economic power. Similarly, the government's recent infatuation with such enormous science programs as in oceanography, or drilling to the earth's interior, or interplanetary explorations, while all interesting and potentially significant, contain a distorting element. Valuation of their costs in terms of competing social needs is dismissed in the name of science and knowledge for its own sake.

The science-technology race has tended to become a substitute for economic reform. In the absence of a political consensus concerning American social and structural problems, the R&D establishment becomes the beneficiary of a substitute consensus which, in the name of science, frees massive funds to create jobs. We are witnessing a sophisticated form of public works, "science being used as a front for technological leaf-raking." [15] R&D has become a magic formula precisely because it has no specified end-product, price, or means of accountability. As pump-priming, R&D spending for defense and space may contribute less to the nation's overall economic health than do the traditional forms of blatant pork barrel, such as the rivers and harbors appropriations, PWA-style programs, and the distribution of post offices and federal prisons.

"The scientific and technological explosion," declared John Rubel, "is not just a matter of more productivity or more product, and neither is industrial growth. Both are part of an evolving revolution in the structure, perceptions, and capacities of modern society." [16] The gap of $25 billion a year between civilian goods the nation produces and its capacity to produce is a rather important matter. But raw and uncritical attempts to buttress or expand the economy are meaningless and self-defeating. "I am not quite sure what the advantage is in having a few more dollars to spend," says economist John Galbraith, "if the air is too dirty to breathe, the water is too polluted to drink, commuters are losing out on the struggle to get in and out of the cities, the streets are filthy, and the schools are so bad that the young wisely stay away, and hoodlums roll citizens for the dollars they save in taxes." [17]

In his 1966 budget message, President Johnson declared that the great society must be "bold, compassionate, and efficient." There are hard choices to be made and no simple guidelines for their making. Nor is there any brand of expertise which can relieve political leaders of agonies of uncertainty and the risks of commitment and leadership. The starting point is to make visible the process of decision-making which has devolved to the board rooms of private power. A consolidation of planning responsibility in the hands of public authority might arouse the pluralistic discourse of democratic politics.

There is considerable promise in the movement of the federal government during the middle sixties. The pursuit of social objectives will as a by-product enlarge the economy and, in the long run, redound significantly to the real prestige and power of America in the world. It is against the national interest for any region of our own country to be so technologically poor and backward that it cannot participate in the mainstream of our social, economic, and political life. Ultimately it may be necessary to extend this recognition to the entire globe.

The challenge in the area of R&D is clear: the need to view the long-term challenges of international diplomacy with realism and calm, to apply the grindstone of economic reality and the principles of cost accounting to both private and public science and technology, to put aside the illusion that in order to maintain both national security and economic health we must undertake all at once the uncritical pursuit of technology for its own sake. National policy must seek to insure that our overall resources are conserved and that explicit choices are made in terms of the great social values; that public R&D is not exploited by private ambitions; and conversely that private R&D is not impaired by the simultaneous pursuit of national security. Government contracting should not become a form of leaf-raking for American industry; rather, the problems of economic growth should be dealt with as such by frank recognition of the legitimate role of government and democratic choice in the allocation of economic resources. Science may by this route be redeemed from its role as surrogate, alibi, evasion, and pork barrel, and honored in its proper place as a method serving human needs and values.

"Life is good in America, but the good life still eludes us," Interior Secretary Stewart Udall told a Dartmouth graduating class.[18] We have yet to solve the problems created by our abundance and to find a way to distribute the good life among all of our citizens so as to call forth the fullest productive capacity of the nation. Science and technology provide the necessary and inevitable, but not sufficient, instruments of our creative evolution as a people and as a responsible part of the world community. We are committed, in the words of AEC Chairman Glenn T. Seaborg, "to an accelerating cycle of knowledge gathering and knowledge exploitation," but we must fashion our knowledge to the uses of man. "We cannot visualize in detail the consequences for the future. We can, however, clearly see that, if we have the wisdom, we have the power to create an environment essentially of our own choosing." [19]

CHAPTER

VI

SCIENCE: PROCESS AND IDEOLOGY

The word "science," which in its root sense means simply "to know," has acquired in our civilization a highly authoritative quality whose unmistakable halo effect is borrowed by many different self-serving groups. The appropriation of the word by all manner of charlatans emphasizes its popular association with all things that are good, true, and beneficial. In short, "science" has become a dangerous word, frequently used to invoke authority. Its features have been worn away by the touch of many hands, and, for this reason, its ideological as well as technical meaning has become blurred.

The average man views science very much as a believer does a revealed faith. For him, subatomic particles are part of an invisible world whose existence he accepts on authority. The mathematical formulas of the physicist have become a species of sacred text. The initiated ones, the scientists, are thought to penetrate these mysteries and can awe the laity by the miracles they perform in modern agriculure, transportation, medicine, and weaponry. For most of the people in the world today, science is the only dominant religion of our times. It is the first universal religion, shared alike by Christians, agnostics, and Marxists.

Science has become an intrinsic part of modern culture, penetrating even into the underdeveloped regions of the world. Its cultural impact is inescapable, and it generates much of the symbolic imagery and metaphysical vocabulary of all our thinking. Its concepts and methods have permeated not only all intellectual life but the customs and tenets of workaday trivia. All philosophies and disciplines now share with science the need to work with such concepts as space, time, quantity, matter, order, law, and verification. It should surprise no one that the symbols of thermodynamics, statics, and hydraulics turn up in odd combinations in the poetry of Walt Whitman or in the writings of Ayn Rand. The success of technology furnishes powerful analogies in all fields of thought. "Guiding ideas—such as conditions of equilibrium, centrifugal and centripetal forces, conservation laws, feedback, invariance, complimentarity—enrich the general arsenal of imaginative tools of thought." [1] In the best sense, science has become the necessary and natural expression by man of his life and the world in which he finds himself. It can be at once poetic and powerful, symphonic or silly, simple or misunderstood, corrupted or pure.

Science and "Scientism"

As an uncritical mystique, the cultural form is recognized by philosophers not as "science" but as "scientism," a vulgarization of the pure discipline. The distinction is not clear-cut even in the attitudes of scientists. The most rigorous scientific investigator is deeply involved in a metaphysic which arises from his culture and the symbols of language and thought. The element of faith is clear. The essential quality of faith is the act of trust, which enables man to rely on the regularity of his experience. The Christian, the agnostic, and the Marxist share a common trust in the technical world whenever they flip a light switch. When a home appliance fails they do not reject science but blame the gadget for not living up to the standards set by science itself.

Many seek to exploit this trust in order to sell a product, to win assent for a favored policy, to make laws, to promote private or public ends—as though each "scientific" interpreter possessed some special truth. The value of a particular cosmetic, the proof that this toothpaste will combat decay better than another, the claims of "scientific" historiography or the behavioral "sciences"—all manifest the ritual genuflection to science which is now so deeply a part of Western culture. The Marxists have gone further, officially denouncing traditional sects and establishing a state religion of economic determinism based upon a "scientific study of history." In non-communist nations the genuflection is part of an all-encompassing ritual which prescribes rules of correct behavior. Modern man is ready to adapt his behavior to the latest theories of child rearing, sexual happiness, and diet. He carefully follows the "directions for use" on a medicine bottle. He is practiced in the ritual of taking a pill twice daily or changing the oil in his car every two thousand miles. There is an inherent ethic in the technical world which punishes any fall from grace and rewards faith and obedience to "scientific" law.[2]

Scientists, while they resent vulgar scientism employed for values other than their own, are perfectly willing to exploit it when it suits them, and with a very few exceptions are victims of the myths that envelop their discipline. Most scientists take for granted the artifacts and notions of contemporary scientific culture, believing in the solidity of the basic concepts and theories that represent the prevailing consensus of their colleagues. But sophisticated scientists no longer believe that scientific laws have an element of necessity and represent the inexorable course of nature. Today the "Laws of Nature" are considered to be a kind of symbolic shorthand which cannot properly be extended beyond specified conditions except tentatively, subject to the test of new experiments and observations. Scientific theories that endure represent a selection from many varieties of ways of symbolizing events.

Often in the history of science, parallel theories based upon the same experimental data have existed side by side (for example, crepuscular vs. wave theory of light). The ultimate choice is based not only on Occam's Razor (that is, simplicity is preferable to unnecessary complexity) but on relative ability to integrate other theories and new data. The ultimate choice also depends upon a theory's imagery, the cultural environment, existing political ideologies, and the ethos of the age. In this respect, scientific theory differs little from creations of art and poetry. Its evocative and allegorical qualities may be more important than its manipulative and predictive power (which with appropriate improvements could be built into other conceptual material).

The study of comparative languages leads to the same insight. Some of the most solid concepts of the Western world, such as "time, velocity, and matter," are by no means essential to the construction of a consistent picture of the universe. "The psychic experiences that we classify under these headings are, of course, not destroyed; rather, categories derived from other kinds of experience take over the rulership of the cosmology and seem to function just as well." [3] The Hopi Indians, for example, have little concept of time in the Western sense;

consequently all those terms which relate to time—such as speed, sequence, cause and effect, simultaneity, past, present, and future, and so on—are communicated within an entirely different conceptual framework.

The myths of science have undergone partial disintegration during this century, but large portions remain intact and are deeply ingrained in education and popular culture. Such notions as: science is difficult to understand and open only to those with special gifts; the real truth of science loses something when translated into plain and understandable language; the belief in the inevitability of progress and the tendency to look to the men-in-white-coats as the source of that onward-and-upward climb toward wisdom, toward better social institutions, toward a peaceful and healthy world where eventually death itself will be conquered.

Science flowered during the period in which the dynastic state declined as a political entity. It was grasped as part of the ideology of revolt against authoritarian systems and as a means of rationalizing the new order of society brought about by proliferation of new technology. This led to the habit of identifying science with the liberal tradition, the values of the democratic state, and a free market economy, a habit fundamentally shaken by Sputnik I.

Scientism is the broad cultural aspect of science and technology as they are assimilated in those activities of life where laboratory conditions cannot be maintained and where environment is rationalized by combinations of cultural symbols. It constitutes a major ingredient of the world-view of the average person.

For example, a typical housewife applies a combination of pagan myth, conventional religion, and modern science to most things in her experience. But for those activities which demand her professional attention (that is, cleaning rugs, polishing furniture, or preparing food), she tends to imitate the scientist at work in his laboratory. She develops hypotheses that have operational consequences, and she tests them in accordance with standards of what constitutes a well-kept house and a satisfying meal.

On the other hand, the scientist believes in the universal validity of the scientific method and prides himself on applying it systematically to all aspects of his existence. But he cannot manipulate his wife or control his children, and his relations with salespeople are utterly unsatisfactory. As a result, in most aspects of his life he behaves remarkably like the housewife. Concerning these relationships, his hypotheses and predictions are frequently contradicted by experience.

This does not disturb his faith in the correctness of the scientific method, rather it merely fortifies his recurrent hypothesis, namely, "people are crazy."

In the laboratory the real nature of his science becomes evident. While he cannot always control the variables of his experiment, each failure forces him to identify a new factor, to modify the experimental design and his hypothesis. He knows that at each step his hypothesis is better and with each experiment he is better equipped to define and predict the behavior of a complex system. The values that motivate and guide him in the laboratory lead to a method objectively very similar to that of the housewife preparing dinner. But there is an important difference. She is unlikely to embody the practical wisdom (acquired through preparing countless meals) in an abstract formulation which would be intelligible to anyone else on earth, which would enable a future generation either to carry on her work or integrate it with that of other isolated housewives. But the scientist must accomplish exactly these goals. He does not satisfy his calling by serving the results of his experiment on a luncheon table or applying it to polish a specific chair. His social product lies exactly in communicating the information acquired by tedious and demanding laboratory work.

The scientist operates within the context of a relationship with other scientists. His work may fill in a gap in the total system, extend it further in some direction, or fundamentally change its direction or challenge the consensus of scientific knowledge already established within the existing context in which he works. The product of his work is information or intelligence which has immediate relevance to a body of theory embodying all the experiments of brother scientists. Such theoretical knowledge is a concise repository of the thousands of experiences of scientists exploring many different environmental boundaries related to the social life of man. This distillate serves as a guide not only for future experiments but for the application of social resources in new or different ways.

Thus the individual and society learn of changes in the physical world which all cannot experience directly, which the scientist is charged with experiencing in the laboratory. Scientific knowledge abstracts, diffuses, and assimilates such change. It becomes the current coin of reality which enables the individual and the group to adapt their reflexes, allocation of energy, and actions in new ways to maximize individual or group values. Whoever can control information concerning a human or physical condition may thereby manipulate the concept of reality to which the group responds. Applied science and engineering ("development") bring into being the materials, sources of power, and forms by which laboratory science can achieve this social dimension.

A great deal of what the scientist does in the laboratory may contribute to the body of theoretical knowledge without having early impact on the doing and making of social institutions. Yet the scientist is tolerated and supported because overall he has an important social utility. Any elaboration or revision of scientific knowledge ultimately influences man's capability to utilize his resources or impress his image upon nature.

Nature is careless of man's place. She can be cruel, wanton, and unfeeling. Her physical dimensions and power render all human emotions absurd. For man, she endlessly renews the old task of survival. When she is gentle, man makes some headway and civilization flowers.

Against this unequal opponent we do not choose science as preferable to other means of action or knowledge; the process of gaining knowledge is simply necessary and inevitable. It is the natural mode of man's address in confronting the boundaries of his existence. Man interacts with nature in all the ways that his life, form, and place allow, learning from his experience. He cannot resist the urge to generalize from his experience in trying to cope with the future.

The method of science is neither new nor unique; it is the natural mode by which men test and record experience. In short, every person is to some degree a scientist, and the rudiments of self-correcting discipline are not only universal but also spontaneous. Who has not observed a small child become suddenly aware of a new sensation or experience and with great delight repeat it many times as if to confirm an implicit hypothesis?

Scientific man is not necessarily any more "intelligent" than his unscientific brethren. Warren Weaver describes the accurate deductions that enabled primitive men to become skilled hunters or to exploit natural herbs as a means of treating ailments. Ephedrine is the active ingredient of an herb employed by Chinese physicians for more than five thousand years. Certain African savages have a custom of taking some dirt from the floor of their old huts when they move to a new village. They do this to avoid the anger of their gods who might not want them to move, but by this process they bring to the new location the soil microorganisms that continue to give them some degree of protection from certain ailments. We honor Fleming and Florey (discoverers of penicillin), but Johannes de Sancto Paulo, a medical writer of the twelfth century, prescribed moldy bread for inflamed abscesses.[4]

The social role of the scientists' activity is important, just as are the other roles by which men direct and concert their energies. But science has not solved and will not solve the puzzle of the universe, the secret of life, the beginning and the end, the intrinsic nature of the arbitrary ideas by which man symbolizes and classifies experience—matter, energy, heat, light. Even if the day comes when men from earth shall visit Mars, shall synthesize self-reproducing proteins in a test tube, manipulate coherent light waves and neutrinos, they will only have deepened the mysteries of life and the universe. Man is, in Pascal's words, "the central point between nothing and all, and infinitely far from understanding either. The end of things and their beginnings are impregnably concealed from him in an impenetrable secret. He is equally incapable of seeing the nothingness out of which he was drawn and the infinite in which he is engulfed." [5] The best he can do is elaborate myths to explain the ultimates and the irreducibles, myths which will reflect his own concerns and his own history more than they reflect a larger truth. Here science once more becomes scientism and is indistinguishable from metaphysical faith. It offers bewildered man highly qualified and circumspect propositions of limited generality, a most inadequate substitute for a world-view which explains, justifies, and gives continuity to personal existence.

Thoughtful scientists have learned, in the words of J. Robert Oppenheimer, "how fundamentally narrow and constrained and accidental" are their views about the world of nature. This is the experience of science just as it is the experience of life itself.[6]

The Role of Technology

Compared with the elemental forces of nature, man is pitifully small and weak. But the accidents of evolution have given him a number of advantages over other species. He can stand upright and move about while his arms and hands are free. His hands are formed so that he can oppose his thumb to his other digits. He may turn things over, feel their properties, and fashion natural materials into tools. When he shapes a bit of clay or chips a stone he takes the first step in augmenting his place in nature through the accretion of technology. He continuously adds new capabilities for further discovery and for adapting the natural environment to his needs.

In this way, man, "the marvelous artificer," extends his senses and his powers. Nature may overcome him in the end, but he makes it a good show. The materials he learned to manipulate have largely determined what man could do at every stage of history.[7] The enormous transition from clumsy hand tools to massive suspension bridges and nuclear reactors has come about through man's ability to discover ways of intervening in the processes of his environment, in order to take advantage of what may be called "trigger actions," where the disposition of natural forces enables the small cause to produce a disproportionate effect.

At the beginning of the American republic, human labor furnished half and animal labor the other half of the energy required for life. By 1900 every man, woman, and child had two horsepower working for him day and night, and today the figure stands at ten. With less than 10 percent of the population of the globe, America controls over half the supply of available energy, a fact largely responsible for a standard of living seven times the world average.

The growth and elaboration of technology provides the hard rib of mankind's history—from the water wheel in the sixth century, the stirrup in the eighth, the horse collar and boat-rudder in the ninth, the windmill in the twelfth, to the sudden takeoff in the fifteenth century when the invention of printing accelerated the diffusion of knowledge. From the seventeenth century onward technological change moved swiftly and utterly transformed the institutions of men.

Each of the artifacts of our world is a museum containing the whole history of our civilization: the lamp by which we read contains a concise record of the history of glass which goes back to ancient Babylonia and Egypt, the discovery of the vacuum, Pascal and the Puy-de-Dome experiment, the discovery of the rare gases which now replace the vacuum in the bulb, the conquest of high temperatures, the search for metals, the perfection of processes by which man creates a wire finer than a human hair.

Science and technology are not autonomous but rather closely coupled and inseparable aspects. The debt of theoretical knowledge to technology is clear in all areas. For example, the idea of universal gravity was recognized in the fourteenth century and the notion of centrifugal forces even earlier (Aristotle asked the question: Why, if the earth is moving and turning, doesn't it explode?). The laws of inertia were also practically understood although they had not been abstractly stated in general laws. Galileo and Newton were the beneficiaries of mechanical devices which began to make possible crude measurement of these forces, thus making a general mathematical theory of gravity and celestial mechanics possible. As technology continues to improve we are able further to refine the hypothetical law. The orbits of satellites around the earth do not entirely conform with Kepler's calculations and have led us to modify our concept of the uniformity of gravity and the shape of the earth. The determination of the structure of large molecules by X-ray examination (for which Watson and Crick recently received a Nobel prize) was possible only because a very high-speed computer was available to process data produced in experiments.[8]

The development of mathematics by Copernicus, Kepler, and Galileo was dependent upon remarkable advances in mechanical engineering in the fifteenth century, especially the development of clockwork mechanisms and mechanical toys of great ingenuity. The telescope and the microscope, extensions of the human eye, opened the world outward and inward, a process which continues today with the electronic microscope and with radio astronomy. The existence of the jet stream in the atmosphere was discovered when B-29's first reached an altitude of twenty thousand feet, and the Van Allen radiation belt awaited the penetration of near space by the first satellite equipped to telemeter such information.

The relationship of science to technology in the developing field of space weapons and research is clear. Since the early 1920's inexpensive mass-produced light metals, highly efficient oxidizers, and electronic equipment have become available for the revival of the ancient art of rocketry. In the 1930's private groups, inventors, and engineers in many countries were working on rocket propulsion; in the U.S., Robert H. Goddard; in Rumania, Herman Oberth; in Russia, Constantin Ziolkovsky; in Germany, Max Valier; and so on.[9] In the words of Wernher von Braun: "Why can we all of a sudden build manned space rockets? We can do it not because somebody has come up with an ingenious idea, but because the number of hitherto unrelated fields—digital computers, propellants, rocket engines, inertial guidance systems, drift-free gyroscopes, and a host of other things—great progress has been made during the last decade or two. . . . We can suddenly accomplish something that previously has been beyond man's reach." [10]

It is something of a conceit of modern scientists to emphasize the dependence of technology upon science. They frequently state as self-evident the proposition that sometime during the nineteenth century tinkerers and mechanics could no longer innovate technology until new scientific knowledge was acquired in the laboratory. This is part of the myth which delineates social status systems of technologists in our time. Scientists think of themselves as freer and more creative and therefore more important than engineers. Even within the scientific disciplines, the "pure" scientist is held to be more virtuous and significant than his "applied" colleague, while both are the precursors and leaders of the regiments of engineers who translate their findings into new or improved industrial processes, devices, and products.

The whole process is intrinsically unitary and its parts inseparable. But historically science traces its origins back to philosophy, cosmogony, and theology, while technology comes from the world of the craftsman, the mechanic, and the blacksmith. There is a clear "mind-body" duality in much of the self-consciousness of today's scientists. Science is identified with "mind" or "spirit," while technology is "flesh," somehow tainted and subject to corruption: the sneer with which C. P. Snow refers to "gadgets"; the scientist's resentment of R&D as tending to conceal how little of the nation's investment is going into basic science.

The myth that laboratory research in electromagnetism and in chemistry in the mid-nineteenth century reversed the historical relation of science and technology is based on a misconception. What happened in fact was that the philosopher-scientist was transformed into the technologist-scientist who, as he became a recognized professional in the educational system, overcame his aversion to soiling his hands and began tinkering in the laboratory with electric cells, coils, Bunsen burners, and Leyden jars. Since that time the role of the scientist as technologist, himself exploring and manipulating new technology, has become obvious.

Scientific theory may suggest new experiments and new ways to manipulate materials—out of this may come new discoveries which fit or modify original theory. But to suppose (as is implied by the view of many scientists) that theory ever has more than a formalistic and conditional correspondence to the environment is very unscientific. Theory does not really anticipate future discovery; it merely offers hypothetical elaborations and extensions as bases for future research. An example of this may be seen in two opposing interpretations of what time dimension may be "deduced" from the Einsteinian theory of relativity. Eugene Sanger takes the view that interplanetary and intergalactic space travel will be possible for man as he achieves the capability of traveling at the speed of light. He holds that the foreshortening time will enable space travelers to travel a thousand light years from the earth and back again in twenty-two years of time relative to the spacecraft itself, while the earth to which the crew would return would be not twenty-two but two thousand years older. William R. Brewster, Jr., on the other hand, interprets Einstein to mean that real time would be the same in the spacecraft and on earth. The foreshortening of time for the spacecraft would be apparent only to observers on earth, and the spaceship would still have to travel at the speed of light for a thousand years in each direction in order to complete the trip.[11]

There are countless examples of the need for technology in science. During World War II Ernest Lawrence sought the shortest route to isolate uranium 235 for the atomic bomb project. He tinkered endlessly with electromagnetic calutrons, improvising magnetic shims, sources, and collectors on a strictly trial-and-error basis.[12] In a classical 1942 experiment, Fermi demonstrated the possibility of nuclear self-sustaining reaction; his major contribution was building a lattice of graphite and uranium into a pile with the characteristics for unleashing adequate numbers of neutrons. The great bulk of work in the Manhattan Project was technological, a scaling-up of Fermi's basic design to produce weapon material, separate it, and design a deliverable weapon. The great bulk of the effort was in chemical, metallurgical, and mechanical engineering.[13] Physicist Robert R. Wilson: "I found all my life that I have gone back and forth . . . emphasis on physics, then emphasis on technic. I often found that doing a particular experiment with a machine involved changing the machine. So the two get very much mixed up, and with it myself as a physicist, myself as a machine builder." [14]

The technologist-scientist is at the frontiers of technology, but he is not alone there; the engineers and mechanics may still provide an independent source of innovation and discovery with or without theoretical formulation, contributing equally to the accretion and complication of new technology.

Scientific knowledge is the subjective aspect of technology. It is the reduction of the bewildering diversity of unique events to manageable uniformity within one of a number of symbol systems. These systems, whether qualitative or mathematical, are as much human and cultural artifacts as technology itself. Much of the reality of the world which these symbol systems represent is closely related to the grammar and structure of the symbol system itself and to the mode of human perception and action. Symbolic logic now takes for granted the fact that not only mathematical but also descriptive verbal systems are essentially tautologies; that is, they are circular. Reasoned conclusions are already presupposed by the definitions or assumptions of relationships between symbols.

The *act* is thus the most meaningful and essential form of meaning, not the subjective knowledge or the symbols which may arbitrarily fuse one or another set of symbols. In short, it is we ourselves who determine the world, who find in nature what we are able to make of it. Arthur S. Eddington expressed this well: "We have found a strange footprint on the shores of the unknown. We have devised profound theories, one after another, to account for its origin. At last, we have succeeded in reconstructing the creature that made the footprint. And lo! it is our own." [15]

This is the reason why great theories of science come to dominate our cultural perception of reality. They are the poetic myths that give meaning to our acts and interpret and dramatize our relation to each other and to our physical environment. No less than art and music, the great theories of science contain a creative and arbitrary element. The habit of precipitate explanation is irrepressible, but the virtue of science is to subdue this tendency by the discipline of empirical testing, universality, efficiency, and economy.

The Role of Political Choice

As with other specialized activities, the role of the scientist and engineer is related to social need and must be evaluated in terms of its relative cost and social utility. It is clear that some of our resources must be expended in basic science, permitting those that are qualified and motivated to study whatever it is that interests them along the frontiers of technology. It is also clear that in achieving other kinds of social purpose, the role of mission-oriented applied science is a necessary adjunct; it is a means of gaining necessary technical intelligence or creating new tools by which certain missions may be more effectively carried out. The basic issues of allocating resources to R&D are never wholly scientific nor technical, but political—identical to those involved in allocating resources among all competing needs and demands which arise in the body politic. They constitute problems of values and are appropriate for consideration in the political arena. Certainly there is a magic in technology, in the parable of frail man rising against awesome elemental forces. The comforts and longevity of life are positive achievements, but new problems continue to replace the old. The equating of science with "good"—part of the myth of progress—ignores the real ambivalence of innovation. Like other human activities which unite and divide humankind, science and technology both help and harm human lives and values. For each new social capability there is a new social problem; for each new convenience there is a price to be paid somewhere in the society. With wonder drugs come unprecedented instruments of mass murder. With increased longevity and comfort come population growth, overcrowding, and insidious pollution. In modern society, science and technology now move man ahead only in isolated instances, barely keeping him abreast of the negative spin-off of past scientific and technological development.

The problems of man have not lessened but merely changed their character. Albert Einstein in his later years concluded that the costs and consequences of innovation were already beyond human control, and that science had become not a great boon but an enemy of man; that if he could do his life over again he would not choose a profession "that has to do with the search for knowledge." [16] The scientific revolution of recent decades has made the direction of scientific research a matter of public policy. It is no longer sufficient to rely upon the ad hoc decisions of private foundations, industry, universities, and individual researchers. Today all are dependent upon the allocation of major resources and organized effort by public authority. The free market no longer provides the mechanisms for technological innovation. Just as our economy is now generally administered by the complex relationship between highly concentrated private corporations and government agencies, so big science must develop the means of operating within traditional democratic institutions to determine how our R&D resources should be allocated.

It is a fallacy to believe "that any problem can be solved by gathering enough scientists and giving them enough money. . . . Just finding out something new is not by itself sufficient justification for research. It needs to mean something when we find it." [17] Technology is not an end in itself, and no single motive, such as the race with the Soviet Union, provides a healthy basis for allocating limited scientific resources. Admiral Rickover assails the "tacit assumption, whenever technology contravenes human desires, that man must adapt himself to technology." [18] A politically and socially mindless technology may evade the process of democratic choice and ultimately destroy the ability of the people to choose, while the military and industrial decision-makers, in the name of science and security, appropriate the power of choice for themselves.

Science and technology, in Rickover's words, are not "an irrepressible force of nature to which we must meekly submit . . . marvelous as they are we must not let ourselves be overawed. . . . We alone must decide how technology is to be used and we alone are responsible for the consequences. In this, as in all our actions, we are bound by the principles that govern human behavior and human relations in our society." [19] It is easier to create a public consensus to visit the moon or to invest billions of dollars in high-energy physics (both efforts of marginal and even questionable social urgency) than it is to attack the immediate and concrete problems that infest life on earth. This is perhaps the basic reason why science and technology is overfed and overvalued during the present period.

We should continue to hold science and its work in high esteem (as we do all professional functions that serve our needs), but we must reject the flagrant overvaluation of science and technology as an ideological slogan for the R&D cult which, supported by enormous public funds, continues to assume a disproportionate economic and political role in our society. Science and technology in themselves offer no guidelines and promise no miracle cures for our anxieties; they cannot relieve us of hard choices with which we must grapple in traditional ways.

CHAPTER

VII

THE SCIENTISTS: HIGH PRIESTS OR VESTAL VIRGINS?

Since Hiroshima, scientists and technical experts have proliferated around Washington like psychiatrists at a juvenile court. The emergence of the scientist-technologist as government adviser, industrial manager, public and private decision-maker, court jester, and scapegoat is part of the contemporary rise of science and technology as instruments of national policy. The conquest of power, distance, and time by new technology has been purchased at the price of an unprecedented arms race in which the complexity of weapons, their early obsolescence, and the geometric increase in costs have caused a major realignment in the process of allocating public resources. As research ideas grow to prototypes, to production orders, to established weapons systems, costs rapidly increase from thousands to millions to billions of dollars, forcing to the highest levels of governmental decision the issues of the laboratory. Since 1957 every major department and agency of government has elevated research and development to the policy level, extending from the Office of Science and Technology at the top through the huge diversified empires of NASA, the Atomic Energy Commission, the Department of Defense, and throughout the executive branch, with equally important countermotifs in the congressional system.

Like the business community in the 1930's, the scientists-technologists find their activities losing their private character and passing under the shadow of public control. With the public office thrust by events upon their specialty come the terrors and temptations of political influence and power. The monastic discipline and the solitary exultations of earlier generations of laboratory workers have been broken by the clamors of the crowd. It has become impossible for scientists to obey Robert Hooke's injunction "to improve the knowledge of all natural things" but not to meddle with "divinity, metaphysics, morals, politiks, grammar, rhetorik, or logik." [1]

Having entered politics as a major constituency, the scientific community, a complex and variegated social order containing as wide a range of conflicting interests as society at large, has become more divided and impotent, the coinage of its special authority inevitably devaluated. The scientists have *never* been a monolithic community with a unique and unifying value system; now, in the public arena, it is inevitable that there should be a fractionation of the so-called scientific establishment. This is not necessarily bad and may reflect an inevitable habit of democracy to defrock would-be priests and force them to assume the undignified postures of political gossips. The growing tribe of official savants could easily become a public nuisance in an age when science is overfed and overvalued as a means of dealing with the contemporary world.

The process of politics is more difficult than the pursuit of knowledge in the laboratory. This is its natural state and the result of its function in creating consensus upon which policy-making proceeds in the face of uncertainty, lack of time, pressure for action, inadequate information, and conflicting interests and values. The personal tragedy often left in the wake of the trial-and-error business of maintaining a social order besets lawyers, economists, businessmen, and labor leaders, as well as scientists, with an even hand.

Who and What Are They?

The scientist, according to one study, is a unique species differing in certain aspects from the common run of mankind: he is "highly intelligent, individualistic, radical, retiring and unsociable, generally unhappy in his home life, and married to an unattractive woman." [2] In another study, eight of ten scientists were found to lean toward the Democratic party; to be disproportionately of Jewish or non-Catholic background, having agnostic tendencies.[3]

A federal judge, sentencing a young physicist for contempt of the House Un-American Activities Committee in 1954, expressed a prevalent myth of the period, denouncing the naiveté of "the younger generation of pure scientists" who have become "a fertile field for communist propaganda." [4] In the backwash of the atomic espionage case involving scientists Klaus Fuchs and Allen Dunn May, this view was part of vulgar mythology. Generalized to the whole scientific community, it played an important role in the Oppenheimer security hearings of 1954 and figured as an ominous background theme of the great disarmament debates in which segments of the scientific community played a leading role.

In the campaign to discredit those scientists who attacked nuclear testing and a continuation of the strategic arms race, the Senate Committee on the Judiciary offered a poignant analysis of an alleged fatal flaw in the scientist's makeup which makes him especially vulnerable to communist influence. "The devices used by the communists," the staff analysis asserts, "for the seduction, misdirection, exploitation, and . . . the direct subversion of scientists," include:

1. Building up the idea that the United States can rely for survival upon purely ethical superiority and that military defense considerations are dishonorable, unethical and ineffective;

2. Building up the fear that further atomic testing, even when it does not contaminate the atmosphere, may escalate into nuclear war;

3. Playing upon the scientist's guilt complex, his sensitivity to his ethical and moral responsibilities, while encouraging disregard for the powerlessness of scientists under the communist dictatorship;

4. Encouraging the idea that the Soviet dictatorship stands for pure science and research, while the United States allegedly stands for commercialized, militarized, and decadent science;

5. Encouraging the American scientists to believe that the American government stands for war and the Soviet government for peace, that the American government stands for armaments while the Soviet government stands for disarmament, that the American government proposes to utilize the atomic bomb for aggressive purposes and continues testing toward that end, while the Soviet government stands for scrapping the atomic bomb and stopping tests, that Soviet scientists are better off and that science is more advanced in the Soviet Union than in the United States;

6. Appealing to American scientists to act against their own government in the name of world solidarity of scientists, while the Communist party of the Soviet Union maintains a rigid control of science;

7. Encouraging scientists to oppose certain domestic and foreign policies of the American government and its duly chosen representatives . . . ;

8. Urging scientists to bring pressure to bear upon the American government in behalf of Soviet-supported policies;

9. Flattering American scientists who follow Soviet policies through the award of honors by communist controlled organizations in the United States and abroad.[5]

The Senate committee offered a number of interesting contributions to psychoanalysis, citing the autobiography of physicist Louis Infield and the testimony of J. Robert Oppenheimer to prove "the childlike innocence and irresponsibility" of scientists. From Infield the Senate publication quoted: "I created a protective layer around me by studies and books. By living in an atmosphere of abstract scientific problems I tried to diminish the impact of my environment and the impact of social problems carried to my world on the waves of anti-semitism. . . . I saw in science an escape from reality, a source of emotion in the glow of which even my grandfather seemed only a harmless and unhappy fool." [6] After World War II Infield returned to his native Poland where he has since been a persistent force for democratization and nationalism.

From Oppenheimer the committee study quoted:

> I was not interested in and did not read about economics and politics. I was almost wholly divorced from the contemporary scene in this country. . . . To many of my friends, my indifference to contemporary affairs seemed bizarre, and they often chided me with being too much of a high-brow. I was interested in man and his experience; I was deeply interested in my science; but I had no understanding of the relations of man to his society.

The report commented: "Certainly Dr. Oppenheimer's background is typical of that of a number of American scientists who attended the various Pugwash Conferences—and equally certainly, a man with this background is no match for a professional politico and conspirator. . . ." [7]

The criticism of the so-called egghead, the scientist who doesn't know anything about baseball or can't make love, is as old as Western civilization. The first philosopher, according to Greek legend, was a fellow who fell down the well when he was looking at the stars, and everyone laughed at him. A number of scientists themselves have laid claim to the converse of this view, namely that scientists possess a "more-objective-than-thou attitude" in human affairs. This claim fosters public faith in the magisterial air of omniscience of the scientist, alleged to be based upon overriding dedication to truth in all things.

So Kenneth Boulding can write: "For reasons which are still not entirely clear, the scientific community in the course of its development created an ethic and standard of value in which the truth took precedence over any individual identity." [8] And C. P. Snow, the best-known exponent of the transcendental calling of science, speaks of the gift of prescience which enables its practitioners to feel "the future in their bones." [9]

So also can the critics of these claims generalize broadly in the opposite direction, calling upon the scientists to awaken to the fundamental fact that the making of public policy "is not a quest for the right answers. . . . There are only hard choices, the consequences of which will be uncertain and the making of which will often seem interminable in time and irrational in procedure." [10] Another can cite "the failure of the scientific skill group to recognize the true character of the political process . . . the compulsions it contains for seat-of-the-pants decisions made with a sensitivity for competing values." [11] The metaphor of a monastic order leads Dean Robert Elliot Fitch (Pacific School of Religion) to describe the scientist "emerging from his cell, with its austere discipline and caste aspiration," being "profoundly shocked to see the way his own truth and power are prostituted to ends with which he cannot become reconciled. He will do like all pietists before him: propose simple solutions to complex problems, see all issues naively and out of context, and make absolute moral judgments where the need is for shrewd compromise." [12]

Few took the side of English literary critic F. R. Leavis when in 1962 he mounted a veracious attack on C. P. Snow's scientific wisdom and literary accomplishments. "The peculiar quality of Snow's assurance," he declared, "expresses itself in . . . a tone of which one can say that, while only genius could justify it, one cannot readily think of genius adopting it." [13] Robert M. Hutchins suffered somewhat less abuse for a similar injury to scientific self-esteem: "My view, based on long and painful observation, is that professors are somewhat worse than other people, and that scientists are somewhat worse than other professors." The narrower the field in which a man must tell the truth, Hutchins declared, "the wider is the area in which he is free to lie. This is one of the advantages of specialization." He agreed that Snow was correct only insofar as the scientific profession imposes its own self-enforcing morality: "There have been very few scientific frauds. This is because a scientist would be a fool to commit a scientific fraud when he can commit frauds every day on his wife, his associates, the president of his university, and the grocer. Administrators, politicians, and butchers are all likely to be more virtuous than professors" in a wider range of their lives, "not because they want to be, but because they have to be." [14]

The Leavis-Hutchins onslaught came at a time when Congress was showing the highest degree of skepticism toward scientists since a hysterical conversion to scientism in the wake of Sputnik I. Spokesmen for the organized scientific community treated the attacks as though they were aimed at rationality itself. The personal bitterness of this reaction can be understood in terms of the political issues of the day and the declining influence of the scientists in public decision-making. The tendency of scientific spokesmen to distinguish between "good" and "bad" scientists underlines the scatter effect that politics is having upon the so-called scientific establishment. The priestly attitude of this emergent skill-group, with a deepening interest in public policy and an occasional opportunity to participate in its making, becomes increasingly ineffective. The turnabouts of politics tend to weaken self-serving ideologies which may in their time have enjoyed great vogue. The scientists are sharply split in all directions today, exactly as are all specialized skill-groups with equally well-qualified and well-intentioned experts on every side of every issue.

In the first place, only a small percentage of American scientists has been involved in national politics, either as insiders or as opposition critics. Roughly 1.2 million Americans are scientists-technologists of one stripe or another, well over half of whom are tied to allocations of the federal budget, either directly in government or indirectly by government contracts with private industry, foundations, or universities. A number of studies have posited "an influential elite" of at most a thousand and "an active elite" of four hundred, with even fewer numbers having been consistently influential over a long period of time.[15] Since World War II this elite has always been divided and diversely aligned with a wide variety of broader interest groups and political factions. Those scientists who have exercised real influence at the centers of power at any given time have never demonstrated the childlike idealism, internationalism, and ignorance of the political process that have been ascribed to them.

Real responsibility and influence impose a discipline of their own which makes ideological oversimplification irrelevant to the practical politics of choice. Vannevar Bush, James B. Conant, the Compton brothers, Oppenheimer, and others have consistently demonstrated a practical and realistic grasp of the national interest, military requirements, and the ways of Washington. It ought not be forgotten that this group favored the first version of the Atomic Energy Control

Bill of 1946 (the May-Johnson Bill supported by the military) which precipitated the first large-scale organization of scientists as a public pressure group.[16] Oppenheimer favored the use of the first atomic bombs on Japanese targets, the decision that carried the day, while large segments of the scientific community in the A-bomb project (led by Leo Szilard) sought to promote other alternatives.

The concern for "truth" and "internationalism" imputed to the scientists is in reality an integral part of the values of Liberalism which attained ascendancy among practically all intellectuals during the 1920's and 1930's. Belief in progress, the perfectability of man, social and economic reform, softening of the "narrow special interests" which pervert nationalism to their own purposes and foment wars—these were all part of the prevailing cultural consensus that emerged after World War I and did not decline substantially until long after World War II. The alleged idealism of the scientists can more easily be found among the rejected and non-influential who are forced into the public forum by exclusion from the inner councils of policy. Nor is this unique. Idealism has always been the universal slogan of out-of-power groups which seek to arouse the decent impulses of the public as a means for enhancing their influence, for obtaining or renewing their mandate of power. A quick glance at the slogans of opposition groups will confirm this observation. But once in the seat of power, any man, be he grocer or scientist, experiences the breakdown of the opinions he may have formed in his days of impotence. In their place he formulates a concrete grasp of the conflicting pressures, interests, and personalities intrinsic in the political approach to any problem. The Labor victory in England has given C. P. Snow cabinet-level responsibility in the area of science policy. Dr. Leavis may find it diverting some day to analyze the impact of this experience on the simplistic scientist-power concepts of Sir Charles's earlier work. The innocence of Snow's abstract concept of political power, endearing as it is in his totally humorless novels, may be transformed in the new Snow who emerges from the humbling uncertainties of real public responsibility.

The persistence of the myth that only scientists can properly make government science policy arises from the prestige which science derives from technology—the impact of the successful atomic bomb project, the shortening lead times of weapons development which mark the postwar arms race, the important technological choices which are intrinsic to the exercise of diplomacy by a totally engaged great power upon whom have devolved great responsibilities, the impact of Soviet leadership in space, and the deliberate use of this prestige by the scientists themselves to claim their new vested interests. This myth is similar to that of the generals who claim to be best qualified to decide issues of war and preparation for war, or like that of the economists who during the Great Depression enjoyed the vogue of an overvalued expertise. Bricklayers may not claim a special gift for making housing policy, but architects do.

The so-called scientific establishment is diverse and divided, containing advocates of every conceivable policy alternative, from solemn to silly, as well as a vast majority who passively join whatever becomes the prevailing consensus of the nation. The parable of former AEC Commissioner Willard Libby is a case in point. In the late 1950's Libby clashed with a large segment of his colleagues over the utility of backyard bomb shelters against nuclear attack. Intent on demonstrating their practicality and confuting his critics, Libby constructed behind his California home a shelter against blast and radiation, built of railroad ties and bags of dirt: total cost, $50. His critics called it a homemade crematorium. Libby regaled the press with the economy and safety of his design until a brush fire left his shelter a blazing ruin.

Seeking to enhance their influence, the opposition atomic scientists have argued their special moral responsibility to warn the nation of the dangerous forces they unleashed with the atom. This is a self-serving claim not unlike those made by other rejected elites who seek by harangues in the marketplace to influence the councils of decision from which they are otherwise excluded. Indeed, they have a moral responsibility as do all citizens, but it is not a special one and does not give their claims special authority any more than, say, the moral responsibility felt by electricians when customers electrocute themselves.

Almost two decades have been transfixed by the nuclear dimension, the self-generating strategic weapons race, and the emergence of the scientist-technologist into the public cockpit. These portents provided an auspicious moment for C. P. Snow's statement on the "Two Cultures," which an apocalyptic decade seized as a cliché. Essentially, the thesis expressed a fear (if not a prediction) that new forces "liberated by science" would be misused by non-scientifically oriented leaders—unless they learned, as the scientists presumably had, the unprecedented dangers of traditional conflict. The Sir Charles pronunciamento tended to cloak political values (concerning disarmament, national interests, etc.) with the authority of "science" itself.

Failure to accept his attitude was ascribed to a communication problem arising from the cultural schism between science and the humanities.

But the 1960's appeared to level the strategic arms race; international pluralism confounded the assumptions of a bi-polar Cold War; government developed a healthy skepticism toward claims made (by generals, contractors, and scientists) in the name of a science-technology race with the Russians; and public responsibility in the areas of basic research and education was recognized (at least in principle). These changing conditions weakened the impact of the "Two Cultures" thesis, and far more subtle and complex disjunctions in domestic and international culture emerged to dominate the agenda. The attack on the concept, originally characterized as anti-science, began to soften into a critical consensus.

Increasingly, as the expertise of the scientist is politicized, the real centers of power in federal agencies or in Congress pick and choose those scientists who will proffer the advice they want; the chosen scientists provide a rationale and a justification for policies arrived at by other means. Scientific advice becomes a new species of lawyers' brief and advocacy, and a cover for conducting politics in bureaucratic corridors or congressional cloakrooms. In all key postwar decisions, scientists have been present in ancillary roles both in and out of government, but they have made none of the major decisions. The most clear-cut case was the decision of the General Advisory Committee of the Atomic Energy Commission to defer a crash program for the development of hydrogen bombs prior to 1949. This was overridden by non-scientists (assisted by other scientists) as soon as the first Soviet atomic explosion raised the issue to higher centers of power. Scientists did not figure importantly in, and in fact have largely opposed, the decision by President Kennedy in 1961 to go to the moon in this decade. The NASA proposal to go by way of lunar-orbit rendezvous was submitted for review to the President's Science Advisory Committee only after the decision had been taken.

The scientific community has tended to coalesce in clusters around the interests of various government agencies engaged in playing a real role in the decision-making processes of the White House and Congress. The independent scientist finds himself used first by one group and then another as each seeks to strengthen its case by rallying public support or joining in the criticism of rival agencies. PSAC and the Office of Science and Technology stand at the head of this new political dimension, seeking to maintain an independent perspective so that the President is not overwhelmed and can maintain administrative control in the difficult balancing act of national and international leadership.

Like ordinary men, scientists have not neglected the opportunities afforded them to advance their personal and professional interests. In 1609 Galileo made a present of his new telescope to the Venetian Senate, accompanied by a letter in which he explained that the instrument would prove of utmost importance in war since would-be invaders from the sea could be seen "so far off that it was two hours before they were seen with the naked eye." [17]

Similarly, today we see the fraternity of astronomers providing the most solid support for the nation's space effort. Dr. Martin Schwarzchild: "I wish there were time, Mr. Chairman, that I could tell you what unbelievable luck it is for a man in my profession to happen to live at the time when these new tools come along . . . Tools that happen to have been developed for obviously quite different reasons, but that, in at least good part, the government has permitted us to use for purely scientific purposes." [18] Or John C. Lilly rejoices in the unprecedented funds NASA has made available for studying the behavior and the language of dolphins on the ground that an attempt to learn dolphin language will enable space travelers to communicate with life found on distant planets.[19] So architect Charles Luckman testifies before Congress in favor of the establishment of a National Science Academy which, if built by his firm, would provide a unique architectural climate for educating scientists. So strategist Herman Kahn testifies in favor of an augmented civil defense program as a new instrument of diplomatic pressure short of war, an instrument whose utility as a low-intensity mobilization step he proposes should be investigated by his Hudson Institute. Wernher von Braun sees "a great deal of just plain envy" toward the lunar landing program in the criticism coming from sectors of the scientic community: "I can understand very well the scientist working in a different area, who hasn't been very successful in raising funds for his effort, may feel a bit envious that space is apparently more generously furnished with funds." [20] Thus, inadvertently, von Braun also characterizes the motivation of those scientists who have been enthusiastic NASA supporters.

The New Millionaires

It has become somewhat of a cliché to speak of the increasing national importance of science as a "revolution," and to cite the emergence of a scientific elite as "the new priesthood." [21] A deeper investigation eliminates this myth, revealing that scientists as a group are doing quite well but are becoming neither wealthy nor able to dominate public affairs.

At most a few dozen scientists have become directors, owners, and managers of companies engaged in advanced technology on the basis of government R&D contracts. Professors at the California Institute of Technology and in the Boston area (Harvard and MIT), mostly physicists, chemists, and engineers, have founded a number of companies during the last fifteen years, conducting research and manufacturing precision instruments as subcontractors to government primes. Most of these companies have enjoyed moderate success and some still pursue an independent existence, but many have been acquired by larger corporations, often the primes upon whom they were dependent. Of these scientists few have become millionaires, though all enjoy a higher level of income than the majority of their academic colleagues.

One of the few to make it is Dr. Arnold Beckman, once a physics instructor at Cal Tech, who began making instruments in his garage and soon moved to a modern plant at Fullerton, California. According to one estimate, he has a net worth in excess of $60 million. Another is Dr. Edward Land, who rode Polaroid photography into a personal fortune and an industrial empire which is becoming increasingly diversified.

Among the scientists who founded companies with sales from $1 to $20 million a year may be cited: Dr. Denis M. Robinson, 57, president of High Voltage Engineering Corporation, Burlington, Massachusetts, a firm he founded in 1946 with two other scientists to build accelerators; Kenneth H. Olsen, 39, president and co-founder of Digital Equipment Corporation, Maynard, Massachusetts, founded in 1957 to build digital computers and associated equipment; Marshall P. Tulin, 38, vice president and co-founder of Hydronautics, Inc., Laurel, Maryland, started in 1959 to do research in hydrodynamics; Isaac L. Auerbach, 43, president of Auerbach Corporation, Philadelphia-headquartered information-systems design and consulting organization he founded in 1957; and Dr. V. N. Granger, 47, president and one of the founders of Granger Associates, a nine-year-old Palo Alto manufacturer of communications equipment. Drs. Simon Ramo and Dean Wooldridge are in a class by themselves, having in less than four years parlayed a few thousand dollars apiece into multi-millions and control of one of the largest corporate empires in the nation (Thompson-Ramo-Wooldridge, Inc.). Their case is far from typical, however: they were officially sponsored by the Air Force, beginning with nothing but their own technical skills plus control over tens of billions of government dollars provided by their sponsor.

Also in a class by itself is the marketing of Nobel Laureate prestige and insider savvy by a corporate holding company, Quadri-Science, Inc. Physicist Ralph Lapp was the driving force in adopting the methods of corporate organization to convert knowledge and reputation into cash. Quadri-Science includes Dr. Harold Urey, University of California chemist and winner of the Nobel prize for the discovery of heavy hydrogen; Dr. Polykarp Kusch, Columbia University physicist who won the award for atomic measurements; Dr. Joshua Lederberg, Stanford University geneticist, awarded the prize for genetics research; Dr. James Van Allen, Iowa State physicist-astronomer whose name is indelibly associated with the radiation belts that surround the earth; and Dr. Samuel K. Allison, director of the Enrico Fermi Institute for Nuclear Studies at the University of Chicago.[22]

The name Quadri-Science was chosen to indicate four broad areas in which this group is primarily interested—earth sciences, oceanography, atmospherics, and space research, all areas in which government interest has led to vastly increased R&D funding and national programs. The formula of the corporation is to exchange its stock for a number of shares in other research companies. In many cases, one or more Quadri-Science founders may be added to the boards of such companies. For the scientists themselves the whole arrangement is strictly a part-time commitment, protecting them against industrial consulting requests at low rates.

According to the Quadri-Science treasurer, its shares have increased in value "about 50 percent every six months." On the other hand, the enterprising scientists do accept government consulting at modest fees. Lapp claims that "we've passed up a number of attractive financial deals because we felt the companies wanted us only for window dressing." They seek to avoid direct conflict of interest between their corporate affiliations and their advisory role to government. This rule thinly veils the most important commodity that Quadri-Science offers: the reputations of its founders among government administrators and in the professional pecking order, and the overview of the direction of government R&D interest obtained by their laboratory work, their role in the scientific community, and their advisory role to government.

Much the same incentives have led large corporations to seek out scientists as employees, advisers, and sometimes (rarely) as board members. As the research director of a major drug firm commented, in terms of overall budget it does not cost very much and ". . . once in a while we get some useful work out of them. I think we can chalk up a lot of the money to public relations." [23]

Most physical scientists do occasional consulting; government fees range from $50 to $100 a day, while the rate for private industry is $100 a day and up. Within the drug industry it is a common practice for major firms to pay annual retainers up to many thousands per year, depending upon the individual's usefulness and the amount of time devoted to the company. Consulting serves many good purposes, keeping the scientist-teacher aware of the problems of industry and government, enabling the academic researcher to disseminate his broader knowledge to specialized commercial laboratories, and maintaining a link between the companies and the universities from which industry recruits full-time technical people. It enables the consultant, if he is one of the successful few, to double his academic salary, thereby in a sense subsidizing the university, making it economically possible for him to remain there in the face of tempting offers from industry.

At the other end of the scale we find the almost extinct independent scientist who is forced to moonlight for a livelihood, is not consulted by either industry or government, and can get neither contracts nor grants from any source. Such a primitive is William Fox, a full-time policeman doing research into the physics of fluids in his Staten Island basement. In 1964 he was invited to present a paper in Brussels but was denied a $526.30 travel grant by the National Science Foundation. Nobel Laureate Polykarp Kusch says of Fox: "He didn't dream up projects to get support, as some people do. He followed his own curiosity, he had a vision of what life could be and that's what led him on. And if somehow or other we can't work out public policies to encourage and help people like Fox, then there's something very wrong with the whole system." [24]

On the basis of conversations with scientists at several universities, this writer concludes that few scientists obtain annual incomes over $20,000, but most manage to take home more than non-scientist colleagues. It is among the managerial elite of major corporations, most of them recruited from careers in law, finance, public relations, and engineering, that the new wealth is to be found. The largest number are found on corporation boards whose members derive their income from investment banking and corporate finance. Salaries in the six-figure range, stock options, and profits from personal investments are the sources of great wealth. For every U.S. millionaire in 1948, there are now seven, a total in 1965 of about ninety thousand with combined assets of $250 billion. The vast bulk of this wealth is in the form of stock (at least 75 percent), the millionaires holding about one-third of all corporate paper individually owned.[25]

As a group, scientists participate in this wealth on an extremely modest scale; they may be "corrupted" to some extent by the interest they acquire in the financial prospects of government contractors, but they are not paid for it as well as are other groups.

Challenge and Response

Throughout our history, when scientific and technical advice was required by government it was forthcoming by one means or another.

In the earliest days, for example, government turned to the National Academy of Sciences (NAS), which from its 1863 founding until World War I gave advice on such matters as the preservation of paint on Army knapsacks, exploration of the Yellowstone, and tests for the purity of whiskey.[26] During the Civil War the NAS was consulted on the magnetic deviation of compasses; in 1871 it furnished guidance in connection with the exploration of the North Pole; its advice was sought in discovering a means to preserve the original Declaration of Independence and the Constitution, and to prevent landslides into the Panama Canal. It assisted in the creation of the Weather Bureau in 1890, the National Bureau of Standards in 1901, and the U.S. Forest Service in 1905.[27] In 1915 Josephus Daniels, Woodrow Wilson's Secretary of the Navy, decided to establish a "Department of Invention and Development" in the hope of securing effective counter-weapons to that "new and terrible engine of warfare . . . the submarine,"[28] and turned to Thomas Edison, the most respected technologist of the time.

When advice has been needed it has been found, generally on an ad hoc basis—until World War II, in whose aftermath the Atomic Energy Commission was created and permanent advisory groups were introduced by all the military services. Henceforth, in the midst of continuous international crisis and a technological arms race, the trend has moved swiftly to formal institutions at ever-higher levels for funneling scientific expertise to policy-makers and for coordinating rapidly burgeoning R&D empires.

Long before the 1957 creation of the post of Presidential Science Adviser, there were *de facto* science advisers in the vicinity of the Cabinet. During World War I both Dr. Robert A. Millikan and Thomas Edison served at high levels, and in World War II President Roosevelt turned to Vannevar Bush as the nation's science czar. Formal statutory creation of the AEC General Advisory Committee (GAC) in 1946 provided a permanent body at the highest level, chaired by Robert Oppenheimer. At the same time General Hap Arnold, Chief of Staff of the Army Air Force, called upon Theodore von Karmen of the Jet Propulsion Laboratory to establish an advisory board to "investigate all possibilities and desirabilities for postwar and future wars developments as respects the AAF." [29]

With its formal role in nuclear matters, the GAC found itself in the uncontested center of the stage where the key issues of postwar strategy were being formulated. In effect, Oppenheimer became Vannevar Bush's successor as presidential scientific adviser. The Soviet A-bomb in 1949 and the GAC's recommendation against a change in nuclear weapons policy led to the first substantial challenge of this central role. The myth of unity of the atomic scientists was broken as AEC Commissioners Lewis Strauss and Gordon Dean, supported by physicists Ernest Lawrence and Edward Teller, endorsed the military protest of the GAC's position, soliciting congressional support to force the issue to the presidential level, where the policy was overruled. Thereafter the GAC was demoted to a fractional part of a divided array of would-be scientific advisers, each supporting different values and political groupings in the government, struggling for the ear of Congress and the President. The Joint Committee on Atomic Energy, some members of the AEC, some non-GAC physicists, and the Air Force were united in countering GAC influence and politicizing decision-making. They emphasized rapid development of thermonuclear weapons and the building of an unchallengeable pre-eminence over the Soviets in all strategic systems. The Oppenheimer-GAC, on the other hand, came to stress the need for more versatile nuclear warheads and civil defense.

The advent of the Eisenhower administration put the Strauss-Teller group on top, with Strauss as official AEC chairman and thus unofficial White House science adviser. But the politics of this period were to change rapidly, and the Oppenheimer-GAC scientists still played a formidable role in the formulation of civil defense policy under the National Security Council. The Lincoln Report of 1952 (calling for elaborate early-warning systems against potential Soviet bombing) and Operation Candor of 1953 (an assessment of Soviet thermonuclear capabilities, pointing toward the inefficacy of escalating the arms race and the need for new disarmament initiatives) represented the final influential work of this group. Strauss and the Air Force managed to convert Candor to "Wheaties," the Atoms-for-Peace initiative coupled to intensification of the new missile phase of the arms race.

The Oppenheimer security hearings of 1954 marked the culmination of all of the forces that undermined and destroyed the last vestige of influence of GAC scientists. Admiral Strauss and Dr. Teller remained the government's *de facto* science advisers, and scientists throughout the country mounted an intense campaign to counter their influence. The growing deadlock between Democratic congresses and the administration aided the cause of the out-of-power scientists, providing them a forum and a means of action for continuing the struggle for power.

Sputnik I crystallized the next phase of the situation, leading to a formalization of scientific advice in the Office of the President under the first official top scientist, James A. Killian. He turned to many of the out-of-power physicists of the old GAC for manning the new Science Advisory Committee (PSAC). During the remaining Eisenhower years, agencies of the government and scientists themselves were deadlocked on every major strategic question, including military strategy, NATO policy, disarmament, and space.

Free from most of the scars of these old battles, John F. Kennedy ascended to the presidency with a free hand, supporting the space and missile strategists but also opening new disarmament initiatives. He gave the GAC scientists of the Oppenheimer period a renewed mission and authority, eventually rehabilitating the reputation of Oppenheimer himself by plans (carried out by his successor) to give him the government's highest scientific medal, the Enrico Fermi Award.

In spite of this formalization of the scientific role at the highest levels, the trend has been away from the acceptance of scientific authority, toward the assimilation of this new tribe of experts into the ambiguities of traditional politics. In essence the scientists have not become high priests but rather part of the universal brotherhood of minstrels, pilgrims, jugglers, faith-healers, and advocates dealing in the marketplace of public policy, using and being used by more praetorian powers. They are neither sloe-eyed vestal virgins of abused innocence nor pristine high priests of a new technocracy. Neither is science one true cross; the so-called sainted brotherhood marches in all directions, and some of the brothers hold beneath their robes profane images and murmur extremely banal prayers.

Influence over public policy is nebulous and difficult to measure, but it is clear that, whatever its constituents, scientists individually or as groups have not possessed inordinate amounts. In evaluating their political influence, Daniel S. Greenberg, one of the best working journalists in the field of science policy, writes: "It is useful to remember that unlike labor, business, or the farmers, scientists didn't crash their way into Washington with voting power and political contributions; they were invited to Washington because their skills were needed by the men who had come to occupy positions of political power." [30] The moral of this tale is that the day is swift arising when the special competence of the scientist will be properly valued and understood. He is neither "the Wizard of Oz nor the benign doctor in the baby-food ads; he enjoys normal activities, hobbies, prejudices, political views, and a love life; he is as well fitted as other educated men to play an important part in the evolution of general culture. . . ." [31]

As a group, scientists can provide intelligence information concerning the physical parameters which must be reckoned with in the process of decision-making. They can offer informed but hardly decisive opinions concerning promising areas in which, through research and technological development, the investment of government funds may provide means for improving the tools of society to accomplish politically possible and desirable ends. They can judge somewhat better than others the credentials of researchers who seek public support for work that promises no direct or early social utility. But they have no special insight into that proportion of public funds that can and should be diverted for this purpose.

Put in proper perspective, science-technology issues are no more abstruse, complex, or forbidding than such issues as balance of payments, tariffs, or the tax structure, subjects which the lay politician has learned to master politically, if not to the satisfaction of the specialists. In modern society there are no public questions not enveloped with the specialized abstract knowledge and claims of experts, ranging from bankers to labor leaders. The genius of the political process lies in the recognition that expertise is useful in boiling out common-sense issues from conflicts among experts themselves. In the area of science and technology, politicians are acquiring self-confidence from witnessing the frequent clashes of scientists on all issues. Influential men in the legislative and executive branches of government know they cannot afford to rely solely upon any one source of advice from any brand of expert.

Congressional committees and federal agencies ignore this lesson only when they are already committed to a course of action and do not wish to weaken their position. This is unfortunate, perhaps, but inevitable. It is much more important that in the congeries of conflicting advice the President not fall into this self-serving systematic error. At the apex of decision, science must be relegated to its proper place by reflecting a broad representation of all conflicting advice in order that the underlying political issues may still be extracted as the focus for public debate and governmental consideration.

In his last science address (to the NAS), President Kennedy said, "As the country had reason to note . . . during the debate on the Test-Ban Treaty, scientists do not always unite themselves on their recommendations to makers of policy. This is only partly because of scientific disagreements. It is even more because the big issues so often go beyond the possibilities of exact scientific determination. I know few significant questions of public policy which can safely be confided to computers. In the end, the hard decisions inescapably involve imponderables of intuition, prudence, and judgment." [32]

CHAPTER VIII

THE POLITICS OF ARTFUL BRUTALITY: OPPENHEIMER AND STRAUSS

In June 1941, six months before Pearl Harbor, President Roosevelt created the Office of Scientific Research and Development (OSRD) under Vannevar Bush with responsibility for mobilizing scientists and engineers for impending war. Bush himself had been the major impetus of the action, supported in this role by a number of leading government and university scientists. The OSRD made a contribution to the war effort in the development of proximity fuses, radar, and minor improvements in weaponry. It aided the launching of the large-scale atom project which succeeded in producing two weapons before the end of the war in the Pacific, but whose real impact was not to come until later years.

At the start the atomic bomb idea was based on purely theoretical calculations and laboratory demonstrations. The great funds required for more solid findings necessitated government aid. Without the help of Bush and James B. Conant, it is doubtful that the early funding would have materialized as quickly as it did, making possible the Lawrence and Fermi experiments which gave substance to theory and rapidly expanded the government commitment.

As the project moved from laboratory to drawing boards and large-scale hardware, the scientists were increasingly robbed of paternity. As the Army Engineer Corps (represented by General Leslie Groves), the War Department, and the engineering and production contractors (represented by Du Pont's Crawford H. Greenewalt) stole the action, anxiety and resentment became the lot of many of the scientists who, under the leadership of Leo Szilard and others, increased their political role as the only independent constituency within the secret bounds of atomic policy-making. When the European war ended and availability of the first bomb by early 1945 appeared certain, the Chicago scientists pressed their views concerning the use of the bomb and the necessity for formulating postwar policy toward international control. This activity forced the government to consider these issues; but it was clear that non-scientific considerations were to play the determining role, and the great decisions were being made in the office of Secretary of War Henry L. Stimson.[1]

The significance of the scientists' role can be seen in the experience of Szilard who, unhappy with the course of the decision-making, sought to utilize the same route of influence which had been so successful in initiating the A-bomb project. He obtained a letter from Albert Einstein to President Roosevelt, which he hoped would give him an audience to argue against the use of the bomb on Japanese territory on grounds that such use would destroy the moral authority needed in the postwar world to lead the way to international control. The letter was found unopened on the President's desk following his death in April 1945. Szilard then called at the White House to bring the letter to President Truman's attention, but he was shunted into the office of an aide who arranged for him to speak with the new Secretary of State, James F. Byrnes. Byrnes reported in his memoirs that Szilard "complained" that he and his associates did not know enough about the policy of the government with regard to the use of the bomb, that he criticized Bush, Compton, and Conant as giving unrepresentative scientific advice to the President. Szilard argued the necessity that other scientists be permitted to present their views at the Cabinet level. Byrnes was annoyed by Szilard's claims but promised that J. Robert Oppenheimer would be brought into the consultations so that a wider range of scientific opinion might be represented.[2] As head of the Los Alamos lab where the bomb had been assembled, Oppenheimer had been close to inner policy and had also retained the confidence of the scientists.

The Szilard-Byrnes encounter convinced the former that the President and his Secretary of State did not grasp the true significance of atomic energy and were misled by those anxious to justify the $2 billion secret expenditure by military use of the weapon. On the other hand, Byrnes acquired a distinctly unfavorable opinion of Szilard, which confirmed his wish to exclude interloping amateurs from cardinal choices in the waning days of war. The significance of the episode is clear. Except for token consultations with a limited number of "practical" scientists, the atomic scientist community was to be excluded from the major role it sought in postwar policy-making, just as it had been during the production and engineering phases of the Manhattan Project.

The legislative battles of 1946 over control of atomic energy proved the opposition scientists to be vigorous, wily, and persistent organizers, lobbyists, and publicists. The battle had at stake the question of what future role certain scientist groups would play in the making of the postwar world. The organization of the Federation of American Scientists, the launching of the unique journal, the *Bulletin of the Atomic Scientists,* and the galvanizing of national opinion behind the scientists were the result of the debate. Much of the subsequent homage paid to the claims of science for participation in the policy process may be traced to this unprecedented 1946 success.

After 1946 J. Robert Oppenheimer became in effect, if not in fact, the leader of the new constituency and the foremost government scientific adviser. He was the chosen successor of Vannevar Bush who saw in him the combined skills of scientist-statesman which the role required. As a brilliant student of Ernest Lawrence, Oppenheimer had emerged in the 1930's as an outstanding teacher of the new physics. His diffident manner, the clarity of his thought and expression, his ability to efface himself in moderating conflicts between others, enabled him to interpret and lead the developing consensus of the postwar period.

These qualities had led Bush and Lawrence to press his appointment in 1943 as director of the nuclear weapons laboratory at Los Alamos. Here Oppenheimer recruited and directed a brilliant scientific team, integrating their efforts in what was certainly an unprecedented demonstration of political tact and administrative skill. In the policy discussions of 1944-1946 he won the respect of military men, responsible government officials, and politicians, while somehow managing to retain the confidence of the scientists. His leadership ability was founded on realism and practicality combined with patience, restraint, and consideration for others. Secretary of War Stimson thought him "the best of the scientists."

In late 1946 the President appointed Oppenheimer to the AEC's General Advisory Committee, which elected him its chairman. At the same time Bush brought him into the Military Research and Development Board, where he headed panels engaged in sorting out potentially useful military applications of atomic energy.[3] Under his leadership the GAC concluded that "the principal job of the commission was to provide atomic weapons and good atomic weapons and many atomic weapons." In addition he pressed for a broader concept of AEC responsibility. "We thought from the first that however remote civil power might be, the commission had an absolute mandate to do everything it could . . . to get on with the exploration of it." The commission "needed to respond to requests from the military and needed to alert the military establishment as to other applications of atomic energy of military use," such as propulsion and radiological warfare. "The third thing that we felt—and it was not really third in our feelings, but simply in a budgetary and practical way—was that the commission had a mandate to stimulate basic science" and "the training of scientists." [4]

His method of working (described in testimony by AEC officials, military men, and scientific colleagues) destroys vapid generalizations about the "idealistic innocence" of scientists. "A scientific adviser," Oppenheimer declared, "has one overriding obligation. It is his principal one in which he is delinquent if he fails . . . to give the best fruits of his knowledge, his experience, and his judgment to those who have to make decisions." It is his task to "study the problems that are put before him, to analyze them, to relate them to his own experience, and to say what he thinks will happen and what he thinks will not happen; what he thinks experiments mean, what he thinks will happen if a program is developed along certain lines." In the early days the GAC had to suggest to AEC members such areas of advice as they should be seeking, but "as time went on and the commission . . . knew more and more about the program, we tended to let the questions come from them." Sometimes the commission would address to them questions not limited to scientific advice, such as the hassle precipitated by military demands for custody of atomic weapons: ". . . in this case we confined ourselves to talking about the technical problems and pointing out that there were much more important political ones which it was not our job to pass on." [5] These were halcyon days of the nascent scientific establishment, and Oppenheimer was increasingly the catalyst of its unity and its spokesman to government.

The shock waves of the first Soviet atomic detonation in September 1949 jarred the nation and set in motion conflicting forces which were to rend the scientific community asunder. The event triggered a great debate on nuclear and diplomatic strategy, a debate which in one form or another continues today. In its most objective form, the transition of 1949 (leading to a crash program to develop the super-bomb and eventuating in 1953 with simultaneous U.S. and Russian achievement of that objective) marked the end of reliance upon one-sided atomic deterrence founded on the U.S. monopoly. It was now clear that the Russians could maintain scientific-technological parity with the West. The central assumptions of American diplomacy and military strategy were broken. Within the wall of governmental secrecy there was recognition of the change, and currents began to flow strongly in two contrary directions: one led toward the conclusion that neither side in the future would enjoy decisive advantages in a continuation of the strategic arms race, and therefore new tactical strategies, diplomatic versatility, and arms stabilization initiatives were required; the other current led to the contrary conclusion that Soviet scientific-technological capabilities were transitory, based upon espionage and relaxation of our own pursuit of the arms race—therefore a crash program for the hydrogen weapon and the multiplication of all strategic systems would restore American superiority and diplomatic strength. After 1953 the race of technological lead time was renewed by both sides with unprecedented fury in the new area of missiles and warheads, leading inevitably to the race in space after missile development endowed both with this new capability.

The H-Bomb Decision

Within hours of the first Soviet atomic shot, routine radiological surveillance by the Air Force brought the news to Washington. An advisory panel under Oppenheimer reviewed the evidence and concluded it was the real thing. Consultations with the State Department followed, in which Oppenheimer urged that the President make an announcement, to which the Secretary of State acceded. The congressional Atomic Energy Committee convened and was apprised by Oppenheimer of the situation. Senator Arthur Vandenburg plaintively asked: "Doctor, what do we do now?" [6] The General Advisory

Committee convened to consider this question. Teller, whom Oppenheimer had convinced in 1947 to stay with work on the H-bomb, urged an all-out effort. After a review by Fermi of the unpromising technical state of affairs, "there was a surprising unanimity that the United States ought not to take the initiative at that time in an all-out program for the development of thermonuclear weapons," but should continue supporting exploratory research in the event that new breakthroughs might justify a future reallocation of short-supply scientific manpower from existing weapons development to the problematical super." [7] After consultations with the State Department and the Weapons Systems Evaluation Committee of the military, the GAC concluded that the super-bomb, even if developed, would not enhance the strategic effectiveness of the existing atomic stockpile which already was capable of inflicting unacceptable damage upon an enemy and doing so with greater military precision because of its more limited impact radius.[8] Oppenheimer later enlarged on his reasoning: "I believe that their atomic effort was quite imitative and that made it quite natural for us to think that their thermonuclear work would be quite imitative and that we should not set the pace in this development." [9]

The AEC approved the GAC advice and forwarded it to the President, who did not challenge it. But the challenge was already building in other quarters: from the Air Force which saw strategic uses for a multi-megaton warhead and feared the Russians would go ahead in any case; from members of Congress who saw in Soviet atomic parity the end of all peace of mind; from minority AEC Commissioner Lewis Strauss, who soon convinced his colleague Gordon Dean—both of whom were persuaded by the Air Force's arguments; from Dr. Teller who convinced them it was technologically feasible; from other scientists with highly respectable credentials, including Ernest Lawrence and Carl T. Compton (then chairman of the Research and Development Board of the Defense Department), who reluctantly became convinced that greater risk lay in failing to learn before the Russians did whether or not the thing could be done; and from leading Republicans who were already engaged in a slashing attack upon the AEC.

The brief three-year unity of the scientific establishment cracked. After the meeting of the General Advisory Committee, Teller arranged a meeting with Senator Brian McMahon. Fermi tried to dissuade him from what clearly appeared to be a political end-run around the AEC, arguing, Teller reported, that "It would be a good idea if the scientists presented a united front . . . that it would be unfortunate if

Senator McMahon would get the impression that there is a divided opinion among the scientists. . . ." [10] In McMahon's office the senator greeted Teller with a comment on the GAC's report: ". . . it just makes me sick." Teller proceeded to pour out his grievances and the senator promised to do "everything in his power" to get a crash program under way.[11] Strauss and a number of Air Force officers simultaneously were briefing Republican congressmen. By the end of November both sides of the case had been pressed upon President Truman and a full rethinking was inevitable. The National Security Council appointed a subcommittee (Secretary of Defense Louis Johnson, AEC Chairman David E. Lilienthal, and Secretary of State Dean Acheson) which rendered the verdict for a crash program that the President announced in January 1950.

The evidence of the Oppenheimer hearings conclusively demonstrates (from the testimony of Teller himself) that thereafter Oppenheimer and the GAC bent their efforts to aid in implementing the decision, especially after the spring of 1951 when Teller and his colleagues stumbled upon a technological breakthrough which promised certain success. The charge that Oppenheimer lacked enthusiasm for the program has an *ad hominum* flavor and gained importance in 1954 only as other reasons for removing his security clearance wilted.

After Truman's declaration the policy role of GAC scientists became a matter of continuous controversy. Center stage was now shared with the politically potent Air Force–congressional alignment, supported by AEC Commissioner Strauss and the Teller-Lawrence scientists, including Dr. David T. Griggs who, as chief Air Force scientist, became the unrelenting antagonist of all the subsequent actions of the Oppenheimer group and the bitterest witness in the 1954 security hearings.

Korea, Continental Defense, and Candor

Between the years of the H-bomb decision and the suspension of Oppenheimer's security clearance, the battle continued between those who sought ways to stabilize the arms race and those who still believed the U.S. could win unchallengeable superiority. Three areas of controversy were involved: (1) A new Army-Navy interest developed in diversified tactical atomic bombs but was opposed by the Air Force which favored accretions to the strategic stockpile and to delivery systems. (2) In the light of new Soviet nuclear capabilities, emphasis shifted to early-warning systems and continental defense against enemy bombers; this was opposed by the Air Force as representing a "Maginot line" mentality, and its leaders sought a larger share of defense appropriations for the expansion of offensive striking power. (3) The most critical question of all, cutting close to the heart of all strategic debates since 1949, was Operation Candor, the report on the effects of thermonuclear weapons, implying the inefficacy of nuclear war as a means of supporting diplomacy, the imminence of a condition of mutual deterrence in the Soviet-American arms race, and the duty of government to inform the people and itself face up to this by exploring the common interest in stabilizing the strategic situation, perhaps jointly adopting means of limiting the spread of nuclear weapons and providing mechanisms of mutual assurance against preventive or accidental war. The Candor Report had a powerful appeal for President Eisenhower but threatened the interests of those who thought nuclear threats still effective if reiterated with determination to accept the consequences. These interests were convinced that credibility would lie with the side which committed itself to this single alternative.

These were the issues that figured importantly in the 1954 effort to discredit Oppenheimer and the views he so persuasively articulated. Witch-hunting and McCarthyism were only proximate causes, exploited by a combination of elements seeking to bend to their coup the otherwise cool-headed Eisenhower.

As doldrums persisted in the wake of the H-bomb controversy, the war in Korea and the problems of creating a viable defense for Western Europe preoccupied the various science advisory bodies. Department of Defense panels and the GAC produced a series of recommendations concerning tactical atomic bombs which led to "the very great expansion in the atomic energy enterprise to support a much more diversified use of weapons, even leading some people to suggest . . . that maybe the atomic weapons on the battlefield would be so effective that it would not be necessary to use them strategically." While participating actively in these deliberations, Oppenheimer decided he should resign as GAC chairman: "The super was a big item on the program. It wasn't going very well, and I wondered whether another man might not make a better chairman." [12]

President Truman late in 1950 called for a study to explore the question: "Is the mobilization of scientists adequate?" Establishment of an emergency office was considered but rejected in view of the already extensive research and development activities in the Department of Defense and the AEC. In early 1951 the President established a new scientific advisory committee which, to maintain independence from the military, was attached to the Office of Defense Mobilization, having indirect access to the President through the ODM director. Among other responsibilities, the committee was charged with studying the controversial area of continental defense. The dilemma on this issue was, in the words of a congressional report, "not only one of funds and resources" but also "the division of labor among the military services. The Army and the Air Force particularly are joined in a bitter contest over weapons and techniques for air defense." The Army strongly favored a system based upon ground-to-air missiles under its control, while the Air Force "would invest heavily in offensive weapons . . . and let the deterrent threat of these . . . guard the peace." [13]

In order to bring the best scientific talent to bear on this problem, plans were made for a two-month summer study to be conducted at the Lincoln Lab of the Massachusetts Institute of Technology. The Air Force viewed the whole subject as anathema and undertook to scuttle it. Air Force scientist Griggs testified later:

> We were concerned by . . . fear that the summer study might get into these things which we regarded as inappropriate for Lincoln and as of questionable value to the Air Force—I refer to the giving up of our strategic air arm, the allocation of budget between the Strategic Air Command and the Air Defense Command—but we were also very much concerned . . . because it was being done in such a way that had it been allowed to go in the direction in which it was initially going, every indication was that it would have wrecked the effectiveness of the Lincoln Laboratory.[14]

Air Force Secretary Talbot asked MIT President Killian to stop the project, but Killian demurred, suggesting that the summer study "would operate to the benefit both of Lincoln and the interests of the Air Force." Griggs told Killian he was afraid the results "might get out of hand, from our standpoint, in the sense that they might be reported directly to higher authority, such as the National Security Council." [15]

Out of this study came proposals for early-warning systems which ultimately were built as a major adjunct of hemispheric defense against Soviet bombers. The attitude of the Air Force toward the advisory committee's influence was revealed in a confrontation between Oppenheimer and Griggs arranged by Isidor Rabi in an effort to overcome Air Force suspicions. Griggs opened the interview with sharp accusations against Oppenheimer, thinly veiling a threat to denounce his loyalty in Congress or in public. Oppenheimer, Griggs reported, "asked me if I had impugned his loyalty to high officials of the Defense Department, and I believe I responded simply, yes, or something like that." Oppenheimer disdainfully called Griggs "a paranoid." [16]

The Air Force took serious umbrage at the third assignment given to Oppenheimer in 1952, which eventuated in the Project Candor document placed before President Eisenhower within a week of his inauguration. It was to be Oppenheimer's last contribution in the magic circle and the most important. Its implications polarized the formulation of high strategy during his long years of exile at the Princeton Institute for Advanced Study where, his face increasingly wrinkled and his frail figure bent, he returned to the quiet agonies of theoretical physics.

In the spring of 1952 Oppenheimer received a letter from Secretary of State Acheson appointing him to serve on a star-studded panel which included such figures as Allen Dulles and Vannevar Bush. The Secretary briefed the group in the presence of high defense officials, charging the Candor panel with the nebulous responsibility of preparing an unrestricted, wide-ranging survey of the problem of regulating armaments: "Was it a feasible goal? Was there any way to go about it? Were there any tricks to it?" Oppenheimer described the committee's work: "We took a look at the armaments situation, getting some estimates of the growth of Russian capability and some estimates of our own as a measure for where they might be sometime in the future. . . . We became very vividly and painfully aware of what an unregulated arms race would lead to in the course of years." [17]

The conclusions of the Candor Report today appear simple and obvious, but within the context of their time and place, and amidst the bitter domestic politics of the arms race, they were explosive. The report called upon government to apprise the people of the full meaning of atomic and thermonuclear weaponry and an uncontrolled arms race. It recommended the loosening of secrecy restrictions under the Atomic Energy Law in order to "work more closely with our allies with problems having to do with offensive and defensive aspects of large weapons," opening the door to new programs for bringing to fruition the peaceful potential of atomic energy. It suggested that the nation undertake "further measures for continental defense as a supplement to our striking capability." [18] The report noted that American superiority in the atomic field had come to an end and that therefore the problem now was to find "a formula for negotiation with Russia which would promote peace by total or partial disarmament."

So great was the impact of the report upon the new President that he embodied it as the core of his first major international pronouncement, the Atoms-for-Peace speech to the United Nations in December 1953. But with its overemphasis on the benefits of the peaceful atom and its underemphasis on ending the arms race, the speech shifted the major impact of the report. A variety of collateral forces were moving Eisenhower and his Secretary of State to ever greater dependence upon massive retaliation as a central strategic doctrine, diametrically countering the tendency of the Candor Report.

Trial, Conviction, and Guilt

The case began with rumbles from Senator Joseph R. McCarthy of impending disclosure of treason in the highest and most strategic places of the Eisenhower administration. Then in November 1953 William L. Borden, former staff director of the Joint Congressional Committee on Atomic Energy and now administrative assistant to Republican Senator Bourke Hickenlooper, sent a letter to President Eisenhower and FBI Director J. Edgar Hoover attacking the loyalty of Dr. Oppenheimer. The damaging information of the Borden letter was old material which had been in the dossier for many years and had passed review more than once, but Borden added to this the construction the Air Force placed upon Oppenheimer's role in the hydrogen bomb decision and in subsequent scientific advice, suggesting that he was a dupe of communist influence if not an outright Russian agent.

Eisenhower, seeing the dossier for the first time, was shocked by the pre-1942 information (membership in three organizations later cited as communist fronts, association before the war with Communist party members, namely his brother, his brother's wife, a woman he once planned to marry, and others) and immediately summoned Strauss. Without further investigation Eisenhower directed that a "blank wall" be placed between Oppenheimer and access to classified information. In a later news conference he explained that the action seemed to him compulsory in the light of circumstances.[19] AEC Chairman Strauss, who had as a member of the commission voted on several occasions to clear Oppenheimer on the basis of the same dossier, did not remonstrate with the President but expressed his disquiet by neglecting to suspend the Oppenheimer clearance immediately. Instead he left for Bermuda, leaving the unpleasant chore to Acting Chairman Dr. Henry C. Smyth. Strauss, however, was not to be relieved of this onerous duty. Smyth refused to act and Strauss, upon returning, was forced to issue the order himself.

The "good man" Eisenhower had not thought to inquire how the Oppenheimer record could have passed scrutiny of responsible men during the previous decade; he did not pause to seek other counsel in evaluating the urgency of the situation; he was not told that Oppenheimer's consultant contract was due to expire automatically in June and that the AEC did not have to consult Oppenheimer at all if they chose not to. Apparently Eisenhower felt his "blank wall" order was a fair-minded way to hold the case in abeyance pending further investigation, serving the purpose also of removing the target of McCarthy's next attack. In memoirs written nine years later, Eisenhower revealed the importance of his concern about McCarthy, citing in justification for his action a letter from historian Robert E. Sherwood:

> I do not know Dr. Oppenheimer, but I do know Dr. Conant and Dr. Vannevar Bush, and it was dreadful to contemplate that the enormous contribution of these distinguished American scientists should provide a series of field days for the McCarthy Carnival merely because, apparently, Dr. Oppenheimer had been guilty of political naiveté in some phases of his career. . . . You have taken entirely proper and wise action which will deprive McCarthy of a great deal of the headline thunder to which I am sure he was looking forward eagerly.[20]

During the same period, administration officials were trying to outdo McCarthy's witch-hunting zeal. In a mid-1954 speech Vice President Nixon crowed over having driven from government Dr. Edward U. Condon, a physicist who had received repeated security clearances, former head of the National Bureau of Standards, and one of the original consultants to the congressional Atomic Committee. "We're kicking the Communists and fellow travelers and security risks out of the government not by the hundreds, but by the thousands," Nixon added.[21]

The Personnel Security Board (under Gordon Gray) voted two to one against reinstating Oppenheimer's clearance. In its report the board took pains to disclaim that it was in any way trying a man for his honest opinions as an adviser to government. The majority report found "no indication of disloyalty despite Dr. Oppenheimer's poor judgment of continuing his past associations into the present." But it declared that "any doubts whatsoever must be resolved in favor of national security. Materials presented as evidence to this board leave reasonable doubts. . . ." While loyal, he might endanger the country by "some involuntary act" arising from "personal weakness." The board rejected the charge that Oppenheimer had sought to sabotage the H-bomb effort but agreed he had been guilty of not lending his enthusiastic support. In effect, the board rejected the whole bill of particulars that formed the basis of the proceedings, substituting its own new charges based on the hearings themselves. He had shown "a serious disregard for the requirements of the security system"; he had "a susceptibility to influence which had serious implications for the security of this country"; his lack of enthusiasm for the H-bomb program raised doubts "as to whether his future participation . . . would be clearly consistent with the best interest of security"; and finally, he had been "less than candid in several instances in his testimony." [22]

Teller testified at great length but refused to be pinned down to a single factual accusation, instead resorting to ambiguous subjective judgments. He provided no hard evidence whatsoever against his one-time friend but managed to convey his full support of the objective of destroying Oppenheimer's prestige. For example, he declared:

> In a great number of cases I have seen Dr. Oppenheimer act in a way which for me was exceedingly hard to understand. I thoroughly disagreed with him in numerous issues and his actions frankly appeared to me confused and complicated. To this extent I feel that I would like to see the vital interests of this country in hands which I understand better and therefore trust more. In this very limited sense I would like to express a feeling that I would feel personally more secure if public matters would rest in other hands . . .[23]

This was the man who (with Lewis Strauss at his elbow) became the potent if unofficial government science adviser as a result of the Oppenheimer hearings. His influence was paramount at least until the creation of the President's Science Advisory Committee in 1957. This was the man who, after Oppenheimer's canonization, was rejected by many leading American scientists and was suddenly to appear at Oppenheimer's side ten years later as the latter was presented the Enrico Fermi medal by President Johnson in a symbolic act of national contrition. At the 1964 White House reception in Oppenheimer's honor, Teller stood for a time at the end of the refreshment table munching hors d'oeuvres. As news photographers unlimbered their cameras, Teller, according to an observer, suddenly dropped a half-eaten cookie and bolted across the crowded room, elbowing his way through a knot of astonished bystanders. He managed to grab Oppenheimer's hand just as the flash bulbs started to pop. The picture that appeared the next day in newspapers throughout the land shows a startled Oppenheimer staring in disbelief at his smiling well-wisher.

The dissenting member of the Gray board, Dr. Ward V. Evans, wrote: "I personally think that our failure to clear Dr. Oppenheimer will be a black mark on the escutcheon of our country. His witnesses are a considerable segment of the scientific backbone of our nation and they endorse him. I am worried about the effect an improper decision may have on the scientific development in our country." [24]

Oppenheimer's attorney immediately appealed the case to the Atomic Energy Commission. In view of the Gray board's rejection of the original specification of charges, it was apparently felt necessary to respecify them, which AEC General Manager Kenneth D. Nichols did in a new document which changed the issues significantly, giving central importance to the so-called "Chevalier episode."

In 1942 Oppenheimer had been approached by Haakon Chevalier, a French writer and translator he had known before the war. The Frenchman asked if he would give information on the bomb project to a certain individual known to be a communist. This Oppenheimer refused to do, but, apparently wishing to shield Chevalier, waited several months before reporting the incident to security officials and, when he did report it, declined to give Chevalier's name. The security office pressed him for three months before he relented. This much of the incident had been known to the AEC for many years and was not found to be cause for action against Oppenheimer. In 1950 Oppenheimer wrote a letter at Chevalier's request to aid his passport application, and the Frenchman spent two days in the Oppenheimer home. In 1953, while vacationing in France, Oppenheimer went to Chevalier's home for dinner. The next day Chevalier took him to meet André Malraux, one of the leading literary figures of France and adviser to de Gaulle. While it was conceded that Chevalier had communist leanings during the war, no proof was ever introduced during the hearings to prove that he was pro-communist at the time of these later contacts.

Dwelling upon the episode, the new document pointed out that Oppenheimer had lied about his relations with Chevalier on several occasions, that he had continued this association, and by not fully reporting these facts to security officers had exercised "arrogance in judgment in security matters." As an afterthought, according to Harry Kalven, the document pointed out that Oppenheimer was "virtually a Communist until 1942," and that as a scientist and governmental adviser he was by no means indispensable.[25]

As irony would have it, the AEC released its decision upholding the Gray board on June 29, the day before Oppenheimer's consulting contract would have expired anyway. Four commissioners (Strauss, Joseph Campbell, Eugene Zuckert, and Thomas E. Murray) concurred with the security board, while Dr. Henry Smyth dissented. (The lone dissenter on the Gray board was, like Smyth, a scientist.) The majority report was based on a careful search of the transcript of the earlier hearings: "The important result of these hearings was to bring out significant information bearing upon Dr. Oppenheimer's character and associations, hitherto unknown to the Commission and presumably unknown also to those who testified as character witnesses on his behalf." The report cited six instances of conflict in testimony, adding that "the catalogue does not end with these examples," all proof of "fundamental defects in his character."

William Borden had written in his original denunciatory letter: "More probably than not, he has [since 1942] been functioning as an espionage agent, and . . . has since acted under a Soviet directive in influencing United States military, atomic energy, intelligence, and diplomatic policy." Majority members of the Gray board and the AEC refused to dignify this charge and even directly contradicted it. But it crept back into the final sentence of the AEC decision, softened and indirect but nevertheless a suggestive seed deliberately planted. The statement declared that Oppenheimer's early associations ". . . take on importance in the context of his persistent and continuing association with Communists, including his admitted meetings with Haakon Chevalier in Paris as recently as last December—the same individual who had been intermediary for the Soviet consulate in 1943."

In his dissent Smyth gave the real question at issue its first clear definition, that is, "whether there is a possibility that Dr. Oppenheimer will intentionally or unintentionally reveal secret information to persons who should not have it?" To this he found a plain answer: "There is no indication in the entire record that Dr. Oppenheimer has ever divulged any secret information." Commenting on the decision, Gerard Piel, editor and publisher of the *Scientific American,* declared: "J. Robert Oppenheimer stood convicted by accusation. . . . If fundamental defects of character were to be found, they were in the judgment, not in the judged." [26]

Giorgio de Santillana has perceptively compared the Oppenheimer with the Galileo trial.[27] In both cases "the accused could not defend himself against the fundamental accusation that was never brought up at the trial." Galileo had no advocate in court and the Copernican theories were not discussed. The only question was, had he disobeyed the Church or not? "Oppenheimer was allowed lawyers, but they had no clearance, and security considerations ruled out any adequate discussion of the facts relating to Oppenheimer's controversial views —which were, after all, the basis of the whole trial." In both cases the scientist was "shown a great deal of official consideration, although in the public consciousness he was clearly branded as one who was either too clever or too scared to commit himself to the major infamy, but whose intentions were sinister from the start." In each case the real purpose was "to inflict social dishonor on the accused in order to deter others from certain kinds of action that the authorities feared."

Santillana notes the similarity in the sentence accorded Galileo, who was convicted

> both for disregard of basic security policy ("thou hast dared discuss . . .") and for lack of candor ("nor does the license artfully and cunningly extorted avail thee"). The actual charges had to be trumped up, but the conflict underlying them was valid. It had to come to a showdown about "who is going to do the thinking around here," and some lack of candor was inevitable on both sides. Loyalty was reestablished at the price of humiliation; we end up exactly where the Gray Board leaves Oppenheimer.

The artful brutalities of politics are manifest in the corrupting influence of the case upon the integrity of all the participants, including Oppenheimer himself. As Santillana notes, Galileo began by challenging his judges but was brought to his knees. Oppenheimer, on the other hand, "is on his knees at the start—as his legal advisers told him he must be—pouring out in public a tale of his past personal attachments and private beliefs, recounting his insignificant indiscretions, protesting that he has learned his lesson, that he can still be useful." Before 1954, while still a highly regarded public servant, Oppenheimer had appeared before the House Un-American Activities Committee and testified on the basis of hearsay and speculation concerning Dr. Bernard Peters, a physicist then employed at the University of Rochester. He was sharply criticized for this performance by Dr. Edward U. Condon (among others), and later wrote a letter to the Rochester newspapers to try to save Peters' job. This incident was used against Oppenheimer in his own security investigation to confirm the ease with which Oppenheimer bowed to influence. The stability of his character was challenged on the grounds that in spite of his criticism, Oppenheimer later stood up for Condon's loyalty.

Haakon Chevalier was later to be heard from, denouncing Oppenheimer as "a hard, ruthless, utterly cynical man," prepared to sell out his friends and former political associates. He insisted the espionage suggestion was completely untrue, gratuitously invented by Oppenheimer in order to overcome his prewar record and win security clearance.[28]

Oppenheimer never openly criticized those who had acted as instruments of his humiliation. He never criticized Strauss, with whom he continued to be associated as director of the Institute for Advanced Study at Princeton, while Strauss was chairman of the board. When Strauss turned to John von Neumann, an old Oppenheimer friend and defender, and secured his appointment to the AEC in an attempt to overcome the whiplash effects of the Oppenheimer case, von Neumann accepted the post and for the remaining brief period of his life demonstrated loyalty to Strauss while maintaining friendship with Oppenheimer.

The Strauss–Teller–Air Force Triumvirate

Even Strauss, whose role in retrospect appears that of villain, was deeply marked by the events. In his book of recollections [29] he exhibits great defensiveness about his actions. Whatever the accidents of life, the man of healthy mind manages to justify his own actions: to cease to be the hero of one's own psychodrama is to choose death. Strauss has lived a long, honorable, and useful life, and in his memoirs he sees himself making the best of hard decisions. In addition to defending his actions, however, he has reflected at times on the moral ambivalence of public life. He was to call the Oppenheimer case "a tragic thing—I shall have to live with it as long as I live." [30] Maintaining a working office in a Washington hotel, he is quick to write to editors and authors detailed letters to keep the record straight. At the same time he has used his great energies and wealth in the enlightened support of young scientists and basic research. (His son is a physicist.)

The moral ambivalence of public life was given immediacy for Strauss in 1959 when congressional Democrats, seeking to embarrass the Republican President, rallied an impressive segment of the scientific community to defeat his appointment as Secretary of Commerce. They subjected him to a public ordeal no less cruel and humiliating than was Oppenheimer's. But where the vicissitudes of politics reversed the actions of 1954, bestowing upon Oppenheimer the apologetic homage of the nation, the decision of 1959 which forced Strauss to leave the Cabinet in disgrace lingers untouched by unseeing history.

There is no one to disturb the accumulating dust or to find in the record evidence of his helplessness to resist the grip of powerful forces that controlled his actions and forced him into the role of public executioner. The men whom he served in his days of honor—Woodrow Wilson, Herbert Hoover, Bernard Baruch—are gone, and Strauss wears out his days writing letters in a desolate hotel room.

In 1960, on the occasion of the death of Ernest Lee Jahncke, member of the Olympic Games committee in 1936, Strauss perhaps composed his own epitaph in a letter to the *New York Times*. Jahncke had protested the holding of the games in Berlin, where the Nazi government had barred participation by Jewish athletes, but was not supported by the U.S. committee. On that occasion, as on many others, wrote Strauss, Jahncke demonstrated that "he was a man of courage . . . although in this instance he was vindicated by history, he later told me he had never received any apology for the indignities to which he had been subjected. Many will sincerely mourn the passing of a brave man who did not hesitate to fight in a good cause even when alone." [31]

Strauss and Teller emerged from the 1954 events as the *de facto* science advisers of the Eisenhower administration. In the same year the Air Force succeeded in winning approval of a crash program for production and deployment of strategic ICBM's. This unleashed a new struggle for survival among the three military services, from which the Air Force emerged in 1958 the victor. For Strauss the period until 1958 continued to be full of tribulation and adversity. President Eisenhower became an ever-stronger advocate of the conclusions embodied in the Candor Report. Strauss had to serve as his agent, yet at the same time he was agent and advocate of Air Force strategic doctrines. He was torn from within and without in all the ensuing clashes of his remaining years of service. In his interior castle, the contest between the requirements of an infinite arms race and those of *de facto* U.S.-Soviet mutual deterrence was enacted, not as theoretical debate but as violent collision between the State Department and the AEC, between the President and the Air Force, between Strauss and the Democratic majority of the Joint Committee on Atomic Energy (which wrote into law ever-increasing hegemony over AEC powers), between the scientists struggling for funds to develop massive weapons systems in every area of advancing technology and those who pushed for a relaxation of nuclear secrecy, who painted grim pictures of thermonuclear effects and accused Strauss of duplicity in reporting the deadly dangers of radioactive fallout.

The job of AEC chairman and *de facto* presidential science adviser was deeply troubling. Discredited in the leading universities of the country (whose faculties competed furiously with one another to invite Oppenheimer for guest lectures), repudiated by most of the nation's leading scientists, driven by the Democratic leadership on the Hill into the unrewarding round of conforming to their wishes and begging their understanding, increasingly at odds with the President's new disarmament initiatives, selected (after the Dixon-Yates case) as the favorite target for Liberal attacks upon the "big-business orientation" of the administration, and hounded by the japes of newspaper columnists, Strauss carried on, a tough but expendable buffer for a universally loved President.

It was as a surrogate for the President that Strauss was ruthlessly cut down and eliminated from government in 1959. In June of that year, by a vote of 49 to 46, the Senate forced him to leave the Cabinet, the eighth Cabinet appointee in the nation's history to be thus rejected. During the Senate confirmation hearings Strauss was painted as a "master of deceit" and as a "sinister and remote control" over the nation's scientific programs. While the Air Force and the President, the two masters whom Strauss had tried to serve, stood invulnerable and untouched in the background, Strauss was brought to bay, convicted in the Congress and in the public prints of "deviousness" and "flaws of character," the same charges that had felled Oppenheimer.[32] The method of the 1959 hearings differed only in detail from that of 1954 or 1633. The testimony of the scientists was full of rancor and bitterness, their charges overflowing the cup. Republican Senator Hugh Scott declared with considerable justice: ". . . the record ought not to be left in this state. I have been around this town for a long time and . . . I have never in my experience heard testimony . . . which is more vindictive or more bigoted or more infested with a venomous desire for revenge against a public official than I have heard in this instance." [33]

Strauss came to the role of science adviser with excellent credentials, a record of varied experience in large-scale scientific undertakings and in government science policy, counting among his longtime friends most of the leading scientists of the nation. His long period of service on the original Atomic Energy Commission had been preceded by a tour of naval duty during World War II, in which he was deeply involved with production of new technology. Having no college degree, he was yet a well-educated man in the best sense and during his long years as financier made it his business to support creative scientific developments. As early as 1937 he was involved with nuclear physics, having financed construction of a surge generator to explore high-energy ranges for Drs. Arno Brasch and Leo Szilard. Convinced of the importance of their efforts to produce radioactive isotopes by bombarding certain materials with subatomic particles, he was faithful in his patronage and, when the need exceeded his resources, sought participation from industry. This was at a time when no government support was available and neither private foundations nor industry had a very great interest in this research. He aided Dr. Ernest Lawrence in finding funds for the first cyclotron which crossed the frontier of 100 million volts, and he personally paid the bills of Dr. Robert A. Millikan in research aimed at producing gamma rays for the treatment of cancer.[34]

After the Oppenheimer case Strauss made strenuous efforts to win back the confidence of the scientific community, playing a major role in revising security regulations, seeking to eliminate unnecessary secrecy surrounding AEC laboratories, continuing to support a substantial amount of basic research and an academic atmosphere in government laboratories. He sought and won the personal support of Dr. Detlov W. Bronk, president of the Rockefeller Institute and later of the National Academy of Sciences, in trying to win a positive relation with the community of scientists, even those portions that rejected him. He intervened to reverse the dismissal of Dr. Allan V. Aston, who had run afoul of a powerful constituent of the Secretary of Commerce when the Bureau of Standards, which he headed, refused to certify a valueless automotive battery additive. Even his bitterest critic, Dr. David R. Inglis, then chairman of the Federation of American Scientists, conceded that Strauss had made "an able administrator," at least "in the narrow sense of the material needs of science—he has brought to you senators a proper analysis of the material, the financial needs of science, and helped to make provision for them. . . ." [35]

But all this proved unavailing. In June 1954 a Los Alamos scientist expressed the consensus: "There'll be no strike of course, but I find my enthusiasm dropping to zero just when lack of enthusiasm has become illegal." [36] In April 1955 the University of Washington was forced to cancel an important scientific conference after most of the invited participants refused to participate on the grounds that the university had recently banned an Oppenheimer lecture. Even four years later, when Senator Paul Douglas polled the members of the American Physical Society in his state, of the 327 replies (41 percent of the sample), 70 percent of those who had an opinion voted *no* on confirmation of Strauss.[37]

The period of the Strauss-Teller advisory role was also the period in which the new ICBM phase of the arms race began. Administration initiatives toward disarmament (beginning with the 1953 Atoms-for-Peace, through the Open Skies proposal of 1955, to the stirrings of interest in a test-ban agreement in 1957) were compromised from the start by the nation's commitment to acquire the new weapon capability.

Many of the same officials who strongly endorsed this commitment also glimpsed the implications of strategic stalemate and were sincerely interested in further explorations with the Russians into disarmament possibilities. The virtual exclusion from governmental influence of the Oppenheimer scientists forced them to seek new routes to the public forum where they could argue the futility of the arms race and the inevitability of strategic parity, pointing to radioactive pollution of the atmosphere and heightened danger of accidental war as products of a policy which could not give either side any ultimate advantage. Linus Pauling, Hans A. Bethe, Isidor A. Rabi, Ralph Lapp, David Inglis, Leo Szilard—a complete listing would be very long. Albert Einstein put his tremendous prestige into the cause, remarking: "I would rather choose to be a plumber or a peddler than a scientist, in the hope to find that modest degree of independence still available under present circumstances."[38] The scientists found their most effective forum in such publications as *Science* magazine, the *Bulletin of the Atomic Scientists,* and the *Scientific American.* In addition, they used the platforms of their professional associations to raise their voices, to call upon people and the government to listen "before it is too late." In mid-1955 a group of internationally prominent scientists called upon all the governments of the world "to find peaceful means of settlement for all matters of dispute." A long list of Nobel prize winners joined Albert Einstein and Bertrand Russell in issuing an appeal for an international conference of scientists "to appraise the perils that have arisen as a result of development of weapons of mass destruction. We want you," the appeal read, to set aside national differences and "consider yourselves only as members of a biological species which has had a remarkable history, and whose disappearance none of us can desire. . . . We have to ask ourselves, not what steps can be taken to give military victory to whatever group we prefer, for there no longer are such steps." The question is: "What steps can be taken to prevent a military contest of which the issue must be disastrous to all parties? The general public, and even many men in positions of authority, have not realized what would be involved in a war with nuclear bombs." [39]

Thus began what came to be called the Pugwash Conferences, named after the Nova Scotia estate of Cleveland industrialist Cyrus Eaton, which was provided for the first conference after the U.S. government refused to grant visas to participating scientists from Iron Curtain countries. The meetings of Pugwash, held each year in a different country, enabled American scientists excluded from influence by the Strauss-Teller school to carry on a kind of quasi-diplomatic probing of Soviet motives and intentions, searching for evidence of the growing common interest of the two combatants in controlling the arms race and codifying the implications of strategic stalemate. These meetings enabled the American delegates to return to their own country and to speak with enhanced authority against some of the misconceptions concerning Soviet intentions which were being used by those who argued in favor of accelerating the American contribution to the self-fulfilling prophecy of the arms race.

As the Joint Committee on Atomic Energy became more and more the advocate and the sounding board of an infinite arms race, the scientists whom it had earlier favored as a means of attacking Strauss and the administration were dropped and replaced by Teller and his confrères. The Pugwash scientists, on the other hand, soon acquired a new congressional forum in the Disarmament Subcommittee of the Senate Foreign Relations Committee. Here the subcommittee chairman, Senator Hubert Humphrey, granted access at the national level to the Oppenheimer and Pugwash schools, thereby creating a counter-forum against the prevailing scientific advice within the administration, and opening the doors for genuine debate in Congress, in the country, and increasingly within the administration itself.

Sputnik I was to bring the voices of these scientists back into official government in late 1957. But that acceptance was quite different from the complacent unanimity of the scientific establishment during the early postwar period. Instead it represented what might be considered a more important breakthrough: the normalization of science politics. From this time onward the cruelty of 1954 began to recede amid recognition of the value-loaded nature of scientific advice. The process began which would assimilate the scientists as other kinds of experts had previously been assimilated to the rigors of politics.

The charge of deviousness imputed to Strauss recalls the testimony of Edward Teller against Oppenheimer, in which Oppenheimer's actions were described as "complicated . . . difficult for me to understand." It should be apparent that all such adjectives applied to men who have played significant political and administrative roles are gratuitous: "complicated and devious" is the nature of the political animal. No one can long serve as the mediator and broker of the conflicting passions and interests of men without the gift. It is only the innocent bystander or one who is futilely involved in pursuit of a blind abstraction who can afford the virtue of consistency. Responsibility for and power over organized social groups and governmental institutions denies this virtue to homo politico, who must instead cultivate the virtue of courage in accepting both the necessity and the risk of choice. The political man learns that he cannot expect justice or understanding for his motives and intentions but must depend upon the unpredictable factor of success. What is more, yesterday's success, however widely applauded, does not reduce the risks of today nor offer guarantees for tomorrow.

Whatever the contemporary image of justice surrounding the Oppenheimer-Strauss tragedies, both situations were locked in the grip of powerful historic forces from which no one has escaped. Those who thought to destroy Oppenheimer, just as those who sought to destroy Strauss, were seeking to do what C. P. Snow did to Lord Cherwell (Lindemann) in his book *Science and Government,*[40] that is, to punish someone for the loss of their own innocence.

Henry Smyth, who stood by Oppenheimer to the end, perceived the truth when he urged his fellow physicists to put aside the emotions aroused in them by the Oppenheimer case and to open their minds to the Teller-Strauss regime. He urged them "not to be set one against the other in the public mind for . . . advantage to one faction or another in or out of the government." While they should retain loyalty to honesty and truth, they should also exercise "a certain humility, and show a willingness to accept different points of view." In regard to those aspects of government science which may be inimical to them, "we must recognize the novelty and difficulty of the situation in which this country finds itself. We must be patient with the inevitable clumsiness of our attempts to meet this situation."[41]

CHAPTER

IX

THE EMERGENCE OF PLURALISM

Sputnik I turned the world-view of most Americans upside down. Before October 1957 a smug complacency prevailed about the superiority of U.S. science and technology. The backwardness of Russia was purblindly thought to be guaranteed by the social and political rigidities of communist ideology. Americans believed that their lead in strategic weaponry and the overhanging threat of massive retaliation would hold the Soviet Union's innate aggressiveness in check indefinitely—at least until internal forces brought about the downfall of communism or softened its militancy to a point at which American disarmament terms, the unification of Germany and Europe, and the pacification of the revolutionary areas of the world would be accepted. Sputnik I forced the West to reckon with the fact that its historic science-technology lead, unchallenged since the beginning of the industrial revolution, was coming to an end. Complacency swiftly moved to the opposite extreme of panic and despair.

All the pent-up frustrations of American society burst forth in a chorus of demands for change and action, each special interest looking for ways to turn the "crisis" to its own use: the academic community rose as a man, calling for reforms and new government programs for support; the Air Force launched the "missile gap" slogan, calling for further redundancy of strategic weapons; the other services argued the implications of strategic stalemate which required a build-up of subnuclear capabilities; social reformers blamed the national mortification upon the "power structure" which had maintained its special privileges and resisted change throughout the postwar years; scientists demanded greater government support of basic research; industry-labor politicians, caught in the wheel of economic recession and stagnation, raised their demands for protection, subsidies, tax relief, and new government contracts to maintain the health of the economy; and so on. No interests neglected to exploit Sputnik I to advance their own ends, or to offer self-serving formulas of all kinds which they claimed would revitalize the nation and save us from the Russians. The clamor was united in demanding change, but on little else, and, as the nation re-geared to the challenge, these forces once more fell to quarreling among themselves.

Out of this mood and the subsequent deadlocks of government came the slogans of the presidential campaign of 1960, in which both candidates promised to get the country moving again. The scientists who had been in eclipse for four years suddenly found themselves swept back into favor by an anxious nation asking to be saved. It became popular to talk about bringing Oppenheimer back into the government, as if he possessed some miracle tonic. The *New York Times* noted that the Oppenheimer case was rising up out of the past "to pose haunting second thoughts like a guilty conscience." [1] Senator Henry M. Jackson called for Oppenheimer's reinstatement to mark "a complete and final repudiation of the era of mistrust."

The shock of Sputnik awakened President Eisenhower to the need for improving the quality of advice he was receiving from self-serving executive agencies. "I think the President became aware," said Dr. George Kistiakowsky, "that the advice he was getting from his statutory advisers—the Cabinet members, the heads of the agencies like AEC, and the Bureau of the Budget and his National Security Organization in the White House—that their advice was somewhat inadequate. . . . President Eisenhower felt himself being sort of crowded by the military demands for ever-increasing R&D money and new weapons systems." He felt "that the channel of command wasn't screening the proposals effectively enough and that therefore he was forced to approve actions which he wasn't agreeable to for lack of technically meaningful advice." [2] Even before Sputnik the government had been torn by dissension on nuclear policy and military appropriations. The events of late 1957 gave impetus to the need for an independent overview.

Lewis Strauss had long since become more of a liability than an asset in maintaining links with representative segments of the scientific community. His parochial attachment to the Air Force had undermined his credibility; White House Chief of Staff Sherman Adams more and more leaned upon Dr. Detlov Bronk, president of the National Academy of Sciences, and Alan Waterman, director of the National Science Foundation, for guidance on science matters.

In November 1957 the White House announced creation of the post of Special Assistant to the President for Science and Technology, and James R. Killian, president of MIT, was named to this responsibility. Primarily an administrator of science and education, Killian had wide-ranging experience with the nation's scientists and quickly undertook appointment of a representative committee (the President's Science Advisory Committee), including many of the scientists associated with the Oppenheimer period. PSAC was modeled after the science advisory group of the Office of Defense Mobilization, in which Oppenheimer had played an early role. It was composed of eighteen scientists and engineers from outside government, appointed for rotating four-year terms, the members to maintain a high degree of autonomy. They were to serve part-time except for Killian himself, making their own procedures, pursuing studies on matters of interest to them as well as responding to questions assigned by the President. Until 1962 PSAC enjoyed executive privilege which protected its members from congressional scrutiny, denying Congress the opportunity to intrude its advocacy of fragmented bureaucratic interests upon their work.

PSAC quickly demonstrated its usefulness as an administrative tool for rationalizing federal R&D machinery and providing a perspective for evaluation of competing technical advice and demands. "They have been a scientific fire-brigade," filling the vacuum created by the inability of the Secretary of Defense to contain or resolve the competing demands of the services; they helped offset "the failure of the Department of State to secure technical competence adequate for dealing with such problems as arms control, nuclear test cessation, international scientific cooperation, and NATO technical problems"; they assisted the President "in coordinating important programs cutting across departmental lines . . . as technical auditors of certain ongoing agency programs." [3]

A scientific coördinating committee already existed among federal departments (made up of bureau chiefs and laboratory heads) but had little influence. It was transformed into the Federal Council for

Science and Technology, directly under Killian's oversight, gaining the prestige and authority of the President.

The impact of these new executive institutions was felt by all the advisory committees throughout the government. Previously these had been used for little more than support of agency special interests; now they became part of a centralized system which vaulted the crannies of the bureaucratic hierarchy. Before PSAC it was "very easy for an executive department . . . to say 'thank you' and proceed to disregard any advice given. The military have sometimes been past masters of this art," according to scientist Walter Brattain of Bell Laboratory. During the Strauss-Teller period, independent scientists summoned to advise lower reaches of government felt that any advice diverging from already firm positions was futile, "since . . . nobody will pay any attention anyway." The advent of PSAC changed this: "We as advisers had an independent channel to the top. Therefore we were listened to and taken seriously, at least part of the time." [4]

The role of technical advice in decision-making is never without political dimensions. From the beginning Killian and his successors found themselves deeply embroiled in the conflicts that divided government and nation. PSAC became the instrument by which the President could grasp the dimensions of these conflicts, even if he were unable to resolve them. Eisenhower now entered upon an active role in exploring opportunities for formalizing the fact of strategic stalemate. But his own weakened leadership and the course of international events prevented him from resolving fundamental issues. Too often, as in the case of test-ban initiatives, PSAC and the State Department (under Christian Herter) were forced to mount inconclusive battles against combined attacks and sabotage by the Air Force, the AEC, and powerful members of Congress. Too often Eisenhower balked at resolving difficult questions, resorting instead to the simpler expedient of budgetary controls in the name of "fiscal responsibility," often serving not to force choices but to intensify the circle of internecine strife.

PSAC played an important political role in strengthening the hand of all those elements (such as the State Department and the Army) that had been losing ground to the advocates of the arms race. It led successful opposition to congressional attempts to close the alleged "science gap" by a proposed Cabinet-level "Department of Science." It was largely responsible for the attempt to undercut Air Force control of space by asserting the principle of civilian supremacy which led to the founding of NASA. It was instrumental in reorganizing the control of R&D scattered among the military services, providing an independent means of evaluation and budgeting. Through the President's Scientific Adviser, PSAC was represented on the Committee of Principals which formulated disarmament policy. It cut through politically inspired funding of competitive weapons systems, forcing a presidential decision to stretch out the B-70 bomber, the nuclear bomber, and the Nike-Zeus anti-ICBM missile. It secured White House support for increased funding for basic research and education, strengthening the National Science Foundation's share of the federal budget. It was responsible for reactivating the scientific attaché program in the Foreign Service and led the move for greater integration of NATO scientific resources. The list could be expanded. In short, PSAC served as primary representative of the whole scientific community, offsetting parochial technical-political advocacy by less representative constituencies. It became a kind of governing body of the scientific community itself, and an instrument for continuous oversight of the place of science not only in defense but in all areas relating to the health of science and its impact on American life.

But PSAC's role was by no means decisive. In fact, it became a new focus of controversy; in many areas its influence was defeated during the balance of the Eisenhower tenure. As a staff agency it could do little more than enlarge presidential perspective and depend upon the quality of presidential leadership. The best Eisenhower was able to achieve in his last two years was a policy stalemate in which the most infantile ambitions of certain interests were held in check but were nonetheless able to keep government divided and block positive policy initiatives.

Disarmament: A Case Study

Disarmament policy furnishes a good example of the political divisions. The Candor Report of 1953 and the evidence of Soviet parity in strategic nuclear weapons convinced President Eisenhower that nuclear stabilization had become a key goal for the national interest. The early role of Harold Stassen as Special Disarmament Adviser to the President, blocked and frustrated by contrary forces and by his own personal ambitions, expressed this preoccupation in the early days of the administration. Renewed emphasis upon missiles and hydrogen bombs in the late 1950's accelerated the arms race into an area of even greater U.S. vulnerability. The failure of Atoms-for-

Peace and Open Skies, and increasing public anxiety about the rising level of radioactive contamination from weapons tests, narrowed the disarmament question. A nuclear test ban became the first concrete objective in exploring Soviet-American common interests in leveling the arms race. Sensationalized in the public prints, Linus Pauling, Ralph Lapp, and other Pugwash scientists captured public imagination in an unremitting campaign against official policies, succeeding in making the test ban a central issue in the 1956 presidential campaign.

In late 1957 the Committee of Principals (including the Secretaries of State and Defense, the AEC chairman, the head of the Central Intelligence Agency, and the President's Science Adviser) provided formal leadership, with a firm majority in favor of a positive approach to negotiations (AEC Chairman John McCone, frequently joined by the Secretary of Defense—first McElroy, then Gates—constituted the negative minority). In Congress, Senator Humphrey and his small band of supporters stood firmly with PSAC and the Pugwash scientists, providing an effective congressional forum for the President's policy. But the Air Force, its huge industrial constituency, the majorities of Congress, and a contingent of scientists and engineers associated with weapons development (under the leadership of Edward Teller) led a powerful bloc working against the Geneva talks. Both groups blocked one another, making American conduct of the negotiations appear indecisive and insincere. Anti-test-ban forces were able to prevent early consummation of an agreement, but their immediate objective—resumption of underground testing—was in turn frustrated by presidential decisions to continue the voluntary test moratorium during 1958-1960. This stalemate drove both sides to intensify domestic debate and political maneuvers, hardening the deadlock of policy and keeping the Geneva talks in rudderless drift.

This is not to say that the contradictions of U.S. policy prevented agreement in Geneva earlier than the actual 1963 treaty. But the inability to project a unified negotiating position prevented an early assessment of Soviet intentions and obscured the problems of changing diplomatic strategy, causing the United States to endure a period of unseemly confusion and uncertainty which contributed to the series of Soviet diplomatic successes and damaged American international prestige. Perhaps the underlying condition that defeated the Eisenhower test-ban initiative was the uncompleted missile phase of the arms race. Dependence by both powers upon slow-reaction, liquid-fueled, and vulnerable (above the ground) missile systems made it necessary for strategists to contemplate the necessity of preventive or pre-emptive nuclear attack (since these weapons could not retaliate if they were committed to riding out a first strike by the enemy). Manned bombers suffered from the same limitation and therefore lent substance to the arguments of Soviet and American Air Force generals that the approach of major war danger should constitute the signal for a massive strike before the same logic compelled enemy action. The "soft" technology of nuclear deterrence was the only force available during the late 1950's, insuring strategic instability and escalating mutual fear. This situation tended to convert the test-ban talks into a tool of fruitless diplomatic probing. Each side sought an escape from the terrible logic of the arms race, but both were anxious to preserve any existing advantage. For the Soviets this meant forbidding any kind of international inspection which might improve American targeting; for the United States, this meant insistence upon inspection not only to deter cheating but to eliminate the most significant, if slight, Soviet advantage in the secrecy of its missile installations, an advantage denied the U.S. by its open society.

This unstable condition was removed during the early Kennedy incumbency, not simply by the stronger executive control of military and foreign policy but by the improved technology of invulnerable strategic systems based upon "hardened" (that is, encased in underground silos of greatly reduced vulnerability), quick-reacting, solid-fuel missiles on land (Minuteman) and those (Polaris) carried by submarines which achieved invulnerability through movement under the oceans.

The congressional Joint Committee on Atomic Energy, its authority as a watchdog of nuclear secrets written into law, became the instrument for blocking progress in the negotiations. In January 1959, as the new 86th Congress assembled, a major engagement was joined. The AEC and the Air Force released the so-called "Hardtack Data" which confirmed Dr. Teller's arguments that secret nuclear testing was possible by detonating bombs in deep underground excavations which "de-coupled" (that is, greatly reduced) seismic effects. This, it was asserted, undermined assumptions of "reasonable risk" upon which the U.S. had predicated its willingness to reduce international inspection on Soviet territory. The Joint Committee insisted that the administration inject the Hardtack Data into the negotiations to prove to the Russians the necessity for vastly more thorough inspection of Soviet territory than the Russians were prepared to consider. The majority of the committee insisted that the talks be broken off if the Soviets refused to accept this new information. However, the Committee of

Principals convinced the President that the talks should continue; in March 1960 Eisenhower conferred with Khrushchev at Camp David. The two issued a joint communiqué calling for continued negotiation and agreeing that once a treaty was signed they would institute a voluntary moratorium on underground tests below an agreed threshold of seismic detection.

AEC Chairman McCone, Teller, and the Joint Committee moved to challenge the threshold set by the communiqué. Teller was the chief witness before committee hearings, arguing that it would not be possible "for at least several years" to develop any kind of reliable seismic system for detecting underground explosions up to at least five times the power of the Hiroshima bomb.[5] McCone arranged a special briefing for the President by Teller (and Dr. Gerald W. Johnson) to convince him that the agreed threshold was far too low and would permit cheating and important weapon improvements.[6] The President was impressed and announced in early May an expansion of the program to improve detection systems. At the same time, American negotiators in Geneva were seeking Soviet agreement for a conference of scientific experts to undertake evaluation of the detection problem and perhaps engage in a joint program. The U.S. delegation quietly informed the press of its displeasure with the President's "jumping the gun." [7]

This denouement provided the means whereby the Joint Committee on Atomic Energy was able to intervene decisively in the negotiations. The Russians rejected the notion that such detection tests were critical but agreed to participate in a conference of experts. The Joint Committee now declared its intention to see that full secrecy required by law be preserved in the use of American warheads for tests in which Russian scientists would participate either directly or as observers. The administration sought to avoid this technicality by suggesting the weapon be encased in a "black box"; but the Russians rejected this gambit, insisting on the right to inspect the device itself, charging that the "black box" raised suspicions that an attempt would be made, under the guise of coordinated research, to test new weapons. The State Department sought a way around this dilemma which would satisfy both the Joint Committee and the Russians, suggesting at one stage that all three nuclear powers contribute obsolete A-bombs to a pool from which the weapon to be used would be drawn. The Russians refused to contribute any weapons to what they called a delaying tactic, and the U.S. had to fall back and offer the Soviets an opportunity to open the "black box." Members of the Joint Committee immediately vetoed this suggestion, letting it be known they were unalterably opposed. There the issue stood, with all the elements of comic opera, buffeted around inconclusively during the remaining months of the Eisenhower administration.

As the conference of experts convened, the President told the press that the U.S. would go ahead with the tests using conventional explosives, declaring himself in favor of a joint U.S.-Soviet program and calling for an amendment of atomic energy laws to make this possible: ". . . everything that they found it necessary to see in order to determine whether . . . this thing is effective, they would see and should see." [8] But the Joint Committee adamantly refused to change the law. The American negotiator (James J. Wadsworth) did his best to reassure the Russians that the U.S. would not resume testing under any pretext whatsoever, taking the occasion of a July press conference to criticize Teller and to doubt "that the scientists who believe the way Dr. Teller does will have any influence in changing the mind of the Eisenhower administration." [9]

During this extended charade, test-ban opponents continued to urge that testing be resumed immediately, citing (as though it were a fact) Dr. Teller's speculation that the Russians were already secretly testing and were achieving significant breakthroughs which had tipped the balance of terror. These charges, endlessly reiterated and clothed with inventive detail (for example, the assertion that the Soviets were developing the cobalt or the neutron bomb), were thoroughly blasted in 1961 when Russia abruptly denounced the voluntary moratorium and conducted its most extensive series of atmospheric tests. This move opened the door for resumption of American testing and devastated the argument that the Soviets profited most from the voluntary moratorium through cheating.

JFK and Science Politics

In a sense, Sputnik I elected John F. Kennedy in November 1960.

Soviet space successes brought to the surface unsolved problems and unsatisfied demands in virtually all parts of American society. Status quo economic and social policies of the Eisenhower administration, drift and indecision, had created a powerful consensus in favor of movement, action, change, and dynamic leadership. Eisenhower policies had enshrined as immutable laws of nature the inevitability of recessions, of increasing unemployment, of incipient social and economic stagnation. For solution to racial problems, growing yearly more explosive, Ike appealed to "the mind and the heart of Americans." Sputnik I followed a long series of Soviet advances in foreign affairs and U.S. compromises or defeats: partition of Vietnam, the immobility of American power during the slaughter of Hungarians by Russian tanks, the loss of the Suez Canal to Nasser as Soviet threats drove our allies (and Western power) out of the Middle East—a situation which our Lebanese landings could not reverse. Associated with verbal clinging to the doctrine of nuclear massive retaliation was a seeping away of confidence in American leadership and a loss of diplomatic strength in dealing both with our European allies and with the Russians. The Soviets toyed with our weakness by means of on-again-off-again crises in Berlin, forcing us to welcome Khrushchev to our shores and making possible a series of deeply mortifying incidents in which governments throughout the non-communist world could not cope with anti-American demonstrations directed at the persons of the President and Vice President on attempted good-will tours.

Virtually every part of American society was shaken by Soviet ventures into outer space. A wave of sharp and widespread criticism arose against the softness, apathy, and decay of American culture. The myth of superiority was shattered, and we were forced to re-examine our educational systems, social institutions, defense organizations and strategy, research and development capabilities, and political leadership. The years of economic slow-motion, unemployment, racial tension, and international mortification focused upon what seemed to be the eclipse of U.S. scientific prowess, creating a consensus in favor of action that elicited only moderate response from the Eisenhower administration. The successive failures of the U.S. Vanguard program and a series of new Soviet space firsts accentuated the "missile gap" cry. Much of the nation's accumulated frustration vented itself on science policy.

As Eisenhower's Science Adviser, James Killian had begun the work of repairing the effects of the Oppenheimer purge. Kennedy completed the job. New disarmament initiatives, a clearer definition of defense strategy, efforts to control Air Force ambitions, and his decision to make an act of contrition to Oppenheimer removed most of the scars, reopening lines of communication between the government and the entire scientific community. This did not mean that scientists were to be placed on political pedestals, nor did it mean the return of the semblance of scientist unity of the early postwar years; on the contrary, it meant normalizing the politics of science as part of the larger mixed constituencies of national decision-making. Public controversy became both more complicated and less intense.

Kennedy showed his tendency early. As President-elect he sent a group of key personal advisers (soon to become officers in his administration) to the 1960 Pugwash Conference held in Moscow. Among these were Dr. Jerome B. Wiesner, to become Presidential Science Adviser, and Dr. Walt Rostow, to become chief policy-planner of the State Department. Ignoring the risks of associating the new administration with an international meeting widely characterized as a tool of communist subversion, President Kennedy used it as a means of opening disarmament conversations with the Russians. As one of the group stated: "Many of us have been busy since returning, communicating and transmitting this information to our own government. We were there as individuals . . . although I think we carried with us the confidence of the government." [10]

Kennedy's personnel selections revealed the trend of his thoughts and intentions. In addition to Wiesner, he named Dr. Glenn Seaborg to head the AEC, a man whose advisory career in government paralleled that of Oppenheimer and who had shared Oppenheimer's views on most critical issues. Army General Maxwell Taylor returned from civilian life to head the Joint Chiefs of Staff, a forum where for four Eisenhower years he had battled unsuccessfully against overdependence upon thermonuclear weapons. The appointment of McNamara as Secretary of Defense laid the groundwork for disciplining the unruly services and taming the Air Force bullies.

During its early months (in the midst of a new Berlin crisis) the administration moved toward a military strategy more appropriate to the changed nature of the Cold War. Early revisions showed a shift from exclusive emphasis on strategic systems toward beefing-up conventional capabilities, enlarging infantry, logistical, and tactical forces, de-mothballing naval vessels and tactical aircraft, and retraining troops in the skills of limited and counter-insurgency warfare. By the second year the doctrine of massive retaliation, in discard in action if not in words for many years, was explicitly rejected. The doctrine was considered increasingly implausible and had tended to deny the United States the wide range of intermediate diplomatic and military options between the extremes of nuclear war and surrender. In its place the doctrine of "flexible response" appeared, based upon limited retaliatory nuclear forces and the build-up of a wide spectrum of capabilities below the nuclear threshold. Deliberate plans were made to phase out vulnerable strategic bombers and soft missiles in order to achieve the stabilizing effect of invulnerable strategic systems which did not depend upon preventive attack for their deterrent value. These actions and comparable changes in Soviet strategy prepared the stage for successful consummation of a test ban in mid-1963, a leveling of the arms race, and an enhanced U.S. capability to deal with the problems of world political change in an atmosphere of reduced danger of surprise nuclear attack. The unresolved dilemmas of international diplomacy were in no way reduced by these adjustments, but—like the politics of domestic policy-making—they were normalized, taken from the shadow of imminent doom and returned to the immemorial patterns of Old World politics.

These actions by the new administration stirred in Congress a faint echo of purge-politics aimed at discrediting the views of the new scientific advisers. The Senate Internal Security Subcommittee continued its two-year campaign to "get" Linus Pauling [11] and issued a study which attempted to prove that the Pugwash scientists were naive and unwitting tools of the Soviet government.[12] But the public merely shrugged, and Wiesner and Pauling emerged unscathed.

Kennedy, who had promised to "get the country moving again," soon discovered the difficulty of moving the Congress toward his economic theories and social programs. After six months as chief executive he declared the moon to be a national objective (against the recommendation of most scientists), almost as a substitute avenue of action. Congress was holding his program in abeyance, the economy was sluggish, a recession was beckoning, and new Soviet space achievements seemed to dramatize America's growing discredit. The first manned flight of Gagarin came in April 1961, followed a week later by the Bay of Pigs—two sharp blows for the ambitious young man who seemed to be a helpless captive in the White House. Vice President Johnson, head of the National Space Council, was pushing hard for an augmented space effort. Key congressmen, all the aerospace industry, lobbyists, and promoters of the Air Force and NASA pushed. Perhaps here was the way. Perhaps on this issue the country could be unified and make the breakthrough Kennedy desired into a positive concept of government responsibility and social reform.

The moon program was hatched during a hectic weekend in May. According to one source, McNamara, Webb, the Vice President, and a few others met around the clock starting on a Friday evening and worked out the crash program which was presented to the President the following Monday. "We were told," one of the participants said, "not to fool around." [13] "Our objective," Kennedy declared in his next State of the Union Message, "is to develop in the new frontier of science, commerce, and cooperation, the position of the United States in the Free World." This nation belongs, he said, among the first to explore the moon, "and among the first, if not the first, we shall be." [14]

The lunar-landing objective is an abstract one, of no concrete utility per se. The objective was chosen because "by the time the Apollo lunar landing has been accomplished, the nation will have developed the scientific and engineering skills and industrial capability to move ahead into other manned space endeavors." The project involved development of the full spectrum of booster and spacecraft capabilities for near-space missions and building blocks for follow-on programs.[15] Apollo was designed to leapfrog the Russians, "to do something they have not yet done, and take a long-enough-range view of it to make sure that it is not the thing that they are already on the way toward doing." [16]

Meanwhile, PSAC's growing importance and political role kept alive congressional demands for a more effective means of access to science policy decisions. To head off adoption of other gimmicks, Kennedy agreed to establish a statutory Office of Science and Technology (OST) which would make the President's Science Adviser available to Congress and more directly subject to its authority through appropriations, therefore more sensitive to congressional influence. In March 1962 a reorganization plan established the OST with few changes in existing arrangements. PSAC remained as before. At the same time, the National Academy of Sciences established a new Committee on Government Relations which would assemble advisory committees for Congress on a basis independent of the executive science establishment.

Kennedy's chief scientist, Jerome Wiesner, primarily an electronics engineer, had worked on electronic components for the first atomic bombs and was later involved in the Lincoln Laboratory air defense studies of 1952 and on various government advisory committees. Killian included him in the first PSAC. A neighbor and friend, Senator Kennedy had relied on Wiesner for many years as his personal science adviser and confidant. As the director of OST, Wiesner raised the level of scientific management, insisting that a policy officer having a science background be made directly responsible for scientific and technical activities of the several departments and agencies. With these officials as its members, the Federal Council for Science and Technology (FCST) began to play the role of a subcabinet for science policy. As the nation's science czar (through 1963), Wiesner appears to have made a useful contribution, demonstrating some of the same qualities of statesmanship that Oppenheimer possessed in full measure. On each issue he tried to present to the President the "big picture," assembling the parochial views of mission-oriented officials into meaningful summaries which sought to separate technical information from political advocacy.

He was often the butt of criticism, but few sought to attack his personal integrity—a mode of politics of little efficacy under conditions of healthy pluralism. His most aggressive scientist critic was Dr. Philip Abelson, director of the Carnegie Institute's Geophysical Laboratory and editor of *Science* (influential weekly of the American Association for the Advancement of Science), who found himself in disagreement with a whole series of administrative decisions beginning with the lunar program. Abelson charged that "too much power is concentrated in a few overworked people," that Wiesner "has accumulated more visible and invisible power than any scientist in the peacetime history in this country," and that he "was a man who had failed to engage in forward planning of science policy." [17] It is true that Wiesner's relation to Kennedy was far more intimate than that of earlier science advisers; but neither he nor his successor, Donald Hornig, seemed to wield much direct influence over final decision-making. The head of OST has no operational jurisdiction over ongoing programs and is but one of many White House sources of advice. He lacks an independent constituency in the agencies, in Congress, or in the nation. Only in the scientific community might he be expected to have bloc support, but on every technical issue in recent years there has been no identifiable scientists' line. The political pressures the Science Adviser can bring to bear are at best marginal, and he is forced to establish his usefulness by the independence and objectivity of his role.

The OST is undercut by the existence of the Federal Space Council, which President Kennedy in 1961 placed under the chairmanship of the Vice President and which became the command post for space policy in an area of mushrooming federal spending. As chairman, Vice President Johnson converted the body from an advisory into a promotional device for larger NASA programs. This arrangement eliminated Wiesner and PSAC from the critical decision to land a man on the moon in this decade. Wiesner's participation was post hoc and perfunctory. His skepticism erupted eighteen months later when NASA announced selection of the lunar-orbital rendezvous route without even the courtesy of perfunctory consultation. Wiesner vigorously protested to the President. The decision had been taken by NASA's Manned Space Flight Management Council after a number of contractor engineering studies; neither the Space Council nor the Scientific Space

Board (set up by the National Academy of Sciences at NASA's request) had been involved in technical evaluation of this crucial determination. The President ordered a PSAC review after the fact, while NASA officials complained of Wiesner's "meddling." [18]

As for Project West Ford (radio-wave reflecting needles in orbit) and the detonation of Rainbow nuclear bombs in space, two projects which became highly controversial among scientists, Wiesner declared: "I believe the government acted in a completely responsible manner" in its efforts to assemble and evaluate representative scientific advice. "Despite this major effort, there was a breakdown in communication with some of the scientific community, and I am afraid a somewhat irresponsible attitude on the part of some scientists. . . ." Such experiments will continuously present such situations, and "for this reason we have undertaken to develop more formal procedures . . . to assure that any agency sponsoring large-scale experiments undertakes, well in advance, to study in depth the possible direct and indirect adverse effects and to assure that these studies are properly reviewed before the experiments are actually carried out. . . . If this process is going to work, the scientific community must develop a better appreciation of the complexity of the judgments involved." [19]

Many scientists have come to Wiesner's defense. J. Herbert Hollomon, a leading government scientist, has said that Wiesner was "critical, helpful, and insistent that the decisions in the Department of Commerce were ours and not his, and that he served only to help the President. . . ." He credits Wiesner with helping the President "open a small path of understanding with the Soviets," having encouraged "the careful analysis of cost and effectiveness that permitted the Secretary of Defense to re-establish civil control over the military," having made "some small steps to connect better the scientific community to the problems of the less-developed nations," and having "unobtrusively insisted on a proper place for science in the affairs of the nation" and for "continued support for free scientific inquiry." [20]

Lyndon Johnson and the Scientists

Donald Hornig, Wiesner's successor, was chosen by President Kennedy but assumed office under President Johnson. A professor of chemistry at Princeton, he had been chosen for PSAC membership in 1961. Wiesner and Kistiakowsky suggested his promotion to the top science post. A younger man, relatively unmarked by the brutal politics of the previous decade, Hornig demonstrated a fresh and independent judgment in PSAC. Unlike the older science-statesmen, Hornig lacked political connections in the administration or in Congress, and his name was previously unknown to the press and public.

The presidential succession of November 1963 left Hornig up in the air; he had yet to assume his duties and he was personally unacquainted with the man he was to serve. The practice of close presidential advisers submitting resignations to a successor did not exactly apply to Hornig who, though named by Kennedy, had not yet been sworn in. Wiesner took up the matter with Johnson and after a get-acquainted meeting Hornig's appointment was confirmed. Since that time the new Science Adviser has performed his duties routinely and quietly. He does not, as Wiesner did, drop into the President's office at odd hours, sitting on the edge of the desk and gesticulating in fervent conversation. He has managed to stay out of the newspapers, a trick in which his predecessor was unlearned, and has yet to appear on late-hour television conversation shows or write argumentative letters and articles in the scientific and public prints. Wiesner was perceptive in his choice: the time may have arrived when the top science adviser can carry out his role in a routine, businesslike manner, permitting the now well-established institutions of government science to function and leaving the politics of choice and the enforcement of presidential decisions to others. It might be said: Killian and Kistiakowsky built and launched the ship; Wiesner was skipper on the shakedown cruise, turning over to his successor a sound and smoothly operating system.

PSAC has changed its complexion since its formation. The preponderance of physicists continues, but there is now a leavening of biology, mathematics, chemistry, and various kinds of engineering, and the Harvard-MIT presence has diminished. Hornig and OST have responded to the President's needs in the traditional manner, providing advice on issues ranging from environmental pollution to tactical aircraft, involving PSAC in the quest for information to satisfy a policy need, frequently requesting individual members to set up expert panels to study specific questions. The President has turned to Hornig and PSAC to help define the requirements of the "Great Society." Apparently Johnson coined the phrase without anything concrete in mind, motivated by rhetorical requirements, compassion for his fellowmen, and vague dreams for mending and uplifting the nation and the world. He has since found it difficult both to translate these into specific programs and build a consensus for action. The President has come to look upon every facet of his administration as a source of ideas and proposals; his faith in the central role of science and technology has endowed PSAC with a special responsibility in responding to the call.

Hornig's performance is current events, not history, and as such cannot be adequately judged. But it is clear that the institutions of science policy have become indispensable as a means of enabling the President to balance the conflicting sources of technical advice received from all parts of government, from Congress, and from scientist publicists.

Johnson's relations with the scientific community have been generally good. After Kennedy's death, he wisely insisted on presenting the Enrico Fermi medal to Oppenheimer personally and held a White House reception, actions which in Oppenheimer's words "required some courage." He put on the Kennedy mantle in an effort to transfer to himself the loyalties that Kennedy had created in his staff, Cabinet, and the nation, striking out at the same time to build his own national image and retain the mixed bag of political loyalties and obligations accumulated during long years of political brokerage. Liberal scientists watched his every move carefully, their attitude saying in effect, "He'd better do right, he has so few redeeming qualities." And in fact he managed to do quite well, maintaining the confidence of a widely representative segment of the public and preserving his own future effectiveness and maneuverability.

In the 1964 election campaign, the Democratic National Committee organized "Scientists and Engineers for Johnson," in order (in the words of its publicity man) to convince "the guy in Pittsburgh in a T-shirt with a can of beer in his hand that the smartest people in the country like Johnson and consider Goldwater unfit." [21] The majority of the nation's leading scientists lent their prestige to the letterhead organization. As a counter move, Senator Goldwater's camp announced formation of a "Task Force on Science, Space, and the Atom," headed by Lewis Strauss and including among its members Teller, Dr. Willard F. Libby, a medical doctor, and a number of retired generals who were currently captains of industry. Johnson had the majority of Nobel prize winners on his side, while Goldwater had one (Libby), if that meant anything. In concern lest either party should cynically exploit "science" for the grubby purposes of winning an election, a third body was formed, the "Scientists' Committee for Information," chaired by geneticist and Nobel Laureate Edward L. Tatum. This group described itself as non-partisan and its intention to "offer impartial and accurate information on scientific issues that may arise in the campaign." [22] In a way, the division of the campaign committees of scientists provided an echo of the battles and alignments of "the bad old days."

The New Pluralism

After the overwhelming Johnson victory, the scientific community resumed its pluralistic role, once more a kaleidoscope of values reflecting its nature as an integral part of the whole society, containing in its midst all the crazy colors and patterns of lay politics. The test-ban treaty, the leveling of the arms race, and the functioning of the new institutions of science policy have tended to take the steam out of the scientist movements that flourished during the Eisenhower years. The monthly publication, *Bulletin of the Atomic Scientists,* founded in the battles of an earlier time, has sought to enlarge its focus as a journal of general public affairs but continues to lose readers at a precipitous rate. The once-militant Federation of American Scientists (FAS) quietly slipped into desuetude, its chairman pathetically writing in the June 1965 newsletter: "I know it is rather futile to ask, but I would like to know who the members are and what they think FAS should be doing. Won't you drop a note to me . . ." [23]

Leo Szilard's last great attempt to organize scientists as an effective independent political force, which he called the "Council for a Livable World," failed to win any broad support.[24] The council was primarily Szilard's personal instrument, engaged in providing him a national forum and a fund-raising instrument during the last two years of his life spent in speaking tours and Washington lobbying. He conceived this effort as "practical" in that it aimed to raise funds for aiding the election to Congress of enlightened candidates who shared some of Szilard's values. In the 1962 congressional elections, the council decided to concentrate its support on the Senate and advised members to give the bulk of their campaign contributions to Joseph Clark, Democratic senator from Pennsylvania, and George McGovern, a two-term congressman running for the Senate from South Dakota. The latter received a reported $22,000 in council contributions, about a fifth of his expenditures, and won by a slim six-hundred-vote margin. Szilard's death in the spring of 1964 denied him the pleasure of a great national campaign into which he hoped to force the issues raised by the council, winning by both compelling logic and practical politics a powerful consensus which would move the world toward rationality and peace. The instrument of this last effort has barely survived his passing and has an uncertain future.

Another political effort to rally scientists foundered from the start in the shallows of pluralism. During the summer of 1962, as the outgrowth of a meeting of the American Association of Social Psychiatry, the first national congress of "Scientists on Survival" (SOS) was held. The majority of participants represented the social sciences and practitioners of medicine, psychiatry, social work, and education. A resolution to protest the Kennedy administration's Rainbow bomb test disrupted the proceedings, many persons demanding that no statements be issued in the name of the congress. Dr. Tom T. Stonier of the Rockefeller Institute declared: "I am appalled by what is going on. This is political action of the most naive sort." There is, he continued, no "shortcut to peace . . . it's a long haul, you stand behind it, you push, but quietly—you don't send telegrams from the hip." [25] In succeeding years, SOS managed to conduct its conferences with greater decorum but without any perceptible movement toward the unanimity and influence that had motivated the founding psychiatrists.

The area of greatest scientist consensus has been opposition to the lunar landing program. For some reason, however, no significant coalition of scientists has invaded the public forum with this issue on its banners. Isolated efforts to use the American Association for the Advancement of Science and other professional groups as platforms have been quickly, and on the whole gracefully, contained by association leaders who do not consider their groups proper forums for controversial political attacks.

The new pluralism can be seen in the fact that on the two issues which in 1965 most sharply split the country, Vietnam and Santo Domingo, the role of scientists *as* scientists was indiscernible.

The pluralism of science in the 1960's has a negative as well as a positive side. Most significant is the tendency to go too far in devaluing the scientist and attempting to use him as apologist and publicist for one or another special interest, or to dismiss his special skills entirely. The political process has a remarkable tendency to learn from experience by substituting new errors for old, to overcorrect mistakes of the past, driven by a kind of restless compulsion on a course that swerves between mania and delusion. The present course of history, however, is not lacking in promising signs that at least the next great bout of collective frenzy will avoid either misguided worship or unwarranted persecution of scientists. No society, whatever its institutions, is secure against the unpredictable conditions of the natural or human environments which unleash the irrational and produce the great dramas of history.

Scientism, the vulgar myth which endows science and its practitioners with pseudo-religious authority, survives in spite of the fact of political pluralism and the growing technical savvy of lay political leaders. The very devaluation of scientific expertise has intensified the myths of scientism as scientists more and more become the hired mouthpieces, advocates, apologists, and special pleaders for a variety of vested interests. The use and abuse of science and scientists is not unlike the way in which secular leaders have always exploited the common faith of their times as an instrument of social control. The dynamics of myth-building are abetted by the scientist's own need to enhance his self-esteem and by the wish of his patrons to exploit the public habit of uncritical faith to advance their own political interests.

On the negative side of scientific pluralism is the practice in all government agencies of picking scientific advisers to support agency interests against the scientific advice of other agencies. For example, right up to the moment of McNamara's cancellation of the Dynasoar program (December 1963), the Air Force conducted "reviews" by groups of carefully selected Ph.D. aerodynamicists employed by Air Force contractors, who not surprisingly recommended continuation of the program, continuously discovering new technical reasons for asserting its feasibility and importance to confute skepticism at higher levels.[26]

Another negative of pluralism is the growing encroachment of consulting duties and promotional activity upon the laboratory and classroom. Too many scientists both in and out of government turn away from the arduous but creative inward-isolation of their special tasks, behaving more and more like lawyers or salesmen—jumping airplanes, hurrying from one consulting session or conference to another, in Washington today, Cambridge tomorrow, and Los Angeles the day after, returning to the laboratory only to check their mail before rushing off to another meeting in Europe, all the while treading the endless round of cocktail parties, committee meetings, and the promotion of new federal grants for their universities or contracts for their corporations. Who benefits from this merry-go-round? Certainly not their students or their research. Perhaps some benefit accrues to the scientists themselves and to the government agencies they serve, but this is questionable. The chief beneficiaries are probably the airlines and the Washington hotels that depend upon this indirect government subsidy.

Encroachment upon the scientist's time, work, and family life is part of the increased difficulty with which a concerned and responsible scientist seeks to preserve his personal and professional independence. Dr. Abelson's attacks on government science policy, though sometimes intemperate, are a useful service to the nation. It may be that under present conditions attempts to maintain objectivity and independent judgment are purchased at this price. The independent critic, according to Abelson, "pays a price and incurs hazards. He is diverted from his professional activities. He stirs the enmity of powerful foes. He fears that reprisals may extend beyond him to his institution. Perhaps he fears shadows, but in a day when almost all research institutions are highly dependent upon federal funds, prudence seems to dictate silence."[27] No bouquets are to be won by questioning the judgment of the enthusiasts. "It would be desirable somehow to formalize the role of the devil's advocate and to give the devil a little bit of due."[28]

NASA has been continuously criticized even by its scientist enthusiasts for the weakness of its scientific advice and for the abuse of its advisers. The advice given by consulting scientists appears to have little influence on the program; the chairmen and secretaries of advisory groups are NASA employees and may not be of the same caliber as the consultants; there is no advance consideration of agenda, and important technical issues may be deliberately excluded; minutes of meetings, if any, are prepared by NASA's secretariat, circulated many months later, and are too brief, emphasizing what NASA administrators wish to have emphasized.[29] Nobel winner Urey, a NASA and Apollo supporter: "It seems to me that the Atomic Energy Commission is better fixed in regard to this problem than is NASA . . . there are committees that meet quite often to advise them but these committees are often part-time people. . . ."[30] The atomic energy program "had a great positive base. Many top scientists worked in the program. But the space program is just the opposite. NASA grew too fast."[31] Of the 32,500 people on its payroll, only fifty-four came from the ranks of university scientists. More damaging is the fact that the few scientists with national and international reputations are mere adornments on NASA advisory committees which are largely composed of technical people who have special interests as NASA contractors.

NASA's principal advisory support is provided by the Space Science Board of the National Academy of Sciences, composed of outside scientists and engineers. Of this relationship, Ralph Lapp writes: "NASA seems to have availed itself of the prestige of the Academy . . . soliciting advice on space science and then deftly transforming the relationship to apply to the whole NASA program." [32] The academy president, Dr. Frederick Seitz, has frequently expressed support of NASA decisions, permitting the implication to stand that he spoke for the academy itself. Challenging this implication, Abelson conducted an informal poll and found the overwhelming majority of the academy opposed to the lunar program. Dr. Lee A. DuBridge, who is never intemperate, has called for a "more official scientific advisory body for NASA," pointing out that NASA "can ask the Space Science Board questions or not as it sees fit. It can take the Space Science Board's advice or not as NASA sees fit." [33]

In 1965 Space Administrator Webb moved to overcome scientists' complaints and give them a greater sense of participation by naming a group of scientist-engineers to be trained as astronauts for a future role in exploring the surface of the moon. Doubtless Webb would like to send all of the scientist critics of NASA on such missions.

Hornig and PSAC have recently attempted to close the gap by establishing two scientific panels (on space science and space technology) to assess the general direction and goals of the nation's space program, in order to provide the President with an independent evaluation. It has been largely through these bodies that the President turned down early large-scale funding of NASA-proposed projects for a manned Mars mission and extended moon exploration.

Congress and Pluralism

Through the use of the Presidential Science Adviser, OST, and PSAC, the executive branch can maintain objectivity and balance, but Congress has no comparable means to protect the integrity of scientific policy deliberations in the face of pluralism and the eroding independence of the scientific community. Theoretically, Congress should obtain objectivity by playing off all the parochial sources of scientific judgment in government and the nation against the President's advisers. This is the checks-and-balances system, but it is often not the way things are done. Rather, congressional committees tend to become enthusiasts and allies of one or another parochial interest, aligning their considerable powers with this or that dissatisfied department, bureau, or agency, attempting as advocates to impose the parochial view upon national policy.

The problem of creating viable mechanisms for improving the legislators' scientific information and judgment may get no nearer solution through new arrangements proposed or adopted. In fact, congressional preoccupation with organizational problems may be merely a reflection of growing unhappiness with executive effectiveness in knocking together the heads of ambitious agencies, reducing their separate pipelines of congressional influence, disciplining the whole federal establishment to the policy-making and enforcing powers of the Cabinet and the President. Struggles for political power are usually hidden behind the jargon of management and public administration.

It is naive to hope that Congress will change its stripes or that the exigent realities can be changed by exhortation or by organization gimmicks to conform to the theories of the separation of powers. If Congress were truly engaged in the search for a balanced consensus, it could summon adequate resources of independent and representative scientific advice to offset the self-serving tendencies of pressure groups, parochial federal agencies, and the President himself. This has never been the nature of the political process and never will be. As a consequence, the President's Science Adviser, OST, PSAC, and the political activities of unattached scientists in the public forum (like Abelson or Szilard) are essential to the precarious balancing act of public policy. Helpfully, pluralism exists too in the Congress, and whatever success the nation achieves in walking the tightrope owes much to those Capitol Hill mavericks who are willing to resist the pull of the inner circle. Such men as the late Estes Kefauver, the durable and sometimes ludicrous Wayne Morse, the restrained and eloquent William Fulbright, the fatherly Paul Douglas, the collegiate William Proxmire—all constitute a great national asset which serves to soften the distorting influence of special-interest advocacy.

Congressional attempts in 1958 to establish a Cabinet-level Department of Science appealed to legislators as a means of enhancing their influence over science policy. In 1960 the House Science and Astronautics Committee made an interesting innovation, appointing a scientific panel to consult with it regularly. Appointments to the panel included some of the most prestigious of the nation's scientists (such as James A. Van Allen and Lee A. DuBridge), promising an interesting experiment. But as the House committee became more firmly committed to Air Force and industry designs, the role of the panel became increasingly ceremonial. By 1964 its annual one-day seminar was adorned by the presence of outstanding foreign scientists, which symbolized the end of the group as working advisers. One of the members of the panel told Ralph Lapp: "The committee doesn't really work us." [34]

In 1963, as scientific opposition to the lunar program swelled, a number of junior senators sought establishment of a "Congressional Office of Science and Technology" (whose initials—COST—would have had a certain pertinence). Bills were submitted to make available "a small, highly skilled, permanent professional staff and a large body of consultants . . . to furnish advice, evaluations, and reports. It would be the task of the staff to ask questions for Congress, to answer questions of Congress, and to assist the members in handling the scientific matters which come before them." [35] The congressional leadership softly turned aside this proposal, pushing instead the establishment of a Science Policy Research Division in the Legislative Reference Service of the Library of Congress.

All these arrangements are useful but modest devices. Within the limits imposed by the art of the possible, they enhance the relevance of congressional information without altering in any way the structure of political power and the advocacy of special interests. Clinton P. Anderson, chairman of the Senate Space Committee, has expressed the realities of the situation: "I don't believe you are ever going to be able to give Congress the information it wants by trying to build into every staff the complete answer to every question." Indifferent to the addition of scientific advisers or, indeed, other kinds of experts, he prefers a few general staff men (on his committees these have usually been lawyers) "in whom you have complete confidence. They go out and gather information." He cites the role of Joint Atomic Committee staff director James Ramey, who under Anderson's sponsorship became a member of the AEC in 1962: "We told him to watch the AEC. If we found they had swept out the office and hadn't told us about it, we had somebody down there to find out why. We got into a few fights but people began paying attention to things and before long we had a staff that really went into these questions hard." [36] This happened during the second Eisenhower term when the Democratic Congress successfully invaded the precincts of presidential power, carving out a dominant role as broker of the competing interests in the government and the nation, blocking or evading executive control over military spending, the conduct of the test-ban talks, the unification of the services, and the molding of the infant Space Agency. During this period Anderson's Joint Atomic Committee exercised quasi-administrative control over the government's entire atomic establishment.

Senator Anderson's realism is compelling and instructive. He dismisses the intrinsic scientific merits of issues as of negligible importance in political choice. What really matters is demonstrated personal loyalty, reliability, and the proven ability of men to serve each other's interests. Anderson: "I think a politician is someone who has learned that people have interests. They want to find someone who can conserve these interests. And when they have found him, they do like it says in the song, they never let him go." The model politician, Anderson says, "doesn't go around doing 'good,' but he at least goes around trying to understand what other men want." [37]

How do you resolve an issue involving scientific choice? "By the time you have been elected to office a few times . . . you find out very quickly which precinct chairman delivers for you and which precinct chairman doesn't; which man will make a good postmaster or which fellow will be an ingrate the rest of his life if you make him postmaster." That is why, Anderson offered,

> Rickover has a great hold upon the Congress . . . I do think we pick individuals by use of a queer yardstick. We pick the ones we trust. If Norris Bradbury came to me and said "I've found a new way to make a bomb. I am going to take three scoopfuls of common dirt, add an ounce of paregoric and an ounce of something else, and shake them up," I would know it would work. . . . I trust him even if I don't know the scientific reason for what he is doing, or couldn't understand the scientific reason if I knew it.[38]

Although he may bruise the myths of sensibility, Senator Anderson speaks with candor and insight. Even the most uneducated man understands the true nature of politics, however it may be obscured by the myths of consensus or by the oversimplified idealism of the politically rejected.

Proposals for organizational innovation in Congress or the executive branch, and the adoption of codes of ethics for scientists, industrialists, and politicians, though sometimes necessary and useful, are empty of meaning except as they acquire significance through the political process; and they will not significantly alter the dynamics of the process. The issues of science "that arise in the comparatively orderly world of physical matter" will always be subordinate to the infinitely greater complexity and disorder of the human environment. "What a trivial exercise to construct a rocket for interstellar travel, compared with the task of explaining the behavior of Jersey City's electorate." [39]

The re-examination of science and the role of scientists that has been going on in recent years in Congress and in the scientific community may produce positive results, or it may be merely pseudo-soul-searching whose chief aim is to provide a substitute for a real confrontation of the issues posed by the ominous and persistent build-up of social deficits in the contemporary world: the stagnation of cities, the caste system which generates violence between teen-agers in cities and suburbs, deliberate creeping inflation, the concentration of economic power, the growing gap between the have and have-not nations, between affluent and impoverished sections of the national economy, the squandering of resources, the pollution of soil, water, and atmosphere, and mindless technological change. The ever-present imponderables of group conflict may, when the process of politics breaks down, erupt into lead pipes in the streets or thermonuclear fire storms in the skies.

CHAPTER

X

THE CONTRACT STATE

American society has shown considerable resilience in adapting to rapid change. When need arises, an existing institution leaps to meet it, becoming quite a different institution in the process. In recent years the industrial corporation, which "began life as a sort of legal trick to spread the ownership of industrial equipment over a lot of people," [1] has evolved into a routine and immensely powerful method of organizing large undertakings. But the relationship between individuals and corporations has been greatly modified by the overshadowing concentration of corporate power. Government has ceased to be merely a passive arbiter of "the rules of the game" and is forced to become an omnipresent force for balancing the competition for values and controlling the dynamics of social change.

Government has become the economy's largest buyer and consumer. The government contract, improvised, ad hoc, and largely unexamined, has become an increasingly important device for intervention in public affairs, not only to procure goods and services but to achieve a variety of explicit or inadvertent policy ends—allocating national resources, organizing human efforts, stimulating economic activity, and distributing status and power. The government contract has risen to its present prominence as a social management tool since World War II, achieving in two decades a scope and magnitude that now rival simple subsidies, tariffs, taxes, direct regulation, and positive action programs in their impact upon the nature and quality of American life. This evolution has occurred quietly and gradually through a series of improvised reactions to specific problems. Its central role has been achieved without public consideration of far-reaching social and political implications. Even today there is precious little consciousness of the trend; political leaders tend to see each contract as an isolated procurement action, overlooking the general pattern. Just as federal grants-in-aid to state and local governments have (since 1933) become principal means for national integration of divided local jurisdictions, so federal contracting with private corporations is creating a new kind of economic federalism.

The implications of grants-in-aid have acquired some clarity: state taxation still takes care of traditional functions, while new and greatly expanded activities devolve upon local bodies through national decision-making, the states operating more and more as administrative districts for centrally established policies. Here, decision-making is nationalized under the constraints of public attention and democratic politics. On the other hand, economic federalism based upon contracts holds implications that are far from clear. To some degree, the forms and effects of contracting evade the forums of democracy, obscuring the age-old conflict between private and public interests. Mobilized to serve national policy, private contractors interpenetrate government at all levels, exploiting the public consensus of defense, space, and science to augment and perpetuate their own power, inevitably confusing narrow special interests with those of the nation.

Explicit authority for the U.S. government to conduct its business by contract is not found in the Constitution but has historically been accepted as a means of achieving explicitly constitutional objectives. There is ample precedent, such as the use of railroads for troop movements, or General McClellan's arrangements with the Pinkerton Detective Agency for espionage against the Southern Confederacy.

What is new is the persistence and growth of government-industry contract relationships under which, in the words of David E. Bell (then director of the Budget Bureau), "numbers of the nation's most important business corporations do the bulk of their work with the government." The Martin Company, for example, does 99 percent of its business with the government. Bell asked: "Well, is it a private agency or is it a public agency?" Organized as a private corporation and "philosophically . . . part of the private sector," yet "it obviously has a different relationship to governmental decisions and the government's budget . . . than was the case when General Motors or U.S.

Steel sold perhaps 2 or 5 percent of their annual output to government bodies."[2] Except in time of war, the government traditionally has not been the dominant customer for any private firm. The contract state of the postwar world must be viewed as a drastic innovation full of unfamiliar portents.

Grandiose claims are heard on all sides for the "unique contribution" that the contract mechanism has made in preserving "the free enterprise system" at a time when it could have been damaged. Atomic energy has been cited as an example of the new collaboration: "Without contracts, it would be government-owned and operated. With contracts, one person in sixteen in the industry works for government; the other fifteen work for contractors."[3] An aerospace journal cites space technology as "the fastest moving, typically free-enterprise and democratic industry yet created," achieving these values "not on salesmanship" (that is, traditional quality/cost competitiveness) but "on what is needed most—intellectual production, the research payoff."[4] Lyndon B. Johnson, while Vice President, argued: "If we want to maintain credibility of our claim to the superiority of a free political system—and a free private enterprise system—we cannot seriously entertain the thought of precipitating now so massive a disillusionment as would follow a political default on our commitments in space exploration."[5]

The government contract has made it possible to perform new tasks deemed essential without direct additions to the size of federal government, thus preserving the alleged rights of private property and profit. But these huzzahs ignore the real ambiguity of the system that is emerging—neither "free" nor "competitive," in which the market mechanism of supply/demand (the price seeking the level which best serves overall productivity and social needs) has been abolished for key sectors of the economy, its place taken by the process of government policy and political influence. Instead of a free enterprise system, we are moving toward a government-subsidized private-profit system.

Key to the Kingdom

Unlike older government-fostered industries, the new contractor empire operates without the yardsticks of adequate government in-house capability or a civilian market in areas where research and development has become *the* critical procurement and the crux of the system. As described in the 1962 Bell Report: The companies involved "have the strongest incentives to seek contracts for research and development work which will give them both the know-how and the preferred position to seek later follow-on production contracts." [6] Favored corporations that win R&D work thereafter exploit a number of special advantages: They may achieve sole-source or prime contractor status, which eliminates competition and dilutes all cost and performance evaluation. The open-end, cost-plus nature of the contract instrument, the lack of product specifications, official tolerance of spending overruns, all of which increase the total contract and fee (in a sense rewarding wasteful practices and unnecessary technical complication), permit violation of all rules of responsible control and make possible multiple tiers of hidden profits. The systems-management or prime contractor role enables favored companies to become powerful industrial brokers using unlimited taxpayer funds and contract awards to strengthen their corporate position, cartelize the contract market, and exert political influence.

In less than a decade the area surrounding Washington, D.C., has become one of the nation's major R&D concentrations. Every large corporation has found it necessary to establish field offices in proximity to NASA, the Pentagon, and Capitol Hill. Most of these new installations emphasize public relations and sales rather than research and development. The Washington area now ranks first in the nation for scientific personnel (per 1,000 population), although the major product is company promotion and politics rather than science.

The gross figures provide an index of the economic impact: the 1966 federal budget called for $23.7 billion in new obligational authority for defense and space—$11.4 billion for Defense Department procurement of hardware and control systems, $6.7 billion for R&D; $5.26 billion for NASA (virtually all R&D), and an additional $272 million for space-related R&D conducted by the Weather Bureau, the National Science Foundation, and the Atomic Energy Commission. Over 90 percent of this flows to the highly concentrated aerospace industry.[7] Another $3.3 billion was budgeted for other kinds of R&D, making a total of $27 billion. The 1967 budget allocated more than $30 billion to aerospace. Space, defense, and R&D together now comprise the single most substantial allocation of federal funds, towering over all other programs. In the mid-1960's, government R&D (excluding related procurement) stabilized between 2 and 3 percent of the GNP. Cumulative missile/space spending in the decade which began in 1955 amounted to over $100 billion (Defense Department, $84 billion; NASA, $18 billion), and the remainder of the sixties will add at least an additional $125 billion.[8] Virtually every department and agency of the federal government is involved to some extent in R&D contracting, although the Defense Department and NASA account for more than 96 percent.

The first result of this staggering outpour has been the artificial inflation of R&D costs which has enabled contractors to raid the government's own in-house resources. Officials in the lower reaches of the government bureaucracy (both civilian and military), charged with administration of contracts, find themselves dealing with private corporate officials who often were their own former bosses and continue as companions of present bosses and congressional leaders who watchdog the agencies. A contract negotiator or supervisor must deal with men who can determine his career prospects; through contacts, these industrial contractors may cause him to be passed over or transferred to a minor position in some remote bureaucratic corner, sometimes with a ceremonial drumming before a congressional committee.

The military cutbacks that characterized the Eisenhower years were accompanied by expanding military budgets, a paradox explained by the systematic substitution of private contractors to carry out historically in-house activities. This trend was heralded as a move back to "free enterprise." Government installations and factories built in World War II were sold to industry, usually at a fraction of the taxpayers' investment. Others were leased at low fees to contractors who were then given government business to make the use of these facilities profitable. In some instances government built new facilities which it leased at nominal fees. Such facilities were permitted to be used, without cost, for commercial production as well.

The splurge of mobilizing private contractors for government work occurred as a part of the unprecedented growth of the Air Force. As an offspring of the Army, the new branch lacked the substantial in-house management, engineering, and R&D capability that the Army had built into its arsenal system. The Air Force sought to leapfrog this handicap in competing for jurisdiction over new weapons systems, turning to private contractors to correct the defect. In its rapid climb during the fifties, the Air Force fostered a growing band of private companies which took over a substantial part of regular military operations, including maintaining aircraft, firing rockets, building and maintaining launching sites, organizing and directing other contractors, and making major public decisions. In the area of missilery, junior officers and enlisted men were subordinated to the role of liaison agents or mere custodians.

This had several bonus effects, enabling the Air Force to keep its military personnel levels down in conformity with Defense Department and administration policies, while building an enormous industrial and congressional constituency with a stake in maintaining large-scale funding of new weapons systems. The Air Force's success over her sister services during the Eisenhower years established the magic formula that all federal agencies soon imitated. It set in motion a rush to contract out practically everything that was not nailed to the floor and, in the process, it decimated the government's in-house management, engineering, and R&D capability; inflated the costs of R&D through futile contests for supremacy among contractors financed by contract funds; and as a consequence reduced as well the scientific and engineering resources available to the civilian economy and to the universities.

The Army learned an important lesson in its struggle with the Air Force during the Thor-Jupiter controversy—that its extensive in-house engineering-management capability was a positive *disadvantage* in mobilizing congressional and public influence to support military missions and budgets. Private industry had provided the Air Force with a potent weapon in Congress for outflanking the Army during all the years of strategic debate over missile development and the role of infantry forces in a nuclear world. In part, the Air Force lobbying instrument of the 1950's contributed importantly to overdependence by the nation on nuclear weaponry and massive retaliation as the primary security doctrine, while the complete range of subnuclear military capabilities was allowed to wither. This lesson was inscribed on the Army-Navy skin by the budget-paring knife of the Eisenhower administration and led to gradual weakening of the arsenal system. In the sixties all the military services and NASA sought to parade bankers, captains of industry, local business leaders, and politicians through the halls of Congress and the White House as lobbying cadres in every new engagement.

The old research triad—government, industry, university—has virtually disappeared. In its place is a whole spectrum of new arrangements, such as the so-called "systems-engineering and technical direction" firms operated on a profit or non-profit basis (for example, General Electric is employed by NASA to integrate and test all launch facilities and space vehicles, while Bellcomm, a subsidiary of American Telephone and Telegraph, is employed for engineering and management of all NASA operations; Aerospace Corporation plays a similar role for the Air Force). In between are the major corporations, universities drawing a majority of their research budgets from government, non-profit institutions conducting pad-and-pencil studies of strategic and policy matters for government agencies, and government laboratories operated by industry or by universities.

Knitting the complex together is an elite group of several thousand men, predominantly industrial managers and brokers, who play a variety of interlocking roles—sitting on boards of directors, consulting for government agencies, serving on advisory committees, acting as managers on behalf of government in distributing and supervising subcontracts, moving between private corporations and temporary tours-of-duty in government. Private corporations have contracts to act as systems engineers and technical directors for multi-billion-dollar R&D and production activities involving hundreds of other corporations. Instead of fighting "creeping socialism," private industry on an enormous scale has become the agent of a fundamentally new economic system which at once resembles traditional private enterprise and the corporate state of fascism. A mere handful of giants (such as North American Aviation, Lockheed, General Dynamics, and Thompson-Ramo-Wooldridge) holds prime contracts over more than half the total R&D and production business. In dealing with their subcontractors and suppliers, these corporations act in the role of government itself: "These companies establish procurement organizations and methods which proximate those of the government. Thus large prime contractors will invite design competition, establish source selection bids, send out industrial survey teams, make subcontract awards on a competitive or a negotiated basis, appoint small business administrators, designate plant resident representatives, develop reporting systems to spot bottlenecks, make cost analyses of subcontractor operations, and request monthly progress and cost reports from subcontractors." [9]

They are in the position of deciding whether or not to conduct an activity themselves or contract it out, and they may use their power over a subcontractor to acquire his proprietary information, force him to sell his company to the prime, or make or break geographical areas and individual bankers, investors, and businessmen. They may themselves create "independent" subcontractors in order to conceal profits, to keep certain proprietary information from the government, or for other purposes. Generally, they can and do use their decision-making power to stabilize their own operations, expanding or contracting their subcontracts in accordance with the peaks and troughs of government business, thus protecting their economic strength at the expense of smaller and weaker companies, seeking to assure their own growth and standing among the other giant corporations by mergers, acquisitions, and investments in the flock of companies dependent upon them for government largess.

The same top three hundred companies that perform 97 percent of all federal R&D also perform 91 percent of all private R&D. Most of the private R&D is a means of maintaining the inside track for new awards in anticipated areas of government need. Since these same companies do all or most of their business with government, the so-called "private" R&D is paid for by the government in the form of overhead on other contracts.[10] For example, the U.S. is still paying for Douglas Aircraft's investment in developing the DC-3 thirty years ago. A congressional committee noted the trend:

> At the moment a small number of giant firms in a few defense and space-related areas, with their facilities located principally in three states, and engaged almost exclusively in the application of existing engineering and physical knowledge to the creation of new products and processes, receive the overwhelming preponderance of the government's multi-billion dollar research awards. . . . Clearly, if the resulting technical discoveries are permitted to remain within these narrow confines rather than be disseminated widely through the society, a disproportionate amount of the benefits will be channeled into the hands of the few and further economic concentration will take place.[11]

Prime contractors are becoming brokers of the managerial elites that control American industry, their power limited only by self-restraint and political necessity. Congressman Carl Hebert, after a 1961 investigation, declared his amazement at "the daring of these individuals," recalling that during World War II "the brokers became so prevalent in Washington that they wore badges," and the only office space they had was "under their hats." That was twenty years ago, "and yet we find the same practice not only not halted but increasingly becoming exposed as we go along these days." [12]

Government itself continues to enhance the power of the primes by accentuating the trend toward fewer and bigger contracts. Peter Slusser, an aerospace securities specialist, noted in 1964 the increasing concentration of procurements within a hard core of big contractors. It appears, he said, that the industry may eventually be dominated by a few firms, much as the automobile industry is today. Further, there will be a concentration within the concentration. "It seems to us," he asserted, "that this concentration ratio will probably continue among the larger companies." [13] The House Select Committee on Small Business reported that while defense procurement increased by $1.1 billion in 1963, small business awards declined by $268 million. NASA doubled its prime procurements while the small business share dropped from 11.7 to 8.5 percent. The congressional Joint Economic Committee reported that a hundred companies and their subsidiaries accounted for 73.4 percent of total federal procurement value in 1964; the number of companies receiving annual awards of more than $1 billion has been steadily narrowing. Six companies belonged to that exclusive club in 1964; their combined NASA and defense work: Boeing, $2.3 billion; North American Aviation, $1.9 billion; Lockheed, $1.4 billion; McDonnell Aircraft, $1.4 billion; General Dynamics, $1.1 billion; and General Electric, $1.03 billion. North American alone, whose vending machine business made Bobby Baker a millionaire, held 28 percent of all NASA procurements.

Erosion of Public Control

The dominant centers of corporate power have largely usurped the government's evaluation and technical direction responsibilities. Frank Gibney, one of the early consultants to the House Space Committee, observed that "the spectacle of a private profit-making company rendering national decisions makes the old Dixon-Yates concept look as harmless as a Ford Foundation Research Project." [14] The government's Bell Report of 1962 expressed concern at the erosion of its ability to manage its own affairs and to retain control over contracting, which ". . . raises important questions of public policy concerning the government's role and capability and potential conflicts of interest." [15] The proliferation of quasi-public corporations, both profit and non-profit, springing from the soil of R&D spending (such as Bellcomm, Aerospace Corporation, or Comsat Corporation), symbolizes the bewildering innovations of the Contract State. Congressmen throw up their hands trying to understand their relations to these new organizations under the traditional dichotomy between private and public enterprise.

Nageeb Halaby, former head of the Federal Aviation Agency, insisted that the private airlines fund at least 25 percent of costs of developing a supersonic transport: "I think we have a half-free enterprise industry now, and I don't want to see it any more under government control than it already is. This is going to take some risk and . . . some ingenuity in figuring out how the government helps a relatively free enterprise to remain so, at a profit." [16] The industry itself insists that the SST will be profitable in commercial service but demurs from investing in a small share of the developmental cost, claiming that government is obliged to do the whole job in the name of national prestige. GE Board Chairman Ralph J. Cordiner worries about the future course of the American system, the secret of whose "drive and creativity" lies in maintaining "many competing points of initiative, risk, and decision." As we move into the frontiers of space, he declares, "many companies, universities, and individual citizens will become increasingly dependent on the political whims and necessities of the federal government. And if that drift continues without check, the United States may find itself becoming the very kind of society that it is struggling against." [17]

Labor joins its voice, decrying those sectors of industry that in the name of preserving free enterprise call for ever-greater farming-out of government responsibility: "It is purely charlatanism to claim that the government is in any form in competition with private industry because the government researches and develops and manufactures products for its own exclusive use. . . . It is competing only with the right of private industry to make a profit at the expense of the American taxpayer. But this is competition of a different color." [18]

The arrangements of the Contract State may, in the words of Don K. Price, avoid the problems that come with the growth of bureaucracy, but "encounter them again in more subtle and difficult forms. . . . If public ownership is no guarantee of unselfishness, neither is private ownership." [19] President Eisenhower, who came to office as a believer in the American business community and with a commitment to get rid of "unfair" government competition, ended his term of office with a forceful warning that private corporations (hired for the government's work) were tending to subordinate public to private weal, using strategic positions and influence for economic and political aggrandizement at the expense of democracy and the public interest.

There is no doubt that the flow of billions of federal dollars into narrow areas of the economy tends to create a self-perpetuating coalition of vested interests. With vast public funds at hand, industries, geographical regions, labor unions, and the multitude of supporting enterprises band together with enormous manpower, facilities, and Washington contacts to maintain and expand their stake. Pork-barrel politics and alignments with federal agencies and political leaders provide a powerful political machine to keep the contract flow coming.

The pattern is already in the process of filtering down to state and local governments. In the name of preserving and utilizing the "unique" systems-engineering and management capability that NASA publicists claim as one of the space program's major benefits to the civilian economy, underemployed aerospace industrial teams are now pushing for contracts in such areas as urban traffic management and water conservation. Governor Pat Brown of California has led the way. The only way to approach such problems, he declared recently, is from the systems-management viewpoint. Waste-management facilities will be built with or without the aerospace industry, he told a congressional committee, but the new capability holds great promise. The state has already retained Aerojet-General for preliminary studies. An aerospace trade magazine commented: "The governor's effort to put aerospace talents to work on these problems is an imaginative one. Other states and the federal government are watching the experiment with interest." [20]

If the method of work follows the pattern cut for the federal government, the aerospace firm will win a major contract by hiring some civil engineers (including some from the state agencies already responsible in the area) to write a persuasive proposal. Once the contract is awarded, the corporation will hire away from the state government, the universities, and other on-going operations all the technical people required, paying them greatly augmented salaries for work they were already doing under the traditional arrangements.

The first product will be an integrated state plan with engineering details, specifications, and cost estimates (always optimistically low at this stage). The company will then move for a prime contract to conduct the work, performing some of it itself and subcontracting some of it to the local builders and architects who possess knowledge and experience because of work they have already done for the state on a piecemeal basis. The contractor will cut across the cleavages of state, county, and municipal jurisdiction, insuring everyone a reasonable return while retaining for itself two tiers of profit: one from its own work, the other added to that of the subcontractors. In time the systems teams, having emasculated state and local technical resources in these fields, will by judicious use of contract money build a political machine in the state legislature—relegating to their own managerial cadre (and the investment brokers who control their corporate paper) a large share of the decision-making authority of the State of California.

Later, as the contract arrangements acquire stability, the prime contractor may begin a process of mergers and acquisitions among its subcontractors, directed at building the resources of the company and in the process eliminating competition and much of the cost yardstick. The result will be a new concentration of economic power in these areas of activity which will increasingly give the corporation sole-source status. Like the conventional political party, the corporation will have acquired the ability to allocate state resources, to dispense job and financial patronage, and to insure profits for its investors from the tax resources of state government.

Aerojet-General completed its preliminary waste disposal study in late 1965, recommending a ten-year program to be conducted by systems-management techniques at a cost of billions.

The California experiment may achieve the positive goal of building or imposing a consensus for state-wide planning and technical, rather than political, logic in solving large-scale problems. The traditional political parties have so far failed to achieve this goal. But it must be asked: Cannot social consensus and rational planning be achieved without the abdication of traditional political and governmental processes? The slower progress of traditional politics is preferable to a system that evades democratic controls and may eventually spread its grip into all areas of public policy.

In essence, the same questions may be directed at the full-blown Contract State nurtured by the federal government. Adherence of the R&D contract cult to the shibboleths of free enterprise may be a cloak to conceal the fact that the sharks are eating the little fishes and that a kind of backhanded government planning, in which they participate and from which they benefit, has come to replace free enterprise. In spite of such temporary stimulants as tax-cutting and the multiplier effect of missile-space spending, the civilian economy maintains a faltering pace of growth. The aerospace industries, on the other hand, ride high on unprecedented profits and diversify their holdings, biting deep into the most succulent portions of the civilian production machine in a new wave of economic concentration. In order that their "unique capability" not be wasted, defense firms are now moving into "systems management" of Job Corps camps and national conservation programs.

The politics of corporate finance have accelerated concentration not only in the government contract market but also in the civilian market, both of which are now thoroughly interpenetrated and interlocked. The aerospace giants have built huge conglomerate empires that span both markets, and the old respectable firms are playing major roles as public contractors. Among the top hundred prime aerospace contractors are such household names as General Electric, General

Motors, AT&T, Westinghouse, Chrysler, Ford, Socony-Mobil, Firestone, Philco, Goodyear, and so on. Many of the aerospace companies are mere façades and legal fictions having no individual existence but representing entities of financial and/or political convenience. In a 1965 House Judiciary Committee report, the five largest aerospace firms were cited as flagrant examples of corporate interlock. Douglas has fifteen directors interlocked with managements of seventeen banks and financial institutions, one insurance company, and twenty-eight industrial-commercial corporations (including Cohu Electronics, Giannini Controls, and Richfield and Tidewater Oil Companies). Not uncommon is the pattern by which each company holds stock in its nominal competitors (McDonnell Aircraft holds a large block in the Douglas Company "as an investment"). A study of seventy-four major industrial-commercial companies found that 1,480 officers and directors held a total of 4,428 positions. The anti-trust subcommittee staff concluded that management interlocks today are as prevalent as they were in 1914 when the Clayton Act, prohibiting interlocking directorships, was passed.

Point of No Return?

The quasi-governmental mercantilist corporations, maintained in a position of monopoly power through royal franchises, were anathema to the classical liberals. Thomas Hobbes compared them to "worms in the entrails of man," and Madison in *The Federalist* dealt at length with the problems of limiting their growth. At the end of the nineteenth century Henry Adams emphasized the origin of the corporation as an agency of the state, "created for the purpose of enabling the public to realize some social or national end without involving the necessity for direct governmental administration."

During the second half of the nineteenth century the corporation proved a powerful vehicle for mobilizing and organizing productive resources to achieve rapid economic growth made possible by burgeoning technology. Its very success, the efficiencies of bigness, and the inevitable politics of corporate empire-building thrust into American skies the spires of monopoly power. Since that time sectional and economic interests have shifted and changed, the social and technological landscape has vastly altered, and government has emerged as guarantor of social interests against the claims of private power. Government contracting on its present scale has added another dimension. Business and industry have always been close to the centers of political power, but never before in peacetime have they enjoyed such a broad acceptance of their role as a virtual fourth branch of government—a consensus generated by the permanent crisis of international diplomacy. Sheltered by this consensus, government has accepted responsibility to maintain the financial status of its private contractors as essential to U.S. defense and economic health. Cost competitiveness, the traditional safeguard against corporate power and misallocation of national resources, has been suspended by R&D contract practices.

NASA and the Pentagon use their contracting authority to broaden the productive base in one area, maintain it in another, create more capability here or there for different kinds of R&D, create competition or limit it. Under existing laws they may make special provisions for small business and depressed areas and maintain contracts for services not immediately required in order to preserve industrial skills or reserve capacity for emergency needs. All of this represents national planning. But without recognition of planning as a legitimate government responsibility, planning authority is fragmented, scattered among federal agencies and Congress, and the makeshift planning that results serves the paramount interests of the most powerful political alignments. In place of forward planning responsible to the broad national community, the nation drifts sideways, denying the legitimacy of planning, yet backhandedly planning in behalf of narrow special interests whose corridors of power are closed to public control.

The result is severe distortion in the allocation of resources to national needs. For almost three decades the nation's resources have been commanded by military needs, consolidating political and economic power behind defense priorities. What was initially sustained by emergency comes to be sustained, normalized, and institutionalized (as emergency wanes) through a cabal of vested interests. The failure of nerve on the part of these interests to redirect this magnificent machine toward a broader range of values denies the nation what may be the ultimate basis of diplomatic strength and the only means to maintain the impetus of a mature economy, namely the fullest enjoyment by all of our people of the immense bounty of equity and well-being almost within our grasp.

The shibboleths of free enterprise perpetuate a system by which, one by one, the fruits of the civilian economy fall into the outstretched hands of the areospace group. The so-called "Great Consensus" assembled by President Johnson is based on the paradox of support from great corporate giants as well as from labor and the Liberals. The civilian economy and home-town industry have been systematically neglected in the vicious circle of government contracts and economic concentration, leading the small businessman, vast numbers of middle-management, white-collar workers, and professional groups to embrace the simple formulas of Goldwater conservatism, directing the anxieties generated by incipient stagnation against the targets of autocratic organized labor and government spending for welfare and foreign aid. The exploitation of the myths of free enterprise have deflected attention from the feudal baronies of economic power and the tendency of the administration to attack the symptoms of growing inequality of wealth without disturbing the steepening slope itself.

The dynamics of the Contract State require close scrutiny lest, in the name of national security and the science-technology race, the use of the nation's resources does violence not only to civilian enterprise but also to the body politic. In place of sensational claims about the ability of the American system to meet the challenges of new tasks and rapid technological change, it is necessary to judge the appropriateness and adequacy of national policies that increasingly raise a question concerning the relation between government and private contractor: who is serving whom?

The R&D cult is becoming a sheltered inner society isolated from the mainstream of national needs. More and more it departs from the reality principles of social accounting, insulated against realism by the nature of its contract relations with government and its political influence. The elementary principle of economics applies: whatever is made cheaper tends to grow proportionately. Massive government subsidies to R&D facilitate its expansion beyond the point of rational response to international politics; it becomes a self-perpetuating pathology, intensifying the regressive structure of the economy and making further pump-priming exertions necessary.

As the arms race slows and is sublimated in space and science, as world politics break the ice of bi-polarity and return to the troublesome but more flexible patterns of pluralism, it becomes important that great nations achieve positive values. Military power, though essential, remains essentially a limited and negative tool. Economic and social equilibrium at maximum resource use may hold the key to ultimate international stability, prestige, and national power. Federal expenditures are a response to national needs and aspirations in all areas of public responsibility. The needs and aspirations are limitless, while the resources to satisfy them are relatively scarce. Many rich societies have withered because they allocated their resources in a manner that precipitated the circular pathology of inequity and instability. "Neither Rome's great engineering skills, its architectural grandeur, its great laws, nor, in the last analysis, its gross national product, could prevail against the barbarians." [21]

The problem of bringing the Contract State under democratic control is but a new phase of a continuing challenge in Western industrial societies. The legal fiction that holds economic and political institutions to be separate and distinct becomes ever less applicable as economic pluralism is swallowed up by corporate giantism. The myths of economic freedom tend to insulate the giants from social control, protecting their private-government status and threatening the political freedom of the majority. The tension between private and public decision-making can be a self-correcting process when its causes are visible and understood, and when public authority is not wholly captive to the pressures of narrow interest groups. The process is delicately balanced, and there are points of no return.

CHAPTER XI

THE RAMO-WOOLDRIDGE STORY

Drs. Simon Ramo and Dean Wooldridge were until 1954 obscure and moderately paid aerospace scientists. Less than four years later they had been transformed into multi-millionaires and major captains of industry, holding substantial control of the Thompson-Ramo-Wooldridge Corporation (TRW), among the nation's largest industrial combines with at least nine wholly owned subsidiaries and wide-ranging interests in the U.S. and Europe, standing among the top fifty government contractors (ahead of such old firms as Goodyear and Du Pont) and continuing to diversify and grow in profits and power.

The story of this transformation is not unique. Founded upon the political favoritism of the Air Force and the hidden profits of loose, cost-plus government contracts, the dynamics of TRW's meteoric rise illustrate what has become standard procedure in the success of the nation's largest and most respectable business firms—not only those primarily engaged in aerospace but also the giants of the civilian market that have discovered in government contracts a magic springboard to affluence and influence. The Ramo-Wooldridge episode reveals these dynamics in a clear-cut and shocking parable, demonstrating the nature of the federal contract as an instrument of special subsidy, political power, and economic concentration. The facts are known but have not occasioned either congressional indignation or public scandal, indicating perhaps the most disheartening dimension of the Contract State and a dismal forecast for the future.

In the first War Powers Act (1941) Congress sanctioned a drastic change in traditional contracting procedures, a change made permanent at the end of the war. This act released agencies from the rigid requirements of fixed-price, competitive procurement which had been built up in reaction to the profiteering scandals of World War I and the investigations by the Nye Committee of the 1930's into the munitions industries.

The rationale behind the waiver was to set the wheels of production whirring at maximum speed for the war effort, galvanizing industry to government command. In many areas there was no clear knowledge of future requirements, production costs, and wages; therefore private manufacturers were encouraged to convert to emergency war status without regard to costs or prices. "On procurement items, specifications were so often unusual that there was no basis for determining fixed prices. On construction projects, the contractor more often than not had to break ground for a new plant before the first sketches were off the drawing boards." [1]

Because of the ignominy incurred by cost-plus contracts after 1918, the new arrangements outlawed a sliding scale of profit determined as a percentage of total cost. Instead, a fixed fee was to be calculated on the basis of estimated cost at the time of contract negotiations: the new form of contract was a cost-plus-fixed-fee (CPFF) arrangement. Experience since 1941 has shown that fuzzy negotiating estimates are almost always overrun by contractors, sometimes by a factor of one or more, leading to the practice of re-negotiation or re-contracting to cover overruns, with an additional "fixed" fee added. The profit (which goes on top of overhead charges, an area of notoriously soft accounting) has evolved since World War II into a device that is technically legal but identical in essence to the outlawed sliding fees of World War I.

During World War II the CPFF contract was the primary instrument for procuring aircraft, ordnance equipment, and ammunition, exceeding $50 billion and constituting more than one-third of all purchases. During the war, it should be noted, some major corporations leaned over backward to avoid charges of profiteering; for example, Du Pont accepted broad responsibilities in the Manhattan Project for a $1 fee, returning to the government excess overhead funds. Some contractors, however, had to be nudged by Senator

Truman's Special Committee on the National Defense Program, which discovered widespread profiteering.

Action was taken after World War II to make the CPFF instrument permanently available for conduct of the Cold War. Manufacturers had always farmed out specialized components of their own product on subcontract to other companies as a method of gaining the advantages of specialization and concentrating their own resources. But R&D and systems-management practices in the postwar period converted subcontracting into a method by which the primes, supported by blank government checks, were elevated to quasi-governmental status in a deliberate shift of public activities to private management.

The Original Ramo-Wooldridge Contract

In 1946 the Air Corps (not yet an independent service branch) sought to enter the area of rocket studies pre-empted by the Army, awarding a $4 million contract to Convair (later General Dynamics/Astronautics) to study stabilization, guidance, and power plant problems for a missile with intercontinental range. At the same time Project Rand, a "brain factory" group created by Douglas Aircraft for the Air Force (eventually to become the non-profit RAND Corporation), described a number of space missions which such a missile capability would make feasible. A year later the Convair contract was "canceled," although in fact it was continued on reduced scale under company auspices and paid for indirectly by the Air Force as overhead on other aircraft production contracts. The cancellation represented both an economy move and a decision to rely on the manned bomber as the principal striking force of the future. But progress in propulsion, design, and control of ballistic missiles made by the Army's Redstone Arsenal showed promise of early operational feasibility, and in 1951 the Air Force reactivated the direct contract status of the Convair work. The decisive turning point came in 1952-1953, when Atomic Energy Commission forecasts of smaller warheads suddenly advanced prospects of a practical intercontinental-range missile.

The Air Force turned to Dr. Clark E. Millikan (California Institute of Technology) to establish an advisory group to evaluate the new situation. Out of this came the Atlas program, based on step-by-step actions to solve the problems of guidance, propulsion, re-entry, and heating, projecting ten years as a reasonable timetable for moving through R&D to production and deployment.[2] In mid-1953 firm intelligence data showed that the Russians were embarked upon a similar undertaking; this prompted an effort to shorten the ten-year timetable. Trevor Gardner, assistant to the Air Force Secretary, established a new advisory group to press the case for a crash program, turning to John von Neumann (then at Princeton) and Charles C. Lauritsen (Cal Tech) who recruited a distinguished panel including Millikan, George B. Kistiakowsky, Hendrick Bode of Bell Telephone Lab, Allen E. Puckett of Hughes Aircraft, Jerome Wiesner, Lawrence A. Hyland of Bendix Aviation, Louis G. Dunn of Cal Tech, Simon Ramo, and Dean Wooldridge. The latter two had been employees of Hughes Aircraft but now organized an independent consulting firm called the Ramo-Wooldridge Corporation, which became the technical staff of the committee. It is not clear from the record whether Ramo and Wooldridge quit the Hughes Company before or after informal discussions with their "long-standing acquaintance," Trevor Gardner.[3] The advisory group was designated the Strategic Missiles Evaluation Committee and placed under von Neumann's chairmanship. The Ramo-Wooldridge Corporation (R-W) accepted a letter contract from the Air Force which formalized its staff role.

The new committee concluded that ICBM's could and should be developed and deployed with the *highest* priority.[4] The "should" aspect of the case was reinforced by a batch of RAND studies which recommended "drastic revision" of the slow-paced Atlas program to bring the system on the line in the period 1960-1962. Secretary of the Air Force Harold Talbot, later forced to resign for using his position to promote a private management consulting firm, won approval from Defense Secretary Charles Wilson for a one-year design and engineering study to formulate the parameters of a crash program. In May 1954 Ramo-Wooldridge was given a letter contract to do the job.

As it transpired, this was the key decision that put Ramo and Wooldridge at the decision-making center of future developments as the Air Force assumed command of the new strategic systems. "Considering the decisive impact of the Strategic Missiles Evaluation Committee report on the future course of the ICBM program, the distinctive management arrangement that evolved from it, and the continuing contractual role of the Ramo-Wooldridge Corporation in the program," declared the House Committee on Government Operations five years later, "the question naturally arises as to the degree of Ramo-

Wooldridge participation in developing the report. To the extent that Ramo-Wooldridge authored the report . . . they became in a certain sense the beneficiaries of their own handiwork." [5]

Later investigations leave it somewhat obscure whether the Air Force "told" or "asked" the von Neumann panel about the contemplated contractual relation with R-W. In any case, General Schriever recommended that R-W be given overall responsibility for systems engineering and technical direction, making the Air Force, in effect, little more than paymaster of the private company. Panel member F. R. Collbohm, president of the RAND Corporation, dissented, charging "the arrangement was unworkable . . . would tend to separate responsibility and authority," and several competent contractors already existed who could take on the work while R-W had "no demonstrated competence." [6] Recalling these events eight years later, Collbohm observed that many felt at the time that the Air Force did not have "an in-house capability to handle what I call the project office functions." But he felt the Air Force did, "if it just brought it in. . . . I felt this would be something the Air Force could handle if it wanted to." [7] Collbohm wanted another firm to be allowed to work in competition with R-W, a choice to be made between the two at a later date. Panel member Wiesner sought an alternative: "We tried very hard to persuade Dr. DuBridge to have Cal Tech accept the responsibility." It seems clear that the unique R-W arrangement was a unilateral Air Force action and that ensuing controversy forced an attempt to win the endorsement of the scientist-advisory committee. Wiesner's later comment reveals that dissenters found themselves confronted with an inflexible Air Force position: ". . . at this stage, as individuals, our decision was merely whether or not to participate rather than to try to change what the Air Force was doing." [8] But the panel was shaken by the dissent and a special three-member committee was appointed under General Charles A. Lindbergh which reaffirmed the Air Force "finding" that the role "marked out and already in effect" for Ramo-Wooldridge was "logical and sound and should be continued." [9]

In the course of this controversy, some industrial politicking was going on behind the scenes which led to an R-W commitment that "Convair or some other company" would be a systems contractor under R-W and might receive other large production work in the program.[10] This was the logical solution, according to General Schriever. As part of the Air Force "family," R-W "would control the associate contractors . . . who in turn would control their subcontractors" in the traditional manner. "Convair would be one of the associate contractors . . . ," he said.[11] Another result of the controversy was that Drs. Ramo and Wooldridge, the one-year contract underway, stepped off the advisory group, "considering it no longer appropriate to retain committee membership while performing contractual studies, particularly since a permanent role for their company in the ICBM program was foreseen." [12]

R-W had started in 1953 "with a desk, a telephone, and three or four good people." [13] The company grew to about two hundred personnel within months and soon thereafter to many thousands—including factory production-line workers. General Schriever and a small military staff joined the R-W group in Los Angeles where they went to work re-directing the Atlas program, adding to the original ICBM assignment a second intercontinental-range missile (the Titan), a new intermediate-range missile (the Thor), and an Atlas-boosted satellite system.

R-W Corporate Growth

According to the 1959 House report, the growth of R-W, later to become TRW, was "intimately tied to and largely a result of its work for the Air Force." [14] The missile budget of the Air Force in hand, Ramo and Wooldridge scouted around for corporate respectability, approaching (intermediaries unknown) the Thompson Products Company of Cleveland, an old and well-established publicly held firm primarily engaged in subcontracting for the automotive industry but with an increasing interest and success in acquiring Air Force subcontracts.

Thompson Products put up half a million dollars, receiving a block of *non-voting* preferred stock with a nominal value of $350,000 and 49 percent of the *voting* common stock of R-W. Ramo and Wooldridge each invested $6,750, receiving a controlling 51 percent of the voting common stock.[15] This risk-free investment constituted slightly more than 2 percent of the initial capitalization and an infinitesimal percentage of the market value of their holdings within a few months. At the same time Thompson Products agreed to guarantee any leases entered into by Ramo-Wooldridge.

R-W immediately demonstrated they did not intend merely "to manage" R&D but had ambitions as well in the lucrative field of hardware production. Pacific Semi-Conductors, Inc., became a wholly owned R-W subsidiary engaged in the production of electronic devices.

In March 1954 Thompson Products enlarged its board of directors by adding Ramo and Wooldridge. In its quarterly newsletter to shareholders (July 1955), the Thompson Company saluted the good fortune of having been chosen as corporate wet nurse to an infant born a giant: "With 250,000 square feet of laboratory space in Los Angeles occupied or under construction, and plans already drawn for a manufacturing plant in the Middle West, Ramo-Wooldridge is moving rapidly into such commercial fields as automatic electronic computers, guided missiles, transistors and semi-conductors, weapons control systems, and advanced communications." [16]

Now big money began to flow, with Thompson putting up $20 million to finance R-W growth. The shuffle of corporate paper was arranged in such a way as not to disturb Ramo and Wooldridge's control: $4 million of Thompson's equity was covered by the issue of new classes of non-voting preferred stock, the remaining $16 million covered by ten-year notes. In addition, Thompson received an option to purchase control of the corporation in ten years, providing it bought 60 percent of the other common stock (owned by Ramo and Wooldridge) in its entirety during a two-year period ending in 1966. A ten-year voting trust was established over a portion of Ramo-Wooldridge holdings, with the two scientists and one board member from Thompson as trustees. This was the pattern of corporate control that prevailed for three years. In light of the unprecedented control over multi-millions granted to Ramo and Wooldridge on the basis of their $13,000 investment and the *sotto voce* Air Force commitment of future contracts, these arrangements amounted to cannibalistic absorption of the smiling wet nurse by her ward, an act later codified by formal merger. Shortly thereafter, Thompson put another $4 million into R-W's subsidiary, Pacific Semi-Conductors.

R-W established two major divisions, the General Electronics Group and the Guided Missiles Research Division. The former encompassed Pacific Semi-Conductors and later acquisitions in the area of electronics R&D and production; the latter was reconstituted in 1957 as Space Technology Laboratories (STL), another subsidiary. Incorporation of STL represented a new stage in R-W's relation to the Air Force. As the R-W systems-engineering and management role became more controversial in industry and Congress, it sought to hold and expand its separate role as R&D and hardware contractor which that position secured for it, in order to realize the lusher contracts and profits while avoiding the technicality of conflict of interest.

Three major buildings of the R-W plant were financed by the

employees' pension fund of the Thompson Company, with Air Force contracts as security for the loan. The money was subsequently returned to the pension fund by Air Force purchase of two of these three buildings, leaving R-W with the profit equity of the third building. Two other buildings were constructed at R-W expense and then sold to the Mellon National Bank and Trust Company at book value, the bank immediately leasing them back to the company and enabling R-W to charge the government for the lease under reimbursed overhead. It would require close accounting to discover who really paid for these and how many times. Another building was acquired by R-W under lease (reimbursable), and construction was initiated on a large R&D center elsewhere, later transferred to Space Technology Labs. Interest on the loans that changed hands in financing construction and leasing was not allowable directly to overhead under government policy; but R-W was granted substantial relief from this burden by "a certificate of necessity which permitted it to amortize in five years 50 percent of the cost of the buildings," retaining in company equity what otherwise would have been paid in taxes, in addition to deducting interest payments from its tax liability.

In the investigations of later years, company officials emphasized the high degree of "risk" involved in these financial undertakings. The "risk," if any, was entirely political. The administration or Congress could have called a halt to these shenanigans, but the politics of missilery managed to contain controversy for four years. The Democrats who commanded Congress had no incentive to look closely at R-W's growth, preferring the role of Air Force advocate. The administration showed little curiosity, accepting the necessity of heroic measures to bring the national missile capability onto the firing line at the earliest possible time without regard to cost or possible irregularities.

The actual dollar value of R-W contracts and profits is difficult to determine because of tortured bookkeeping devices, careless cost estimates, and the roles of associate companies and subcontractors. Subsequent growth and empire-building suggest that hidden profits would swell official earnings many times over. Growth was marked: according to the 1959 congressional report, sales increased from $2 million in 1954 to $43 million in 1957; annual net profits increased in the same period from $237,000 to $1,904,000; net worth from $623,000 to $9,145,000. R-W's financial position was improved by a tax offset for the operation of Pacific Semi-Conductors, which showed a book net loss of over $5 million for the first four years, more than covering official taxable R-W earnings; no federal income tax was paid at all during 1953-1957.[17] In this same period, ballistic missile work showed a total dollar value approaching $100 million, which constituted 60 percent of the business. Most of the balance of dollar income (37.1 percent) was diversified in other military contracts, mostly for the Air Force.

On the two basic contract instruments which governed R-W's relation to the Air Force, the average "official" fee was more than 10 percent of the contract cost,[18] unusually high and in no way reduced even after congressional investigation in 1959. At the beginning of the contract the fee had been set at 14.3 percent (15 percent is the legal limit on CPFF contracting). Testimony before the House Appropriations Committee shows that the average fee on R&D contracts let by all military branches falls between 6 and 7 percent, which the Department of Defense considers high enough "to achieve high-level performance."

The chief justification offered by the Air Force for the high fees was the so-called "hardware ban," which was (in the words of the congressional report) "an inevitable but unstable element of Ramo-Wooldridge's intimate association with the Air Force." The ban was written into the original letter contract, forbidding the corporation to "engage in the physical development or production of any components for use in the ICBM contemplated herein except with the express approval of the Assistant Secretary of the Air Force (materiel) or his authorized representative." [19] The company did in fact receive hardware contracts in closely related areas, one from the Lockheed Corporation (which in turn received a number of hardware contracts in ICBM hardware through R-W), and another, approved by the Air Force Assistant Secretary, for production of data-processing equipment for the ICBM program. But these are of no importance in view of the fact that the ban was rendered ludicrous by a decision to permit all ICBM contractors freely to subcontract hardware production to the Thompson Products Company,[20] which now drank deeply of the rich concoction of new contracts that began to flow in its direction. Since R-W substantially allocated ICBM work, the "pay-off" function was obvious and reduced the hardware ban to a thin glaze over a glaring conflict of interest.

The Crisis of 1958

By 1958 the controversies engendered by these arrangements reached a climax and change was mandatory. The timing was providential and was to prove a boon to R-W rather than a source of embarrassment. By the time R&D missile work had been supplanted in magnitude by the more nourishing contracts for general space systems and production of space and missile components. The controversy led to purported reforms of R-W's role which in reality granted the company an even freer hand at a time when its unique role in missile development had already been thoroughly milked of advantages. As controversy grew, R-W and Thompson Products reorganized as a new corporate entity, Thompson-Ramo-Wooldridge, Inc. (TRW), the merger effective October 1958.

The shuffle of corporate paper: All preferred and common stock held by Thompson in R-W was summarily canceled. The stock held by Drs. Ramo and Wooldridge was converted into Thompson common stock at the rate of 13 to 1. In terms of market value, the holders of R-W common stock received approximately $16.9 million in TRW stock as their equity in the merged company. This was a tax-free exchange. Ramo and Wooldridge were able to turn initial investments of $6,750 into paper profits of over $3,150,000 apiece.[21] Space Technology Laboratories, which had been separately incorporated in September 1957 (with retired Air Force General James Doolittle as board chairman), became a wholly owned TRW subsidiary, and $25 million of assets were assigned to it, along with 12 percent of TRW common stock (donated to STL's treasury for the purpose of granting stock options to management, a well-established method of concealing profit and maintaining control). In its merger notice to shareholders, Thompson Company declared that R-W was "presently engaged in a program for the separation of the activities of STL which, it is expected, will permit the merged company to compete for production contracts in the Air Force ballistic missile and space weapons program." [22] The plan was to leave STL intact as the systems manager of Air Force ICBM programs.

In the process of this reorganization, "the Air Force lost the services of Drs. Ramo and Wooldridge," who withdrew from STL completely, maintaining their role through directorships and substantial holdings in TRW. The House committee commented: ". . . the Air Force had believed and relied on the apparent assurance that it would continue to deal with these two men. Thus, through the merger, the government lost the services of the men who were the bases of the original source selection"; even before the merger "Dr. Ramo has been devoting less and less time to Air Force business." [23]

The Air Force cooperated with TRW hopes, maintaining the technicality of a "hardware ban" but gutting its effect completely. Under the policy ruling, TRW was allowed to compete in the normal unrestricted manner for any Air Force prime or subcontracts which did not originate from projects under STL responsibility, except "where it is found to be the sole source for the work under consideration or so uniquely qualified that a determination of ineligibility would materially prejudice national defense interests." The company would also be free to sell to the Air Force off-the-shelf products "'previously sold or offered for sale" by TRW regardless of use, and might compete for work related to the role of STL "but only after review and approval by the Assistant Secretary of the Air Force (materiel)."[24] Even the attenuated hardware ban that remained did not apply to production contracts on behalf of NASA. Of these arrangements the congressional report said: "The financial ties exist, and in a business environment it is too much to expect that the parent and the child will consistently work in isolation and to each other's possible disadvantage. The burden of suspicion cannot thus be cast aside, and certainly the hard-headed business competitors of TRW will not let it be."[25]

The reorganization and merger was one element of the vast train of events set in motion by Sputnik I in October 1957. We have already seen (Chapter Three) how the Air Force surged to power and influence concurrently with the rising fortunes of R-W. The relationship was to remain close through succeeding years, with the reorganized company continuing to play an important part in the internal politics of the Defense Department and the administration. The efforts of the President, his scientific advisers, and the Army to check the imperialistic rise of Air Force power in aerospace weaponry and contracting coincided with the rising chorus of protests in industry against the privileged position of the TRW-STL combine. The result of this convergence was the creation of NASA with full authority over the "civilian" aspects of space R&D, and the creation within the Defense Department of the Advanced Research Projects Agency (ARPA) which was intended to centralize R&D contracting at the level of the Secretary of Defense. Both moves were designed to eliminate Air Force predominance and jungle tactics forced upon her sister services by raising administrative control to higher levels. The odor of the R-W/Air Force arrangement, kept from erupting into a major public scandal by the political interests at stake and a muffled chastisement administered by Congress, led to a demand for fundamental reform. As a result the Air Force agreed to the creation of a non-profit systems-engineering firm (Aerospace Corporation) to assume STL's duties in the missile field.

The Air Force, however, found ways to subvert each of these three organizational moves. As previously described, the appointment of NASA's chief officer, Keith Glennan, was linked to TRW. Air Force personnel and its contractor agents were spread through the network of NASA's management. During this period NASA became in effect an instrument by which the Air Force evaded the forces bent on reducing its influence. After its establishment in 1958 ARPA operated not as an integrating factor but merely as a clearinghouse, retaining nominal control over some minor new projects that came along, such as "Project Vela" (to develop a satellite system for detection of clandestine nuclear weapons in outer space), for which STL was named prime contractor.[26] ARPA in fact became an empty shell with little real use to anyone; when Robert McNamara learned his way around the Pentagon, he chose administrative methods other than ARPA to control the Air Force/industrial empire. Aerospace Corporation did absorb some of the systems-engineering and technical advisory functions of STL but, though a non-profit organization, it was staffed by engineers and managers representing a cross section of Air Force contractors. It came to occupy the role of tribunal for allocating the contract market among the elite industrial coterie of Air Force primes, including TRW whose position was reduced from preeminence to equality. The process of contract-award became one aimed at cartelization of the contract market in the interest of major companies, at the expense of those not members of the club. In 1965 major scandals arose involving Aerospace Corporation too, revealing reckless management, Air Force use of its so-called non-profit creature to escape Defense Department fiscal controls, and other abuses which will be examined later.

The 1958 change in TRW's position did not end the upward trajectory of its growth. Rather, like its confederates among the major contractor group, it launched a great surge of diversification, acquiring as subsidiaries a variety of companies to extend its competence in the government as well as in the commercial market.

By 1965 TRW had acquired (in addition to Pacific Semi-Conductors and STL) the Good-All Electric Manufacturing Company, the Magna Corporation, the Marlin-Rockwell Corporation, the Radio Condenser Company, Radio Industries Incorporated, the Ross Gear and Tool Company, Inc., Bell Sound, and McNeil Machine and Engineering.

TRW has proposed the purchase of Clifford Motors Component Ltd. in England for about $28 million cash, and it has joined a French company (Engines Matra) to form a jointly owned firm to bid on systems and missions analysis and engineering for European space projects.

According to Dr. Ramo, the company (in 1963) was a $500-million-a-year operation, "about one-half government . . . and one-half non-government, purely commercial ventures." [27] It was engaged in a wide range of prime and subcontracts, including engineering design, systems management, and production, ranging from communication satellites and advanced space probes to Gemini, Apollo, and Mars mission analyses, and design and production of the engine for the lunar descent vehicle (all for NASA). It continued its major role in strategic missiles as well as winning new Air Force work; it won a number of Army contracts, including $15 million for production of the M-14 rifle ($6 million of which was to pay for construction of facilities and tooling of equipment to give the company a capability which it totally lacked for this work); [28] for the Navy, it won a $6 million contract for systems analysis, integration engineering, test support, technical support, and engineering evaluation of anti-submarine programs and equipment, thereby achieving in a new and burgeoning area a key staff role in some ways comparable to R-W's original Air Force role.[29] In the words of a recent full-page ad: "Come join TRW. We have something going in nine out of every ten U.S. space launches. Moreover, we have built and orbited more kinds of spacecraft than anyone."

So successful has the company been that in the first quarter of 1965 it broke all previous earnings records: official profits were $6 million on quarterly sales of $155.1 million, up about 10 percent over the first quarter of the previous year.[30]

One of the factors that precipitated the 1958 investigation was industrial complaints to Congress that STL used its role to invade the proprietary information of its competitors, appropriating this information for its own use: "STL would move in to look at their documents and write up a critique which was based on their material and take credit for it." [31] In 1961 a House Armed Services Subcommittee held extended hearings into reports by the General Accounting Office on waste and profiteering under "sole-source" government contracts.[32] So frequently did the TRW name pop into the proceedings that it prompted this exchange:

COLONEL TREACY. I call your attention to item No. 11, up there. That is an armature.

MR. NORBLAD. Yes.

COLONEL TREACY. And this is an item that we bought from Thompson-Ramo-Wooldridge.

MR. HEBERT. Oh, they are our old friends. [Laughter]

COLONEL TREACY. I knew we could get into controversy.

MR. COURTNEY. There we go. [Laughter]

COLONEL TREACY. And they bought it from Westinghouse Electric Corporation, Small Motors Division.

And this item was charged to the government at $35.28. And the subcontractor's price, including about $3 of packaging, was $16.69.

MR. GAVIN. What did the government pay for it?

COLONEL TREACY. The government paid $35.28. And this is a mark-up of 111 percent.

MR. HEBERT. That is par for the course for Ramo-Wooldridge.[33]

The hearings revealed that "sole-source" status was often used for flagrant profiteering. Not only was there no special competency involved in many so-called sole-source items, but the contractor procured the items from other companies at a fraction of the charge made to the government. Further, when items so procured proved defective, the sole-source contractor was not only *not* held accountable but was rewarded with additional cost-plus contracts to correct the defect.[34]

Congressman Porter Hardy declared: ". . . on the face of it, it is an unconscionable mark-up. Unless there is something I don't know about it, this old nasty word of 'profiteering' would seem to apply." The Air Force defense for this and similar cases was that in some cases of secondary procurement, the sole-source contractor charged the government less than cost and that the overall profits were adjusted later to take up "unreasonable" windfalls. (No significant concrete examples were provided to prove this.) In the meantime the company was reimbursed for reported, not actual, charges. If auditing procedures in these instances were no better than those applied in areas about which more is known, the official Air Force position is not reassuring.

MR. BAILEY. Well . . . we think the basic reason for this situation is the ease of procurement in this manner.

MR. GAVIN. Ease of procurement?

MR. BAILEY. Yes, sir, the ability to get what is required with the least expenditure of effort. The process is already set up. It requires only the issuance of the orders . . .[35]

TRW has continued to be a target of General Accounting Office reports. In late 1964 GAO recommended financial sanctions against it for deliberate violations of contracting patent provisions, indicating that "eighteen inventions made under defense contracts up to three and a half years earlier were never disclosed" to the Defense Department as required by law.[36] Compliance with the law would give the government ownership of patents developed exclusively on government-funded R&D, enabling these to be made freely available to other contractors for use on government work. By its evasion, TRW put itself in the position to claim sole-source status for new contracts, or could charge other contractors and the government royalties for the patents' use.

Evaluation

The 1959 investigation led the Government Operations Committee of the House to find positive as well as negative factors in the Air Force/R-W arangement: "On the positive side, there were savings in overhead costs, better integration of research, more simplified designs and specifications, prompt engineering solutions of bottleneck problems, and adoption of components and test facilities for common use in several weapons systems." By contracting with Ramo-Wooldridge for services that otherwise would have been performed separately by several prime contractors, the Air Force "gained a measure of objectivity, continuity in programing, and savings in time and money." [37] This evaluation, however, simply reiterated Air Force and R-W claims for which no convincing evidence had been advanced, giving it an *ad hominem* flavor, part of the tactic of gentle reproof which the committee adopted in avoiding scandalous potential. The committee noted "an element of post hoc argument" in the presentations of the Air Force and the corporation. "Success . . . and unexpected achievements always invite a bit of self-indulgence and the taking of specific credit for actions which have diffuse origins." The congressional report pointed out that "the technical competence" of R-W "was not brought full-blown to the Air Force job. In the beginning, the company was just a high-level recruiting agency. . . ." As for vaunted money savings, the committee declared, "these are difficult to pinpoint in huge costly weapons systems which put urgency before economy and which inevitably are afflicted with costly errors because of the unknowns and risks which must be incurred." [38] R-W claimed to have saved the government more than $100 million, approximately equal to the fee cost of its services. The company cited "its conservative technical approach to keep programs on schedule (and hence save money)," exemplified by selection of the heat sink rather than the ablation nose-cone principle. In the Army view, on the other hand, this was an "unnecessary and costly error involving the expenditure of $200 million or more."

The problems created by the R-W arrangement, the report noted, "go beyond money matters," reaching to the conflict-of-interest issue. Army testimony described the company as "not objective at all" but only "another private contractor pursuing its stockholders' interests first, and the government interests second." Army General Medaris charged that R-W had a vested interest in its own weapons when it came to evaluating the comparative merits of Air Force or Army systems. The committee observed: "There is by no means unanimity of opinion, but a fairly common pattern emerges. Missile contractors generally acknowledge the scientific and technical competence [of R-W] even when they find fault with particulars. . . . STL representatives are regarded as performing those supervisory or directing functions which Air Force personnel themselves otherwise would perform." Most company spokesmen said their men had to spend a good deal of time "showing STL representatives around the plants and keeping STL informed of technical developments." Complaints were numerous that STL tried "to penetrate too deeply into the details of the contractors' own operations. There was considerable sensitivity to the possibility that STL would acquire company technical information for its own exploitation on . . . other contracts." In a few cases, "associate contractors withheld information from government through fear of its exposure to STL. . . ." [39]

"More fundamental," the report stated, "is the privileged position that STL enjoys as a close member of the Air Force family . . . the strict standards and the legal safeguards attending government conduct are considerably weakened. Moreover there is a question as to whether the conflict-of-interest statutes apply to the STL personnel who have responsibilities similar to those normally exercised by government personnel." The Air Force used STL as a tool to overcome restrictions on government salaries not only of professional personnel but "of secretaries, chauffeurs, switchboard operators, custodians and the like." Of all the contractors, "this contractor alone sits at the very seat of government, three or four thousand strong, and wields an enormous influence on the course and conduct of multi-million-dollar missile programs. This influence—there is nothing invidious about the term as used here—is the more powerful because it is exercised in the name of the Air Force." [40] Turning from the thought of censoring either of the parties, the report stated that the arrangement "should not continue. . . . An unhealthy situation has developed; government and private business values have become intermingled to the detriment of both, and there is increasing concern on the part of Congress and the public over the current status and position of STL." [41] The committee called for complete divestiture of STL from TRW, a recommendation never implemented. It recommended that the Air Force itself "retain or acquire the vital technical and managerial resources STL provides. After five years of operations, the Air Force lacks this kind of competence in its own organization and has not taken systematic steps to prepare for the future." [42] Also not implemented.

A blunter judgment was rendered by President Kennedy's Budget Director, David E. Bell, who headed a 1962 task force on R&D contracting which found itself confronting the same problems in even more exaggerated forms. Bell declared: "The kind of situation . . . exemplified some years ago by the Ramo-Wooldridge position is improper"; no organization should "be simultaneously offering technical advice to the government on an activity and in a position to benefit from that same technical advice." [43] The Bell Report pushed vigorously in behalf of a stronger in-house management and technical direction capability for R&D contracting, an intention which up to this time has been only weakly implemented and, in some cases, completely sabotaged.

The 1962 Bell Report led the Kennedy administration to issue a strong conflict-of-interest rule imposing a stricter hardware ban against follow-on production by companies serving as technical advisers during the R&D phase of new programs.[44] The vigorous campaign of Secretary of Defense McNamara to impose rationality and control upon defense and space contracting began to make progress in late 1964. TRW itself accepted a tougher hardware ban in its contract with the Navy for anti-submarine work, the Navy making it clear that "STL will be a technical consultant and no more." [45]

But experience indicates that a "hardware ban" does not reach the real basis of "conflict-of-interest" problems inherent in the contemporary pattern of R&D contracting. It will be difficult for the Navy or the Department of Defense to exclude other TRW subsidiaries from seeking and obtaining contracts in related areas with other agencies of government. It will be impossible to prevent a trade-off of subcontracts between primes who seek to avoid the appearance of conflict of interest but who nevertheless are collaborating in using their positions for private aggrandizement. Attempts to control profiteering and the concentration of economic power have in effect forced even greater collusion and cartelization among major contractors, especially since all are interlocked in one way or another, directly or indirectly. In most cases, the identity of a single company is a legal fiction.

CHAPTER XII

THROWING AWAY THE YARDSTICK

The traditional arsenal system, by which government conducted a large share of military research and development through civil service agencies, is now practically dead, victim of the Contract State and of the revolutionary size, scope, and pacing of the public interest in technological change.

The old system had its legislative roots in a World War I law, still on the books, which declares that the Secretary of the Army shall have supplies needed for his department "made in factories or arsenals owned by the United States, so far as those factories or arsenals can make those supplies on an economical basis." [1] A similar statute applies to the Air Force, except that the optional "may" is substituted for the mandatory "shall" of the Army statute.[2] The naval equivalent of the arsenal system has been maintenance of large Navy yards where R&D and the construction of a portion of the fleet permitted a public "yardstick" by which to measure the efficiency and performance of private shipyards, as well as to give the Navy adequate "in-house" technical capability to manage and technically direct private contractors.

This has been the rationale of the traditional system. Teams of engineering and production personnel held civil-service status and were assured stability and protection of government employment; the system fostered independence from the interests of private industry in evaluating and managing contracts and in advising government decision-makers. Government teams performed all the technical functions associated with developing and maintaining combat weapons, including, in many cases, the design and construction of prototypes of new systems which were then turned over to industry for quantity production.

The in-house function has always been defended as essential. Government could not legitimately be accused of competing with private enterprise since, in the case of armaments, it was like a barber cutting his own hair. Committed to find the least expensive source of optimum quality material to meet defense needs, government requires a means to prevent inordinate private profiteering at taxpayer expense. Private industry could not be expected to invest its own money to carry on expensive R&D aimed at meeting the needs of this customer, so government had to assume this work, taking pains to protect both public and business interests by making civilian spin-off freely available to all private producers and, by open competitive bidding, minimizing economic distortions induced by its own procurements.

"Some of our government laboratory work is outstanding," according to Dr. Walter Brattain, senior scientist of Bell Laboratories, such as in the Bureau of Standards, the Naval Research Laboratory, and the Army's Electronic Laboratory in Fort Monmouth, New Jersey.[3] But in recent decades, under the extraordinary stresses of World War II and the subsequent arms race, the arsenal system has been increasingly by-passed. The Air Force, child of the technological frontier, has seen no virtue in it. General Schriever: ". . . it is generally true we do not go to the service laboratories. We know what they are doing, but our development, our whole philosophy has been one of going to industry and having industry develop and produce for us."[4] It was largely the growth of Air Force contracting during the 1950's that eroded the in-house capability of other government agencies and forced them to follow the pattern, inflating the technical manpower market, depleting the in-house pool, and, as Air Force contracts financed a powerful private constituency to advance its interests, becoming a political tool recognized and exploited by every part of government, not excluding the White House.

Congress has served these constituencies well, providing the legislative umbrella under which government technical activities have been increasingly farmed out to private industry, providing a new and richer brand of patronage. From 1941, when peacetime contracting procedures were set aside "for the duration," through the Eisenhower administration, which sought to eliminate "unfair competition" to free enterprisers, the new system evolved to become no longer an emergency improvisation but rather a mature and expanding institution based upon a broad coalition of interests.[5] The President's Science Advisory Committee warned against this trend in 1959:

> . . . some government researchers have tended to lose heart as the number of contract-operated laboratories has grown over the last decade. Government laboratories are vital national assets whose activities will need to keep pace with the growing magnitude of federal research and development programs. Undue reliance on outside laboratories in placing new work of large scientific interest and challenge could greatly impair the morale and vitality of needed government laboratories.[6]

Congress did not take the initiative in defending the new system while Eisenhower Cabinet members were willing to do the job. But when the Kennedy administration began to look for means of controlling and disciplining the abuses of the Contract State, it precipitated an increasing number of congressional expressions aimed at countering administration tendencies. In May 1961 the House Committee on Science and Astronautics passed a resolution aimed at informing the new NASA administrator on which side his bread was buttered:

> Resolved. That it is the sense of the committee that the free enterprise system in the United States has demonstrated extraordinary capacity for meeting the needs of the Nation, and that the ingenuity of the American business community should be harnessed and utilized to the maximum extent practicable in the Nation's space programs.[7]

In each year's authorization hearings, NASA's Webb has emphasized the increasing percentage of appropriations which flow to private contractors (now about 95 percent of its budget), giving implicit recognition to the preoccupation of Congress. When NASA after 1963 was threatened with the kind of contract controls that McNamara was bringing to bear upon the military, the House Space Committee intervened: "Considering the long-range aspects of the space program the committee believes the best interests of the nation will be served by placing with industry as much of the research and development as possible, for it is private industry . . . which must produce the end product." [8]

An extraordinary reversal of roles between Congress and the executive occurred in 1965 as a result of the administration's efforts. The General Accounting Office, established by Congress as its instrument for auditing executive agencies, was increasingly called upon by the

Budget Bureau to audit R&D contracting as a means of outside oversight, a function which Congress had abandoned in its advocacy of the Contract State. As a result, GAO reports forced Congress into hasty hearings which were more critical of the accounting office than of contractor peccadillos, leading to the resignation of GAO Director Joseph Campbell "for reasons of health" at an embarrassing moment.

The first great Kennedy reform effort was the Bell Report of 1962 which concluded that "there are certain management functions which should under no circumstances be contracted out. Basic policy and program decisions respecting the research and development effort—relating to the types of work to be undertaken, when, by whom, and at what cost—must be made by full-time government officials. Such officials must also be able to supervise any work undertaken and to evaluate the results." [9] The government finds itself, the report said, delegating decision-making and control tasks by default to private contractors. But the R&D coalition managed deftly to side-step the implications of the Bell Report, and the Contract State continued its phenomenal expansion. The next major confrontation occurred in the Department of Defense where McNamara sought to put teeth into the Bell Report; he engaged in inconclusive skirmishes with Congress, with which his boss, the President, often preferred to compromise.

The softness of the Space Agency's budget and its loose contracting practices provided a means by which constituents of the Contract State continued to baffle the administration and the public interest. In an age when R&D occupies the center stage in the whole elaborate structure of government procurement, the yardstick function becomes the *sine qua non* of government control and economic realism, and therefore the most serious threat to pyramiding profits upon which is based the present-day tempo of concentrated and cartelized economic power.

In World War I Secretary of the Navy Josephus Daniels discovered the existence of a shipbuilding cartel. This prompted the Navy to foster the yardstick of in-house shipbuilding in order to determine actual costs. It was discovered, according to AFL-CIO lobbyist William H. Ryan, that "the cost to the government was substantially lower than the lowest bid ever submitted by the cartel." The World War I experience is indicative of today's well-established trends, but on an infinitely larger cash basis. The impact of the Contract State (where the great bulk of government billions is distributed by negotiated and/or sole-source contracts on a cost-plus basis, where millions of components are produced, and where R&D unlocks the door for prime contractor status, systems management, and follow-on hardware production) means that the number of "competitive" major contractors is increasingly reduced, their holdings in both government and commercial markets ever more augmented, their relationships increasingly interlocked. The flow of virtually unscrutinized patronage in government contracts envelops all segments of the body politic. It involves the prosperity of local communities and the status of elite labor unions, imposes upon universities, local business, state and congressional leadership the need to look to federal contracts as the only sure route out of the incipient stagnation of the civilian economy.

Small businessmen become, in Ryan's words, "jackals at the heels of the lions. The lions take the sure money prospects and then farm out the expensive, involved, low-profit return aspects of manufacture to small companies. Competition is extremely keen at this point, and the prime contractors keep sharp watch over the government's money in open bidding for such work." By holding down costs outside the magic circle of major prime contractors, "it is possible to cushion costs on their own activities," even to the extent of procuring the material on cost-plus contracts in quantities adequate to the needs of subsequent contracts "by charging the excess material to scrap or to experimental waste. . . ." [10]

Even the most diligent surveillance by government inspectors cannot control the labyrinth of corporation devices which, by the fiction of "private enterprise," are excluded from public control. It has become virtually impossible to discover what percentage of total funding is spent on substantive line work in research and development and how much is spent on multiple overlays of administrative cost, facilities and sites, construction and acquisition of equipment, all forms of overhead, inflated salaries, and benefits for management (in the form of stock options, bonuses, and such perquisites as corporation-owned country clubs, golf courses, executive gourmet dining rooms, private aircraft for private as well as company use) chargeable to government. The so-called fixed fee—profit added on top of all this (usually around 6 percent)—does not include the overhead on overhead and the fees on fees of multiple-tier subcontracting. No one really understands to what extent hidden profits and the dependence of subcontractors upon primes facilitates the trade-offs, mergers, and acquisitions of concentration and cartelization.

Systematic destruction of the government yardstick opens the door to the runaway problems of the Contract State which, in the name of national security and science, are bringing social and political changes which even the simple yardstick of Josephus Daniels may no longer be able to reform. The yardstick issue today is the tip of the iceberg of accelerated concentration of economic and political power in American society.

Dislocation of Technical Manpower

Companies operating with reimbursable contracts have very little limitation upon what they may offer as salaries. In addition to extravagant rewards for management, the contract cliques can corner scarce specialized manpower which might otherwise be available for government yardstick, for basic research and graduate education in the universities, and for R&D oriented toward civilian economic growth and improvement.

Since salaries for scientists-technologists are a small fraction of the total cost of corporate operations, and since they are subsidized as direct charges or overhead on CPFF contracts, the industrial R&D clique has developed the practice of "stockpiling" manpower with special skills. This stockpiling, the House Select Committee on R&D reported in 1964, "is practiced by some companies or agencies that keep on their payrolls more scientists and engineers than their current work requires, to be in a more advantageous position to bid on and undertake new projects." [11] Dr. Wernher von Braun, faced with holding together the Army's Redstone Arsenal (prior to its transformation into the Marshall Space Flight Center under NASA), complained of the "tremendous amount of high-level scientific personnel hoarded in these big companies," making it virtually impossible to maintain a career in-house capability. "The companies won't let go because they say they have to hang on to the people, otherwise they lose their future." [12]

The best-qualified scientists and engineers are engaged by corporations in writing proposals; when these are successfully sold to the government, they are put on new proposals, and "second-rate scientists" are given the job of performing the actual work. The more highly regarded people continue to scout for new business.[13] According to Dr. Martin Schwarzchild: "We are employing an unbelievable fraction of our very best engineering brains in preparing proposals. . . . I have very much the suspicion that much more thought should go into developing a bidding system that does not require so much of the top engineers as it does at the moment. . . . The government can have enormous influence because they in fact set the scene under which the bidding is done."[14] In the trade this practice is known as "brochuremanship" and is illustrated by the case of the Sperry-Rand Corporation which organized a "systems group" to bring in business by concentrating on the "program-definition phase" of Defense Department and NASA procurement. According to an aerospace trade magazine (which commended the practice to all would-be contractors), the Sperry-Rand group worked on "an expansion-contraction principle in which a small core of experts in diverse technologies can operate relatively inexpensively and efficiently in the pursuit of new business." Since its organization, the group is credited with bringing in $135-140 million in new business.[15]

In effect, the R&D industrial contractors place a premium on scientists-technologists with sophistication in the special market of government contracting. They become a species of salesmen and wildcat promoters rewarded for sales volume, not for achievement in their skill. High-quality R&D performance is secondary, since the government cannot evaluate it anyway, and since grossly expensive mistakes and overrun costs may only mean new business.

Recently the French counterpart of NASA turned to U.S. industry in the hope of buying space-related equipment. The awakening was a rude one. The French discovered "that the glittering advertising claims for much of the equipment far exceeded performance. They expected superior quality and superior reliability. What they got was something different." Telemetry equipment purchased by the French arrived improperly wired and was shipped back to the U.S. in the midst of "protracted long-distance negotiation to bring it up to standard." Three recorders were purchased from three different U.S. firms, each contending for the larger order: "None of the three met specifications." Like U.S. government agencies, French officials lacked the ability to specify exactly what was needed. "Naively, they put themselves in the hands of U.S. salesmen who not only sold them the equipment, but wrote the specifications."[16]

Upon his resignation in mid-1965, Dr. Eugene B. Fubini, the Defense Department's key R&D manager, attacked the tendency: the contractors "do not always face up to their responsibilities" in deciding upon the feasibility of technical proposals in their desire to obtain research contracts. He attacked the tendency, once contracts are obtained, to "gold-plate" specifications and performance, overcomplicating and increasing the cost of the work.[17] There is strong reason to suspect, said Dr. Sterling Livingston, "that many defense contractors are submitting cost proposals that are only half as good as they know how to make," and in fact are encouraged to do so by a collusion between themselves and sponsoring government agencies seeking to get a system approved at higher levels.[18]

Economists, educators, and congressmen frequently point to the personnel ads that stuff the financial section of the *New York Times* as proof of the "enormous" job opportunities available for people with appropriate training and skills, arguing that unemployment is the fault of the engineers and others who have failed to train or retrain themselves in the "thriving" advanced technology. They tell the jobless to stop complaining, and they urge programs for upgrading skills. They overlook the fact that most of these help-wanted ads represent corporation raids upon each other's most successful technologists-salesmen, raids fueled by creeping salary inflation funded entirely from R&D contract overhead. Another but less significant element is the part of the advertising which takes the form of expensive display ads (usually half- or full-page) vaunting the advantages of Southern California, Florida, or Texas and making grandiose claims for the company's products and the wisdom of its military clients. These have become the primary method of aerospace public relations, subsidizing the trade journals at government expense and conducting campaigns in behalf of NASA and Air Force ambitions. McNamara sought to control the old direct form of this activity by excluding such advertising from reimbursable overhead, permitting charge to the government only of personnel-recruiting ads. The old ads continue but now have a few "help-wanted" lines somewhere in the copy.

The raids upon high-level manpower are made not just back and forth between corporate giants, but more importantly in a one-way flow which depletes scarce specialized skill groups from government management and in-house, pirates basic research and educational manpower from the universities, and steals from the civilian economy much-needed innovation forces. This process leaves untouched and in despair the growing pool of unemployed engineers oriented toward the stagnating areas of civilian technology, whose very stagnation is further aggravated by the resultant artificial inflation of R&D costs. Government agencies, funded by Congress without regard to manpower or other resource allocation, are forced to enter the competition themselves, loosening the financial controls over their own contractors to make it easier for them to acquire adequate manpower to do the agency's job. Dr. Philip Abelson, one of the sharpest critics of this practice, declared: "The left hand doesn't know what the right hand is doing. . . . The government pays big salaries to one outfit to take people away from another outfit." [19]

Decimation of government's control and evaluation function follows close upon the heels of this servomechanism, augmenting private power and systematically eroding sectors of the economy that could provide a yardstick. The government is forced into the self-defeating round of hiring systems-design, engineering, and management skills from the same contractor groups, and establishing so-called "non-profit" brain factories such as Aerospace, MITRE, and ANSER, largely controlled by personnel hired from the same contractors (slightly augmented by a sprinkling of university scientists). These organizations become in effect not the objective and independent replacement for government yardstick, but juntas representing the prime contractors in cartelizing the procurement market. Less and less is the civilian economy, the university laboratory, or the university-administered laboratory capable of providing a comparative index of costs to enable government to set standards for effective public control of R&D spending.

It is true that the numbers of scientists and engineers engaged in R&D have vastly increased in absolute terms throughout the economy. Over the twenty-year period from 1938 through 1958, there was almost a tenfold increase, a doubling in the course of World War II and a more than fourfold increase since 1946. The growth of R&D manpower between 1954 and 1961 was something like 85 percent, the most intensive growth occurring from 1961 to 1964, after which there was a slight leveling. Before 1961 academic institutions held their own with industry and non-profit organizations, while the rate of government growth fell to half the national average. Since 1961 the depressing influence has been felt by universities as well.[20]

The impact on the civilian economy was indicated by Assistant Secretary of Defense John Rubel: ". . . government expenditures for R&D may be more hurtful than helpful from the standpoint of industrial or economic growth . . . at best they represent a distortion of emphasis to worsen rather than improve." They "seem to inflate" costs without "increasing the supply of its most vital resource: trained, competent people." The "vast programs and huge organizations sponsored by government" may be attracting many of the best people into the military-space sector and thus "eroding a pillar of national strength in the private sector." In the long run, the "increasing costs" and "shrinking product base" may "erode our sources of economic vitality and our capacity to compete with other industrialized nations who are not allocating their resources in this way." [21]

As for the universities, the impact is both more obscure and more indisputable. Faculty quality and morale, and basic research of all kinds, especially in genetics and biology, applied areas of medical, consumer, and industrial research, all the soft sciences and the humanities are strained by the undercurrents of public R&D. The whole system of values is distorted, creating pressures for professors, like their industrial counterparts, to become wildcatters and salesmen writing proposals and brochures in search of government or foundation contracts. The most successful tend to be drained off by industry interested not in their scientific skills but in their prestige, their inside savvy of government contracting, and their ability to write persuasive proposals couched in the prose of faddish jargon at the "frontier" of their discipline.

In order to avoid these untoward effects, the universities have been establishing institutes and laboratories nominally separated from the university departments. Through these, some faculty members can augment their salaries without facing the objections of their colleagues in general faculty meetings. Examples of this are the Lincoln Laboratory maintained by MIT and the Livermore Laboratory maintained by the University of California. In addition, many universities have facilitated consulting work by considering it a private undertaking of the individual professor, although his normal classroom teaching may be delegated to a graduate student or reduced by the university, in effect subsidizing his industrial or governmental work.

Efforts by the Kennedy and Johnson administrations to reform the Contract State have had an extraordinary effect upon Congress, whose leaders by some curious reverse-twist of psychology (which can only be characterized as reflecting ambivalent complicity) have undertaken a vicious series of assaults upon the marginal abuses of government contracting and grants to universities. As if acting out its guilt feelings about the abuses of contracting, Congress fastens upon the universities standards conspicuously lacking in contracts with profit firms.

This peculiar congressional displacement has appeared in at least three areas: (1) excessively detailed and skeptical review of basic science projects conducted with funds of the National Science Foundation; (2) morbid preoccupation and magnification of minor irregularities in the grants systems adopted by the National Institutes of Health to support medical education and research, including attacks upon the reputation and integrity of Dr. James A. Shannon, NIH director; and (3) shoulder-shrugging indifference to the pleas of universities for overhead allowances on government contracts which would be adequate to cover actual indirect costs.

The attack on the National Institutes of Health was embodied in a series of congressional investigations in 1962. They provided an ominous inversion of the recommendations made by the administration's R&D task force headed by Budget Director Bell. The hearings focused all the proper implications of the Bell Report against NIH practices with a vengeance reminiscent of the heyday of McCarthyism. The yardstick, management, and evaluation capability of NIH has been eroded by the inflated earnings of medical doctors in a cartelized profession, with effects upon the universities quite similar to those of the R&D clique in other scientific fields. This has led NIH over the years to support various programs to subsidize and upgrade the salaries of medical teachers and researchers. For the same cause, NIH itself has had difficulty in recruiting professionals to supervise these programs, relying upon part-time panels made up largely of those who are recipients of grants and have a vested interest in the advice proffered. Abuses were bound to creep into a system forced to ignore all conflict-of-interest precautions in order to operate at all. In addition, NIH has been caught increasingly in a dilemma brought on by the lack of advanced equipment, facilities, and manpower in university labs and by the pressure of Congress to spread research funds into the coffers of profit-making research organizations. As a result of these factors, NIH in 1960 began for the first time to grant funds to profit-making companies "when it became evident that certain skills that we required could only be obtained in industry." [22] By 1962 this activity constituted only a small fraction of NIH funds. The overall budget, it might be added, had been augmented beyond NIH requests by congressional anxiety about death and disease, and perhaps a hope for breakthroughs which could extend the longevity of truculent congressmen themselves.

In the course of the 1962 investigation, science and scientists of the university variety were mercilessly flayed for venality, greed, and fiscal irresponsibility. Never mentioned was the infinitely greater scale and routine nature of these practices by industrial contractors.

In regard to the issue of university overhead allowances on government contracts: while government contracts placed with private industry provide full reimbursement, including all overhead (often ranging above 200 percent of direct costs) and an added "profit" fee, universities have been held to an overhead limit arbitrarily set as a fixed percentage of costs from 15 to 25 percent. Actual overhead for university management of large government laboratories or internal research programs generally runs between 30 and 45 percent of direct costs, well below that claimed by private industry for similar work.[23] The increase from 15 to 25 percent in recent years was largely brought about by an attempt to make some concessions to university pleas. As stated by the President's Science Advisory Committee in 1958: "Over the years the government has properly compensated industry for the full cost—including capital costs—of scientific work performed under contract, but it has not often done so for the universities." As a result, PSAC reported, "Universities have had to siphon funds from elsewhere in their own budgets to support federal research. . . . Departments of humanities have had to suffer in order that science departments could do research for the government." In addition, the report continued, the schools have been unable to provide for "the economic renewal of capital equipment for science as instruments and facilities wore out or became obsolete." [24]

In 1966 this situation was only slightly improved. The universities' overhead limit was replaced by a more equitable formula. But the general imbalance between profit-making and educational institutions was perpetuated and universities continued to face the same erosion as did government labs.

NASA claims that 8 percent of its expenditures is devoted to in-house activities; but a close examination indicates that a large part of this is administration and construction of test facilities and equipment at government sites, such as Lewis Laboratory in Cleveland, for use by private contractors rather than as part of a real yardstick activity.[25] The bulk of surviving in-house is concentrated in relatively few agencies that command a small portion of the R&D budget. Only 3 percent of the Navy's R&D and procurement budget goes to in-house today, a drastic reduction from the days when (under the Vinson-Trammell Act) the Navy was required to maintain a substantial yardstick. Most in-house exists in the Army, which has tried to maintain the health of the arsenal system, or in the Weather Bureau, the Bureau of Standards, and the Department of Agriculture.

NASA Administrator Webb admitted in a moment of candor that "where some 90 percent of the funds appropriated to us are spent with non-governmental entities, you have a fairly tenuous control of these activities through normal contractual relationships." [26] Jerome Wiesner added a detail which dramatized the erosion: "I once visited a laboratory that was judged by how many coke bottles there were on the filing cabinets, because the people who had the responsibility for judging couldn't judge the technical program." [27]

The Out-House Coup

The growing political influence of the Air Force during the rise of missilery from 1954 to 1958, and the technique of its success, irreversibly set in motion the trend by which the government in-house yardstick was broken and thrown away. The climax came after Sputnik I. With the help of its contractor constituency and the Congress, the Air Force penetrated NASA's ranks and the Advanced Research Projects Agency was gutted of real authority, both agencies becoming new instruments of Air Force pre-eminence. In 1958 the die had been cast which was to lead to the transfer of the Army's Redstone Arsenal and the Jet Propulsion Laboratory (operated by the California Institute of Technology) to NASA, where they were systematically stripped of their yardstick role. This step was consummated during the early 1960's when the Redstone Arsenal, reorganized as the Marshall Space Flight Center, was forced to contract out the bulk of its activity; the Jet Propulsion Laboratory became the target of an elaborate plot which used the early failures of the Ranger program to batter JPL into acquiescing in the farming-out of its responsibilities to private contractors.

In short, the erosion of government yardstick was the immediate and practical result of the politics generated by changing military strategy and service roles, held in turmoil by rapidly changing weaponry. The Air Force's successful use of industrial contractors as its powerful advocate in Congress compelled the Army, and to a lesser extent the Navy, to emulate the pattern, both as a means of self-defense and to secure replacements for the technical competence stolen from their ranks by the private contractors. The outward manifestations of these changes were rationalized by Congress in terms of "unifying the services" and resolving "interface problems." The real situation was instinctively understood as a bloody engagement whose tangible stakes involved the fortunes of federal agencies, members of Congress, and centers of industrial power.

The Redstone Arsenal Story

Shortly after V-E Day the U.S. Army initiated "Operation Paperslip," an effort to woo the loyalty of German rocket scientists. After a series of discussions in Germany, more than a hundred Germans accepted five-year contracts with the Army. For their new employer the rocketmen supervised the dismantling and crating of V-2 rockets, components, equipment, launching sites, blueprints, and their own personal effects and furniture. In December 1945 Wernher von Braun arrived with his party at Fort Bliss to supervise reassembly of the material. Experimental projects began and the program grew by 1949 to the point where a special field organization was needed. Mothballed ammunition plants at Huntsville, Alabama, were reactivated and called the Redstone Arsenal. With von Braun as the driving force, the Redstone missile was developed (sort of a V-3) and successfully launched in 1953.

When the U.S. entered the missile race in 1955, the Army was authorized to proceed with development of a 1,500-mile vehicle to be called Jupiter. Redstone Arsenal soon commanded the services of 3,600 personnel, including five hundred engineers and scientists added to the original German team, the largest and most experienced group of rocket technologists anywhere in the country. This facility was soon to be torn apart by Air Force contractors out to overtake the Army's lead and make the Air Force the nation's missile king. At the start of the Army–Air Force dogfight, the Redstone Arsenal had already been building shorter-range missiles for several years, providing the roots and stem of most future developments. The Navy soon abandoned the liquid-fueled concept and undertook pioneering work in storable solid fuels more adaptable to ship or submarine deployment. Air Force contractors, under the direction of systems manager Ramo-Wooldridge, Inc., borrowed heavily from the technology of Redstone, using virtually unlimited "crash-program" funds to raid the Arsenal's manpower as well. The Air Force later defended its controversial relation to Ramo-Wooldridge by stressing the successful development of missiles within "a highly compressed" lead-time cycle. These claims were somewhat dubious at best; without the work of the Army group (which also exerted a corrective impact upon very serious Air Force/contractor errors) the development of Atlas, Titan, and Thor would certainly have been considerably delayed.

Again, the climax came in the wake of Sputnik I. The Air Force was assigned sole responsibility for development, production, deployment, and control of strategic missiles, but the Army's Redstone Arsenal remained as a threat to Air Force pre-eminence, a reminder of unpaid debts, and a basis on which the Army could continue to claim a share of military and scientific space missions. In the year of decision, 1958, the Army commanded a substantial in-house yardstick and management capability; the Air Force had only a handful of contract officers dependent upon Ramo-Wooldridge's subsidiary, Space Technology Labs, and contingents of aircraft and missile contractors. It was no match. The Redstone Arsenal was devoured.

In December 1958 the Jet Propulsion Laboratory was transferred to NASA. General Medaris chose to fight for Redstone, strongly supported by the von Braun team whose hostility to the Air Force was second to none. Decisions higher up had decreed that the Army be removed from the strategic missile field entirely, and Medaris faced great difficulties in making his case. In bitter memoirs published after his retirement, Medaris charged President Eisenhower, the Army's finest product and bright hope, with lack of comprehension of changing strategy and with allowing himself to be used by his Big Business friends.

Faced with the choice of NASA or the Air Force, von Braun chose the lesser of two evils. The President at first permitted the Army to retain control of the arsenal, which was assigned to NASA's work. By giving up JPL, the Army won concessions for its view that Redstone could best serve NASA "if left undisturbed in its military environment." According to Space Agency Head Glennan, this concession "was conclusive." [28] But like many "conclusive" Eisenhower decisions, the arrangement was conclusively reversed within a year. Redstone and von Braun were transferred lock, stock, and rocket to NASA control, and Air Force General D. R. Ostrander was transferred to NASA "to superintend" the arsenal, whose name was changed to the Marshall Space Flight Center (MSFC). Having acquired responsibility for developing the Saturn family of heavy boosters, MSFC began a period of expansion. Von Braun was assured the right to build in-house the first nine (of a projected thirty) Saturn I vehicles, providing the classical yardstick for measuring contractor cost and performance while retaining real managerial authority in government hands.

These assurances, however, did not avail against the rapid erosion of the civil service cadre of von Braun's team, forcing him increasingly to employ the subterfuges of the Contract State: arranging with contractors to recruit technical people who were placed on inflated, government-reimbursable company payrolls while assigned to work at Huntsville side by side with civil servants. This could no longer be described as a yardstick activity at all, but rather became a mélange of commingled loyalties such as the Bell Report of 1962 was to condemn. Huntsville personnel came to play an increasingly administrative role. More and more, von Braun was torn from the details of design and engineering, forced to parade about the country and through Capitol Hill committee rooms, defending the moon shot against its scientist critics and against congressional budget-cutters. He became ever more deeply involved in political brokerage among the gang of contractors that enveloped every aspect of Marshall Center tasks.

By 1963 NASA officials no longer felt reticent about their objective of eliminating the last yardstick vestiges at Huntsville, converting it almost entirely into a contractor-operator facility with a thin layer of management by NASA employees. Associate Administrator Robert Seamans declared in 1963: "Our overall objective is to get Marshall away from the arsenal-type operation that they have been involved in in the past, and get them more involved in the management of these large developments, using contractor personnel." [29]

NASA contractors had felt the sting of the yardstick and didn't like it. This was demonstrated, for example, in the Centaur (liquid hydrogen) program for which von Braun had systems responsibility. Conflict between Marshall and its contractors led to a congressional hearing in 1962, which revealed the discomfiture of General Dynamics/Astronautics at having its engineering mistakes discovered. Grant Hansen, vice president of the company, sought to explain the irritating episode: The Marshall Space Flight Center design philosophy is "a somewhat more conservative one" than the company's. "Structural design criteria for safety margins which we have normally used in our past design practice and which were used on the Atlas Program" and were "satisfactory to the Air Force" were based on a 25 percent margin of safety. "Marshall . . . design criteria require a minimum of 35 percent margin of safety. . . ." [30] The sharp-eyed technical evaluation of von Braun proved to be especially irritating after repeated failures proved him correct. The extra 10 percent safety factor which the company had resisted, it was admitted, would have prevented these failures: "We have definitely in this particular case, if I might say so, gambled at a very low weight and found we have to correct ourselves." [31]

The dismantling of Marshall's in-house capability became virtually total during 1964 when remaining systems-management and integration tasks were placed under private contractors and great chunks of the Huntsville civil service cadre were transferred to the newly established Michoud Operations and the Kennedy Space Center. For the

Saturn V launch vehicle, von Braun was given responsibility for only three prototypes. He himself announced (with a resigned mien) that Boeing, Chrysler, IBM, and General Electric were assuming broader responsibilities in the program. This was, he said, "a continuation of the Marshall policy to transfer research and developmental work, as well as fabrication, from in-house to contractor plants as soon as practicable." [32]

Responding to the importunities of House Democratic Whip Hale Boggs, NASA de-mothballed a World War II plant and leased it to Boeing for production of NASA space boosters. Michoud Operations, Boggs exulted, had become "the single most important industry that has come to south Louisiana since we were admitted into the Union in 1812." [33] By mid-1964, ten thousand people were at work at Michoud, of whom only 281 were NASA employees, most of them from Huntsville. At the end of the year Congressman Boggs announced that all top-management personnel would be moved there from Marshall, truncating von Braun's in-house center into a largely paper-shuffling facility like other NASA centers. A protest from leading Huntsville citizens yielded Webb's lame explanation that it had become difficult to hire top-level people willing to work in the government-operated facility.

The JPL Story

Perhaps the clumsiest instance of breaking the yardstick was the case involving the respected Jet Propulsion Laboratory (managed for the Army and then NASA by Cal Tech), which holds a record of distinguished accomplishment.

Industrial contractors had for years thirsted for a chance to straddle this proud beauty, and had been restive and unhappy in the role assigned to them by JPL-administered contracts. Before 1964 contractors were limited to the construction of components, with JPL performing all the R&D and engineering in-house. The laboratory would then assemble the equipment (such as Ranger or Mariner spacecraft) and conduct its own testing and operational programs with very minor participation by profit-making firms.

Tight financial and technical management was the JPL tradition and, in spite of personnel raids by contractors for other government agencies, succeeded in holding together a solid core of scientific and managerial talent which performed a yardstick role and made a pioneering contribution to the technology of rockets, jet propulsion, and control systems for unmanned spacecraft. Contractors were impatient with JPL's controls and the glaring cost/effectiveness differences between contractor and JPL work, which could not be explained away simply by the tax-exempt and non-profit status of the university-operated facility.

Webb's assault on the laboratory in 1964 symbolized the failure of President Kennedy's efforts to enforce the implications of the Bell Report. It symbolized the pending task that still confronts President Johnson, that of asserting real public control over the contractor community.

JPL got its start prior to World War II, carrying forward Goddard's early rocket work. Dr. Theodore von Karman, director of the Guggenheim Aeronautical Laboratory at Cal Tech, foresaw potential rocket applications to aircraft. The approach of World War II made funds available for this work beginning in 1939 with a small National Academy of Sciences grant for the study of auxiliary rocket engines to assist take-off of military aircraft. During World War II JPL became a large R&D establishment, a role codified and maintained after the war by a long-term Army contract.

As knowledge of rocket propulsion advanced, JPL emphasized guidance and control systems, developing the first series of tactical surface-to-surface rocket weapons (the Corporal, the Sergeant, etc.), providing the core of scientists-technologists upon which missile contractors depended in the crash program of the mid-1950's. Louis G. Dunn, who played an important role in JPL's Corporal program, was typical of those netted by Ramo and Wooldridge.[34] As twin Army installations, JPL and Redstone worked closely together, launching as a joint effort the first successful American satellite (Explorer I) after the Russian Sputnik I.

After its 1958 transfer to NASA, the laboratory's primary interest focused on developing the technology of unmanned scientific missions to the moon, planets, and deep space, including tracking and data acquisition and responsibility for creating and managing instrumentation networks around the world. JPL, of course, is best known to the public for the spectacular close-up pictures of the moon and Mars taken in 1964 and 1965, whose success was built upon a series of earlier failures, including consecutive failures of the first six Ranger launches. The success of Mariner IV in taking the first photos of the Martian surface represented a superb JPL achievement. Abelson noted: "The accomplishment required the proper functioning of 134,000 parts after seven months in space. The magnitude of the success is highlighted by the failure of others to attain the goal. The Russians, who have some first-class engineering talent, have not succeeded in their dozen or so attempts at attaining close-in data from Mars or Venus." [35]

Four of the earlier Ranger failures were due to malfunctions of the launch vehicles, and only two could be ascribed to the spacecraft itself. These difficulties do not appear atypical in advanced development programs; yet they provided the pretext for JPL's "reorganization." The NASA case against JPL was weak, and the show-trial (staged by the House Space Committee) tended to refute the very charges that prompted the inquiry—but did not prevent the committee's turning every available scrap of evidence against the lab.

The industrial contractors and their trade press had for several years directed a practiced sneer at the "academic" climate of JPL (from which had come some of the most creative technologists in industry and a substantial portion of the pioneering breakthroughs of the space age), which in their view lacked the "sound and hard-headed" business and engineering techniques of private industry. In 1964, as aerospace lobbyists circulated outside the congressional hearing room where JPL Director William Pickering was testifying, sneers became smirks and smiles.

Webb's efforts to dismantle the JPL role began in 1962 after the failure of Ranger V. A board of inquiry (headed by Dr. Robert J. Kelley) conducted a one-month survey and concluded that JPL had not applied to Ranger "the high standards of technical design and fabrication which are a necessary prerequisite to achieving . . . high probability of successful performance." From an analysis of "technical" deficiencies, the inquiry leaped to the "management" area, implying that technical problems could automatically be solved merely by shifting responsibility to private industry. The Kelley study recommended that "an industrial contractor assume responsibility for systems management, including re-design, fabrication, assembly, and test . . . for subsequent Rangers." [36]

NASA moved to implement this, placing under contract the Northrop Corporation for Ranger X and subsequent spacecraft in the series. The change in effect reduced JPL's role after Ranger IX to the clerical capacity of handling the contract paper. More importantly, JPL would have to educate Northrop personnel (and thus in reality do much of their work) while Northrop sought to hire away key JPL people. The transition thereby built private profit into the system, eliminating whatever objectivity had been afforded JPL by its lack of stockholders, financially interested directors, super-salaries, stock options, and industrial empire-building.

NASA's real intentions were inadvertently revealed by the curious habit of congressional hearings to elicit bits of information in far-removed contexts; the mosaic reveals intentions and policies otherwise denied or obscured. So, the real thinking in regard to JPL slipped out during 1963 budget hearings as a NASA official discussed new construction budget requests: "The facilities program with respect to the JPL operation is tied in with a master plan for their operation which will limit the number of personnel that they have physically on the JPL payroll." The facilities "we are developing for them" will be tied to a strict personnel limit which "puts a strong pressure on the part of JPL management to do everything it possibly can out of house." At the time JPL contracted out 60 percent of its budget: "If you watch it over the years I think you will find that percentage goes up rather than down." [37]

Webb used the 1962 Kelley Report to persuade JPL to seek outside prime contractors for the Surveyor spacecraft (unmanned lunar soft landing) and to give up the lunar-orbiter spacecraft (photo reconnaissance of the moon's surface) development. Hughes Aircraft was named as prime for the former with JPL supervision; Boeing was charged with the latter under NASA's Langley Research Center in Virginia. This was the situation that led to the 1964 maneuvers. JPL did not willingly agree to be its own executioner, and its Ranger experts proved to be unusually resistant to the blandishments of Northrop, preferring to stick with the laboratory. In early 1964 the failure of Ranger VI left Webb no choice but to smash JPL once and for all.

Earl D. Hilburn, NASA's Deputy Associate Administrator for Industry Affairs, headed an "independent" inquiry into the failure and concluded that there had been "a number of deficiencies in design, in construction, and in the testing of the spacecraft." [38] Once more the leap was made from "technical" to "management" deficiencies, this time with a determination to go all the way. The stage was carefully set by a letter from Webb to the chairmen of the two congressional space committees and the calling of a press conference. In both, technical faults were quickly passed over in order to emphasize the real tendency. The laboratory, with its background of academic research, Webb indicated in the press conference, was held by some as unable to handle the "engineering and administrative" problems of building highly complex spacecraft. The contract with JPL would therefore be realigned, along the pattern existing in other NASA centers, particularly, he emphasized, "for handling relations with industrial contractors." [39] Two weeks later, at the congressional hearings, he reiterated that the laboratory "resisted" proposed changes designed to improve the reliability of the Ranger.[40] The nature of these "proposed changes" was made clear in his testimony: that full systems management be farmed out to private contractors.

Trying to pin down the specific "technical deficiencies" (upon which the whole NASA position stood), the congressmen saw the charges evaporate into ambiguities. Webb refused to make available to the committee the detailed technical analyses upon which the charges were based, claiming that the Kelley Report was classified because military systems were involved, and that the Hilburn document was only a working paper. NASA officials went further and dissociated themselves from the Hilburn conclusions, which were called "not a definitive Agency position." [41] Homer E. Newell, NASA Associate Administrator, contradicted the key charge of Webb's letter: "It is not correct to conclude from Mr. Webb's letter that the Rangers were not thoroughly and rigorously tested—they were." Oran W. Nicks, NASA's Director of Lunar and Planetary Programs, took issue with several of the Hilburn conclusions mentioned by Webb.[42] Webb himself expressed doubts about the "technical deficiencies" and said he "wasn't absolutely sure that NASA could have produced a better situation by being absolutely insistent on having things operated its way." [43]

Webb's case was even more confounded when the spokesmen of private industry denied that Ranger's technical shortcomings could have been avoided by some kind of management magic. Barton Kruzer, general manager of RCA's Astro/Electronics Division, challenged the assertion that JPL had refused to accept RCA suggestions: "At the time of Ranger VI, I would not, Mr. Chairman, have done anything different than JPL did." JPL's testing program, he indicated, was fully abreast of the state of the art.[44] This evaporation of the underpinnings of NASA's case, however, did not deter the committee from awarding the plumes to Webb. The case against JPL went like this in the committee's final report:

Since the "technical deficiencies" had evaporated, the committee excused itself from considering them: "there is no realistic way" for the committee "to evaluate the various factors. . . . Suffice it to say" that it "does not underestimate the technical difficulties involved in this project." It declared itself "prepared to accept as fact the reputation that the personnel at JPL possess outstanding scientific and technical capabilities." Having dismissed the technical basis, the committee then transferred attention to "management" on the grounds that "technical problems arise from management problems, and are very closely interrelated, and vice versa." [45]

"In view of the history of the Ranger Project, and the facts established during the current investigation," the committee "cannot escape the conclusion that serious management deficiencies have existed. . . ." The evidence which led to this "inescapable conclusion" was twofold: first, that Dr. Pickering's enthusiasm for the "matrix-type" organization had not been shared by NASA headquarters, which recommended a vertical "project-type" arrangement; [46] second, ". . . the apparent reluctance on the part of JPL to accept management direction from NASA headquarters," both in revising its chain of command and in other matters not specified. "If there was a single thread that consistently ran through the testimony of NASA officials regarding the management of Project Ranger, it was JPL's resistance to NASA supervision." [47] The unspecified "other matters" emerged clearly in the report as having been JPL's insistence on maintaining management control over contractors, resisting NASA's repeated urgings that virtually all systems-engineering and management functions be contracted out.

The report does not lack resourceful logic: "Virtually the entire Ranger development project from conception of the spacecraft through design, subsystem procurement, fabrication, testing, and flight has been carried out by JPL. Such all-inclusive in-house responsibilities place JPL in the position of being a prime contractor to NASA rather than a NASA center." [48] Since it was neither an industrial prime contractor nor an administrative center, it could not be adequately supervised, being responsible only to NASA headquarters whose importunities it resisted. This line of reasoning completely omitted the notion that JPL might be acting as should all NASA centers—retaining real authority, competence, and responsibility "in-house" so as to exercise more than fictional control over contractors.

The congressional report concluded that "the orientation and strengths of the Laboratory are such that it is better suited to act as a NASA center than to do the work of an industrial contractor," and recommended appointment of "a general manager, experienced in industrial operations and supported by an adequate staff" to manage such projects as Ranger and Mariner. Upon the expiration in December 1964 of the existing three-year contract, it was recommended that a one-year contract be executed [49] in order, in Webb's words, that there "be a trial run of the tighter controls to see if there can be a meeting of minds." [50] One observer commented: "Here was a bit of comedy relief. It was like having the security guards who check your badge at the Cape Kennedy gate supervise a Saturn launching." [51]

Declaring that JPL needed "a hardheaded business type management," Webb named retired Air Force Major General Alvin R. Luedecke (who had been general manager of the AEC) for the job. He was duly named deputy director at the end of June.[52]

Coinciding with the climax of the 1964 hearings, the Space Agency announced development plans for two new unmanned spacecrafts to explore Mars, indicating that JPL would have no part in the plans. Both would be developed, built, and systems-managed by aerospace firms through the standard pattern of NASA contracting. Both projects proposed to use JPL-developed spacecraft types (Mariner and Voyager) and would be funded "with a total price tag of hundreds of millions of dollars," according to Donald P. Hearth, NASA Director of Advanced Lunar and Planetary programs.[53] This was a surprising cost estimate compared with JPL budgets during the years in which these and other systems were pioneered ($200 million in 1963). The breaking of the yardstick was reflected in a drastic upward revision of costs; the Ranger program went from $6.8 million to $15.5 million in 1964-1965, and estimates of Surveyor during the years 1963-1965 jumped from $495 million to $760 million.

Dr. Homer Newell, Associate Administrator, declared in 1965 budget hearings: ". . . in the immediate past, JPL has had both Ranger and Mariner in-house. This we think has perhaps been too much and in the future we would look to JPL's having a smaller in-house effort than has been the case in the past, and a larger fraction being done by contractor." He suggested that JPL *might* be permitted "to do a subsystem of Voyager . . . to keep them on their toes. . . ." [54]

The passivity of Pickering and Cal Tech appears to have been the result of the adroitness of Webb's maneuvers. The change of JPL's function precipitated strong resistance from influential Cal Tech President Dr. Lee A. DuBridge. Webb's letter served the purpose of denying Cal Tech initiative in appealing the decisions to Washington, forcing DuBridge and Pickering to choose between defending themselves against charges or running the risk of an open break with Webb, who still held the power over JPL's future role in the space program and who (as an experienced politician) was prepared to make concessions to Pickering's primary interest in holding his scientific team together and salvaging a role in advanced research and development.

JPL had always been reluctant to accept the role of managing a growing band of contractors. Its responsibilities for NASA had in three years vastly expanded its oversight of hardware production. It could not be unhappy to relinquish this role, but it feared that the pattern established in other NASA centers—of appointing private industry in a dominant role—would, if applied to JPL, destroy its creative function in advanced R&D. The uniqueness of this function, arising from JPL's history, traditions, and academic ties, had been chiefly responsible for its survival in the inimical atmosphere of the Contract State. In addition, Pickering had to protect the morale and honor of his people, who had resisted the flashy siren song of industrial contractors who were anxious to put them to work writing brochures and wildcatting for new and more profitable government contracts.

Webb conceivably might have re-negotiated the JPL contract quietly. He did in fact do everything in his power to prevent the congressional show-trial that his letter made inevitable. The letter was a symptom of anxiety at a time when NASA and Apollo were already under critical fire from the university community, the military, and the Congress. If JPL had chosen to break openly and had stolen the march to the public forum, possibly cataloguing its own complaints of NASA's inept, inadequate, and wasteful management, the repercussions might have been serious. Now Webb could play the conciliationist role, resisting congressional hearings, reassuring JPL of respect for key values, and acting as JPL's defender before the public and Congress. This is what actually happened.

Committee members were aware that the basic revisions of JPL's role were an accomplished fact, that JPL would lose its independence as contract manager and yardstick producer, that the case was already closed and JPL conciliated. There would be no open protests, countercharges, attempts to inject the JPL issue into the larger political alignments and divisions surrounding government contracting and the space program. This reading of the situation behooved the committee to convert the hearings into a perfunctory and genial waltz around a thinly veiled play of anxiety, humiliation, and anger.

Commenting on the strange spectacle, *Missiles and Rockets,* which for months had been calling for JPL's blood, characterized the hearings as "almost a conspiracy to soothe the wounded feelings" of JPL. The editorial wondered how the committee's concentration on "managerial problems," dismissing interest in "technical competence," could be squared with the charges in Webb's original letter "about the deficiencies in design, construction, and testing." Noting that the probe had stopped short of personality clashes, the editorial said "it was evident that there was much more to these than showed." [55]

In 1966 NASA and industry spokesmen publicly discussed plans to divest the California Institute of Technology of all connection with JPL, removing Pickering and other Cal Tech "academics," converting the facility into purely a NASA field center. The House Space Committee was once again looking into charges that slippage of the Surveyor Project (instrumented spacecraft for soft moon landing) was traceable to faulty JPL supervision of Hughes Aircraft. There were hints that Surveyor tardiness could defeat the national goal of landing on the moon in this decade, and suggestions that JPL might prove a useful scapegoat.

In an earlier moment of hyperbole, Webb had praised the nation's space effort for ennobling NASA's leaders and technologists. "It takes courage," he declaimed in the grand manner, "to build these giant machines such as the Saturn, which can boost 220,000 pounds into earth orbit or send 90,000 pounds outward to the moon. It takes courage for men like Bill Pickering to develop a Mariner and send it out to Venus. . . . It takes courage on the part of Congress, the elected representatives of the people . . . to look to the future of the nation in these matters. May I say that those of us in the Executive Branch appreciate the courage with which this committee has faced up to these problems." [56]

JPL's Bill Pickering showed courage too in 1959 when he wrote an article warning of the dangers of overemphasis on the science-technology race. He warned against blindly pushing development ahead of engineering experience or the state of the art, declaring that the nation may be creating a "cult of failure" because "we have tried to run before we can walk." Crash programs are unnecessarily large gambles entailing heavy fiscal and psychological costs, he wrote, pleading for realistic planning and technical progress with adequate technical staff work and engineering analysis to permit proper testing and reliability programs.[57] These views were a sharp and direct attack upon those very forces that were to profit most from the public humiliation of JPL in 1964. By sticking with the laboratory and sheltering its creative heart, Pickering may have given substance to Webb's glib invocation.

Like the tale of Redstone Arsenal, the story of JPL demonstrates with what consistency and celerity the R&D contract cult is defeating the concept of "yardstick" as a tool of management and accountancy which is responsive to national values and the democratic process. Through the vicious circle of private economic and political power, the methods of R&D contracting create a privileged inner society controlled by the managers of industrial empires, the leaders of banking and investment circles with a stake in corporate affairs, and the key congressmen and federal bureaucrats whose committees and agencies control the allocation of public contract funds. These men come to wield increasing power for a narrow constituency whose names are unknown to the public and also frequently to congressmen and the press.

To ascribe a like influence to university scientists, as have the clichés of the age, is to follow the scapegoat-seeking habit of Congress. Aroused to righteous ire by the intangible and evasive quality of research and development spending, the legislators suffer a blindness which causes them to find targets in basic science, the universities, the National Science Foundation, and the National Institutes of Health. They rationalize this odd displacement in terms of a commitment to limit growth of government and protect the "free enterprise" system. In practice this doctrine serves to insulate and fortify the real industrial power centers of R&D contracting.

The belief that only private industry and the profit system can perform the miracles of modern technology that will protect us from the Russians (and, as an incidental bonus, enrich our lives) dies hard against the crowded record of the technical failures, venality, inefficiency, and cupidity of private contractors, the traditional patronage of science by aristocracy, and the striking technological achievements of the Soviet system. Dr. J. Herbert Hollomon cites his experience with industry which leads him to conclude that ". . . to put it bluntly, the research and development activities of the federal government are probably managed as well as those of private industry generally." [58]

Redstone Arsenal and JPL did not deserve to be sacrificed to the politics of military missions and budgets. Their role is needed as never before in an age committed to the swift pace of technological change. The conversion of laboratories into office space provides a dismal symbol of the new public dedication to science.

CHAPTER XIII

THE NEW BRAINTRUSTERS

The institutions that embody the national response to the science-technology race are no longer improvisations. The classical period of institutional experiment is coming to an end, and the innovations have acquired a stability of law, custom, and usage which resists critical inquiry. The mobilization of intellectuals, scientists, technologists, and managers for public purposes is not new, but the ubiquity and number of private organizations (both profit and non-profit) that work primarily for government are unique in American life. Besides industrial systems managers and advisers, there are literally hundreds of institutes, laboratories, and "think factories" whose main commodities are R&D and technical services. Over 450 non-profit R&D corporations alone have been set up in the past decade and, with educational institutions, absorb close to a billion public dollars each year.

The Jet Propulsion Laboratory provided an early model of a university-operated facility, while the RAND Corporation was the model for the great proliferation of non-profit independent centers. Among the university-operated, government-owned facilities are the Livermore and Los Alamos laboratories operated by the University of California; the Argonne National Laboratory operated formerly by the University of Chicago, now by a group of midwestern universities; and the Brookhaven National Laboratory run by a group of eastern universities. There are also independent, university-linked organizations which do business with private firms as well as government, such as the Stanford Research Institute (Stanford University) or the Armour Institute (Illinois Institute of Technology). In the RAND category, there are such corporations as Systems Development, MITRE, Aerospace, and ANSER. Extending the catalog are the industrial firms that perform systems engineering and technical management through a company division or a subsidiary (such as General Electric's system integration and check-out role with NASA, or that of Space Technology Labs for the Air Force), and the general management role exercised over vast groups of subcontractors embodied in *prime* contractor status. A large number of government facilities are managed by industrial contractors: Oak Ridge by Union Carbide, Hanford by General Electric, the Savannah River Facility by Du Pont, and Cape Kennedy by Pan American Airlines.

These innovations have been both praised and criticized. James A. Killian, the first Presidential Science Adviser, believes that this variety "has countered tendencies toward a centralized bureaucracy, and has kept a great body of scientists engaged in work for the government free of political commitments and outside the discipline of the government's administrative organization." [1] On the other hand, these new organizations have been called "scientific slave traders" and mere "hiring halls" to enable government to evade its own civil service system, allowing public management and decision-making to devolve by default into hands that are anonymous and beyond public control. The organizations are accused of being closely tied to their bureaucratic contracting agencies and major industrial contractors, therefore laced with conflict of interest, serving not in an independent and objective role but as advocates of narrow special interests and as brokers of the contract market.

Defenders claim that their private status permits a high degree of flexibility for government in enlisting special skills to meet quick-changing problems without undertaking long-term commitments. The detractors, however, argue that postwar experience belies this claim, that the system demonstrates a self-perpetuating tendency fortified by pork-barrel patronage prerogatives which have been diminished in government thanks to civil service and public scrutiny.

The brain factories and systems managers have largely supplanted the traditional in-house capability of government to evaluate and technically direct the activities of contractors. These new organizations must be judged by their degree of pluralism, independence, and objectivity in evaluating contract proposals, integrating military and space systems, defining and planning future technological requirements, and formulating "technical" policy questions for public decision-makers. They should be subject to sufficient public control and able to maintain standards of professionalism that protect the public interest at least as well as traditional in-house establishments. With few exceptions, the record does not inspire complacency and raises sharp issues for the future.

The Non-profits and "Objectivity"

RAND, the non-profit archetype, was established by the Air Force after the wartime success of "operations research" in applying mathematical analysis and simulation techniques to the solution of military problems. Starting from an initial base of 255 employees and a budget of $3.5 million in 1948, RAND grew by 1962 to 1,150 employees and an annual business of $20 million.[2] About 70 percent of its current activity is for the Air Force and includes brainstorming future generations of weapons and "paper" analyses of new strategic and tactical doctrines.

As the youngest and most aggressive service branch, the Air Force tied its future to technological innovation. After World War II it sought to maintain a partnership with the technical innovators to overcome the low salary scale and poor repute of peacetime government service. Douglas Aircraft agreed to establish (under Air Force contract) an "independent" scientific group which came to be known as Project RAND (the name deriving from "*R*esearch *an*d *D*evelopment"). To superintend this input, the Air Force created the position of Deputy Chief of Air Staff for R&D and appointed General Curtis E. LeMay to the post. Because other contractors resented the special position of the Douglas Company, RAND was separated and reorganized in 1948 as a non-profit firm with its own board of trustees and with initial funding from the Ford Foundation.[3]

RAND has no laboratories in the usual sense and does not engage in design or manufacture of hardware, systems management, or specific project evaluation. Air defense and missiles provided motives for the spate of new organizations that followed the RAND model. In 1951 the Air Force requested MIT to establish the Lincoln Laboratory for electronics R&D. Out of this (in 1958) came the MITRE Corporation to take over the job of integrating actual production and deployment of an air defense system. The Aerospace Corporation was established in 1960 to assume the duties of technical direction and systems engineering for Air Force missile and space programs. RAND itself has spun off at least two other non-profits: the Systems Development Corporation set up in 1956 for design and programing of a computerized air defense system, and Analytical Services, Inc. (ANSER), set up in 1958 as technical support for the Air Staff, the secretariat of the Air Force's Chief of Staff.

The tendency of such groups to serve as agency advocates led the Defense Department itself to establish the Institute of Defense Analysis (IDA) in 1956 under a group of universities, to screen R&D proposals from the separate services and originate studies of its own. In 1958, with the establishment of the Advanced Research Projects Agency (ARPA), IDA was assigned to provide its scientific and technical capability. The Army and Navy got into the act soon thereafter, the former creating the Research Analysis Corporation in 1961 under the aegis of Johns Hopkins University, the Navy establishing in the following year the Center of Naval Analysis at Philadelphia's Franklin Institute.

All of these RAND-like operations are structured with boards of directors, presidents and all the accoutrements, but are not publicly held. RAND has been described as "a university without students" and goes to lengths to maintain the relaxed and free-wheeling atmosphere of a campus, broadening its expertise beyond the hard sciences into economics, sociology, and psychology. Its imitators, however, have tended increasingly to don the appearance and mode of operation of industrial engineering organizations.

In the rush to develop strategic missiles in the 1950's, conflict-of-interest issues were shunted aside. The most aggravated case was that of the Air Force and the Ramo-Wooldridge Corporation, whose impropriety was obvious, arising from R-W's double role as profit-making firm and government insider. The non-profit organizations sought to remove this embarrassment but often created the same conflct of interest in a disguised and more insidious form. Their avowed service in behalf of "objective" engineering, management, and technical advice (which would theoretically serve a general national interest) confronts tendencies that conflict with the requirements of objectivity: they tend to represent and serve the narrow interests and aggrandizement of the contractor community, the contracting agency, and highly placed individuals in their own ranks. When such motives conflict with executive policy and the general interest, the special-interest incentives tend to color, if not dominate, their role.

The contractor influence. The Scientific Analysis Office (later

ANSER) was assembled by the Air Force in early 1957 and put under a continuing contract to provide engineering studies for development planning. This group was absorbed by Melpar, Inc., a large electronics firm with major Air Force contracts. There was no difficulty with this situation until Melpar's competitors, like those of Ramo-Wooldridge, complained to Congress and refused to provide the firm with their own proprietary information. Nudged to a greater sensitivity to the conflict-of-interest issue, the Air Force requested a RAND study which led to incorporation (in July 1958) of ANSER on a non-profit basis. Whether this change resolved the situation may be judged by the fact that twenty-five of the thirty-eight-man professional staff of the Melpar subsidiary stayed with the new corporation, including the president, and the Melpar Corporation continued to be represented on ANSER's board, which was broadened to include representatives of certain Melpar competitors.[4]

The RAND Corporation itself went through a comparable transition in 1948 which moved it out of Douglas Aircraft. The objection to the Douglas role arose from the company's competitors; the issue was resolved by creating a board of trustees composed of top officials of the big four of the aircraft industry—North American, Boeing, Northrop, Douglas—and eliminating political heat by placing RAND under a consortium of major contractors. This transition protected the interests of a broader segment of the industry and thereby achieved an "objectivity" previously lacking. But all consortium members have common interests not necessarily shared by the public. "Objectivity" in dividing up the contract market does not represent "objectivity" in protecting other constituencies, including smaller contractors, civilian industries and geographical regions, other federal agencies, presidential power and policy, and the general public. The non-profit firms engage not merely in engineering and technical management but also in an essentially political role. Evaluation and control of contract awards and performance tend to be infused with conflict-of-interest problems. This pattern provides the recurrent theme: conflicts of interest not challenged are ignored; when challenged, they are converted into consortiums of the most interested contractors.

The Systems Development Corporation was created for the Air Force in 1956 with a heavy representation of defense contractors,[5] including General Donald L. Putt (USAF retired), president of the United Technology Corporation (subsidiary of United Aircraft, the country's seventh largest defense contractor). Putt was also a member of the Air Force Science Advisory Board. A congressional committee forced him to resign from this board in 1961 when it was revealed that his firm held over half a billion dollars in Air Force contracts related to areas in which he or the Systems Development Corporation had rendered "objective" advice.

The Aerospace Corporation provides a well-documented study of the political role and conflicts of interest of a non-profit systems engineering organization. Designed to correct the abuses of the R-W arrangement, Aerospace was created with a president from the Raytheon Company and a board of directors amply mixed with directors and executives of other Air Force contractors.[6] It has proven a strange creature indeed. From its establishment in 1960, the "non-profit" grew within four years to become one of the nation's largest Defense Department contractors, ranking forty-fifth from the top of the list (over $300 million in contracts during the period). It quickly outgrew its projected size, acquiring (by 1965) 4,306 employees, including 1,752 technical staff, to become the largest of the non-profits.[7] In 1965 an exhaustive investigation by a subcommittee of the House Armed Services Committee found Aerospace operations to be seriously abusing the public interest; the non-profit was in effect an Air Force device for evading administrative controls by the Defense Department and legislative oversight by Congress, and for conducting propaganda in behalf of the Air Force and its industrial contractors.

In government dealings with industry, "Aerospace sits on the government's side of the table," the report noted. "Any non-profit organization created by the government, especially one with such an intimate and privileged relationship as Aerospace, is the government's child, and the government's moral responsibility for the child does not end, but just begins, when the child is born." [8] Evidence indicated that Aerospace's management "wanted the benefits and prerogatives of private industry without any of the risks, plus the protection and security of the government services without any of the restrictions." Aerospace spokesmen answered that its management was responsible "only to its fiduciary trustees for its fee expenditures and assumption of mortgage debts. But these trustees could only be answerable to the public through the Air Force . . ." At the same time, Aerospace claimed to be "independent of the Air Force and did not have to justify to the Air Force, or any public agency, their use of fees . . ." [9]

The House subcommittee blamed the Air Force for "inadequate fiscal control" and accused it of "insincerity" in its dealings concerning the role of the non-profit corporation. "This calls into question the ability of the Air Force to cope with the management problems attending its organization of space programs and will require continual congressional surveillance." [10] Conversely, the evidence raised questions "as to the degree of control exercised by the corporation over essentially military decisions and the wisdom of continued reliance by the Air Force on the management concept which Aerospace represents." [11] An internal paper of the corporation which irked congressmen dealt with finding the best euphemism to describe the company's exercise of public decision-making responsibility, recommending that the corporation "affirm the Air Force ultimate responsibility" while gaining for the corporation "the authority for specific action, since the Air Force is used to the concept of delegation of power." [12] The House subcommittee recommended that the management concept "be reappraised with a view to determining the need for using outside organizations to carry out the basic planning and subsequent management of military missile/space programs." [13]

In 1958 the Institute of Defense Analysis (IDA) became the primary braintruster of the Defense Department's new R&D organization, the Advanced Research Projects Agency (ARPA). In spite of its nominal university auspices, IDA tended to become a consortium of interested contractors, undoing Defense Secretary Neil McElroy's plan to superimpose departmental review over the parochial and self-serving technical bodies and contractors of the separate services. The original concept (told by McElroy to the House Appropriations Committee in November 1957) was for ARPA to take over "all of our efforts in the satellite and space research field . . ." It was necessary, he said, "that the vast weapon systems of the future . . . be the responsibility of a separate part of the Defense Department" in order to provide thorough screening of "will-o'-the-wisp" project proposals.[14] ARPA was to be a top-level management team having its own budget, able to concentrate on the new and unknown without involvement in immediate requirements and inter-service rivalries. It was to have authority to contract directly for R&D, thereby setting the keystone of overall defense procurement beyond the reach of the ambitious, quarrelsome services.

This plan ran aground as Congress denied claims that the Defense Department possessed sufficient authority to create the agency.[15] Air Force General Schriever bluntly recommended that ARPA "be immediately disestablished," a view which the House Government Operations Committee described as arising from the concern that the agency would "become a serious obstacle to Air Force ascendancy in space as well as a disruptive force in weapon developer-user relations." [16]

Since its founding, ARPA has served as a mere staff addition, screening and to some extent coordinating R&D projects, but lacking any permanent, independent in-house capability.[17] According to Dr. Harold Brown, ARPA and IDA mixed together civil servants and the representatives of private contractors in "what I think . . . was an inappropriate manner, in such a way that you could not tell whether the decisions were being made by government officials or whether they were being made by contractor personnel." [18] Air Force Secretary Donald Quarles, former AT&T executive, played an important role in IDA's staffing. In response to congressional queries concerning potential IDA conflicts of interest, Quarles said "great care was taken to select and instruct the IDA employees" to assure objectivity.[19] Originally most of them were on company leave and carried on company payrolls, a situation accepted with equanimity. But after the 1959 investigation of Ramo-Wooldridge, IDA transferred its personnel to its own payroll. An IDA official admitted to Congress the possibility that employees might have standing offers to return to their former employment.[20]

The House Operations Committee touched the heart of the matter: as the reviewing authority for major R&D contracts, ARPA "invites intense curiosity from industrial firms and an understandable desire to have a direct pipeline to the inner sanctum. The issue is not so much the obtaining of immediate contracts but of getting valuable information for future business opportunities." [21] The committee noted that "since ARPA designates the specific companies with whom the military services are to place contracts, the conflict-of-interest potential is present." [22]

The pattern of non-profit braintrusters is further aggravated by the fact that, like profit corporations, their boards of trustees are interlocked with one another. Ralph Lapp reports that he found the same man on the board of five different organizations.[23] In general, the non-profits tend to represent their contractor-patrons in lawyer-like roles, using insider powers of government to adjust and limit competition that might injure their common interests.

The agency influence. Another species of conflict of interest arises between the non-profit's technical work and its potent role as advocate of agency interests. Contract funds are used to finance activities on behalf of the government client which the client cannot undertake for himself, such as attacking official Defense Department policies.

We have seen elsewhere how RAND and other industrial contractors have actively pursued the advocate's role. In this day of scientific expertise, each administrator, admiral, and general requires his own contingent of leashed intellectuals who will perform on command like trained monkeys, writing persuasive analyses in highly sophisticated and quantitative language to support the whims, hunches, value judgments, and opportunism of agency officials, converting scientist and engineer into a new kind of lawyer of technology.

In 1965, for example, the House Armed Services Investigation Subcommittee discovered that the Aerospace Corporation had spent more than $1 million on public relations conducted by its own staff and had hired a New York firm (at a fee of $2,000 a month) for PR work.[24] Claude O. Witze, senior editor of *Air Force and Space Digest,* was carried on the payroll of the non-profit for $150 a month. On top of these costs, Aerospace averaged over $200,000 a year in expenditures for advertising, which, though officially aimed at recruiting, served as an Air Force propaganda tool. The corporation maintained an office with nine staff members in Washington, their purpose given as "technical liaison between the company and its client, the government." With asperity, the investigating committee commented that the corporation's client, "of course, is not the whole government, but a division of the Air Force Systems Command, which happens to be located across the street from Aerospace's El Segundo office" in California. The committee found it difficult "to see any justification for a Washington staff to do legislative liaison and why any congressional interest in the company could not be served through the proper channel of the contracting agency." The Washington staff "spent a great deal of time buying lunches for members of the press and others . . . ," and the office head, who was assigned a private chauffeur, received a salary equal to that of the Secretary of Defense.[25] "In view of the fact that Aerospace's activities are restricted to performing for the government and, consequently, the company is not in competition with private industry, we believe that the need for a public information office is questionable," the committee concluded.[26] Not to mention that "this public relations network appears to run counter to the provision in Aerospace's Articles of Incorporation not to engage in propaganda or otherwise attempt to influence legislation." [27]

During the course of the 1965 House inquiry, a contracting officer newly assigned to the Aerospace account showed a candor that delighted the committee: "In an inquiry so filled with cover-up . . . Colonel Hamby's performance was like a breath of fresh air." Colonel

Hamby had tried to bring Aerospace's operations into line with established regulations and publicly protested lack of support from Air Force Headquarters. His reward was summary dismissal from his post in July 1965, in the middle of the congressional inquiry: "One would assume that the Air Force would have welcomed such vigorous and capable stewardship and it was therefore surprising when the Air Force informed the Subcommittee that . . . Colonel Hamby was relieved of his position . . . ," thereby calling into question "the sincerity of the Air Force in its dealings with the Congress and in its ability to cope with the day-to-day management problems attending our military/space programs." [28]

An interesting case of advocacy and counter-advocacy developed after McNamara's appointment as Secretary of Defense. Having early pre-empted the braintrust field with the RAND Corporation, the Air Force during the 1950's exerted a powerful influence in the National Security Council and on the Joint Chiefs of Staff. It was supported by RAND's closely reasoned, firmly documented technical papers, embellished with mathematical symbols and the impenetrable armor plate of tables and graphs spun out by giant computers whose tireless peregrinations could barely be followed, much less disputed. RAND served its patron well, building elaborate steel cobwebs of obfuscation in support of Air Force positions in the critical disputes concerning strategy, weapons systems, mission assignments, and budget allocations. But an amusing turnabout came in 1961: McNamara, a product of computerized, operations-research-style management, assembled around him leading RAND gray eminences (such as Charles J. Hitch as Defense Department comptroller and Alain C. Enthoven as assistant comptroller). Previous secretaries had been helpless in the web of RAND logic, but now at last the battle of computers was joined with the RAND Corporation pitted against itself—Air Force and Defense Department eyeball to eyeball, computer to computer. The result was foreordained. The RAND magic quickly lost its power and McNamara's hand unflinchingly cut through the gobbledygook to the cost/effectiveness heart of service ambitions and weapon claims.

Self-serving and perpetuation. Vested interests also adhere to the non-profit corporation itself. They are symptomized by extravagant and careless use of public funds and possible accumulation of large capital assets which could be used indirectly to serve the private interests of its managers and as a form of uncontrolled political patronage for others.

In spite of non-profit legal status, these companies "demonstrate all the eagerness of a profit-minded company in expansion and in getting new business." [29] The contracts under which they operate are largely comparable to those with profit-making firms, including a fee on top of costs and overhead (RAND gets 6 percent of estimated costs; Aerospace's fee has averaged 5 percent), indistinguishable from profit taking. A 5 percent fee for a non-profit is roughly equivalent to a 10 percent fee for a profit-making firm because of the former's exemption from taxes. The 1965 investigation of Aerospace noted that on R&D contracts with profit-making firms the average fee was usually below 10 percent, and therefore "with a fee that is at present 5 percent and in some years has been higher, Aerospace is receiving a higher percentage of fee in terms of retained income than a profit-making firm." [30]

The only difference is that a profit firm may distribute its cumulative assets as stockholder dividends, stock options to management, holdings of stock in other firms, etc.; the non-profit has no stockholders. What does the non-profit do with the profit fee it receives? First of all, it can provide perquisites for its personnel (large salaries, private dining rooms, aircraft, Turkish baths, country clubs, and all the items that have become necessities to major profit-making corporations). Second, the non-profit firm can acquire realty, which represents equity accumulation and net worth for the company, and which it may readily convert into credit to support additional realty acquisition and growth. Similarly, it may hold its earnings as a cash reserve for the same purpose or deposit them in interest-bearing accounts. Third, it may build-in a large component of job patronage for retired government officers, their relatives and friends, not to mention members of the political establishment. Fourth, it may subcontract with profit-making firms at artificially high prices, firms in which it's personnel have an indirect interest or from which they receive a rake-off. This is probably illegal but difficult to prove in court, and perfectly legitimate if the costs are reasonable even if collusion is involved in creating a new company or supporting an old one by these means. Another aspect of this, one which never can be completely controlled, is the exploitation by key personnel of opportunities to use the role and influence of the non-profit company to enable somebody else to make a profit, the debt being ultimately redeemed when the person leaves non-profit employment. Large numbers of RAND alumni have traveled the route to important positions with major Air Force contractors, or have set up related profit-making ventures of their own.

The Aerospace Corporation recently has been a target for General

Accounting Office criticism. In its first five years (1960-1965) Aerospace received fees amounting to $16 million which had been repeatedly justified to Congress as enabling the corporation to engage in independent research. But the GAO discovered that only $411,000 had been expended in this manner, most of it after the inquiry began. Comptroller General Joseph Campbell charged that the company had acquired new buildings when government-owned facilities were available, forcing the government into an unnecessary expenditure of more than $17 million. Congressman Porter Hardy indicted the Air Force for deliberately padding Aerospace contracts as a dodge for acquiring for itself facilities that Congress would not otherwise have approved.[31] GAO noted the 20 percent incentive bonus paid top executives on already high salaries (with annual increases of more than 15 percent) and the unusual policy of unlimited sick leave which might have converted "full-time" employees into "moonlighters" compensated on a full-time basis. "Unusual" relocation expenses were also allowed personnel.

The House subcommittee concluded that the fee system was used "as leverage to support many of these practices not supportable for reimbursement," involving "unusually high starting salaries, unusually sharp increases in pay after short periods of employment, very high salary scales for management-level personnel . . . and a pattern of needless and frivolous expenditures from fees for such things as subsidized meals for executives, country club memberships for executives, and elaborate and frequent entertainment." [32] The same kind of gratuities were extended to Air Force officers as the left hand served the right hand and was served again in turn. Air Force interests were served through the use of fees to acquire unauthorized real estate and facilities (the public relations aspect of the non-profit was directly reimbursable under contracting rules). Fees were used for at least two additional purposes: to enable the Air Force to maintain a reservoir of independent funding power to support R&D deleted from its budget by McNamara; and as a means of cost-reimbursement for items not legally reimbursable as either overhead or direct costs.

In effect, the fee arrangement with Aerospace (and by implication with other similarly favored non-profits) undermines attempts at fiscal responsibility or control, even if the contracting agency is seriously dedicated to these ends. The congressional report declared the non-profit fee to be "a self-defeating process . . . not only in preventing unjustified expenditures . . . but in controlling reimbursable expenditures. The contracting personnel are only too aware that if an expenditure is disallowed as reimbursable, it well may be paid for anyhow out of government funds allocated as fee." [33] This goes to the heart of the matter, revealing that contract policy with non-profit corporations creates a fiscally unmanageable institution. Because the government accepts the responsibility for the corporation's financial stability and provides fees "according to need" to safeguard that stability, the responsible contract officers must cover in fees whatever cannot be admitted under reimbursable costs. Fiscal irresponsibility is positively encouraged, and government becomes a helpless pawn of the most self-serving, unreasonable, uneconomical, and unjustifiable excesses. Its non-profit creatures revert to financial infantilism in a cushioned, cost-plus nursery.

Throughout the hearings, Aerospace was unable to give satisfactory information on its use of fees accumulated in five years. The House Armed Services Committee wrote into the 1966 Military Construction Bill a provision forbidding non-profits to undertake "the construction of any facility or the acquisition of any real property" without express authorization.[34] The Senate, however, rushed to the corporation's defense, striking out the clause from its version of the bill.

There is no incentive for the non-profits to retain their capital assets, for their charters provide that upon dissolution such assets revert back to the government—or in the case of RAND, to the Ford Foundation, or in the case of Systems Development Corporation to RAND (if RAND is in existence).[35] Under its charter, Aerospace Corporation's assets will revert to the Secretary of the Air Force in case of dissolution. But the charters do *not* compel the corporations to dissolve if their contracting relation is terminated. In spite of the fact that all its assets represent government endowment for public purposes, Aerospace's charter leaves it free under an "independent" board of directors to engage in similar activities for private business, providing only in the vaguest terms that such work serve "the interest of the United States." Even in the event of dissolution, the government has no control over the prior disposition of assets the corporation might choose to sell to others, or which would devolve in the first instance to the firm's private creditors.

NASA's Bellcomm and GE Contracts

At the very moment the Bell Committee was formulating new guidelines for R&D contracting, NASA was breaching these guidelines spectacularly by signing contracts with AT&T and General Electric under which the government abdicated major management responsibilities for the nation's space program. They were handed over on a profit-making basis to two huge industrial complexes, both already deeply involved in defense/space contracting. GE ranked as sixth largest contractor in 1964 (NASA and Defense Department), with a total for the year of more than $1 billion in contract value. AT&T (and its twelve subsidiaries, not including Bellcomm) was the nation's seventh largest Defense Department contractor ($635.6 million) and fifty-sixth largest NASA contractor ($4.5 million) in 1964; the world's largest corporate enterprise, it reported net earnings in 1965 of over $1.8 billion.

The NASA awards were made on a sole-source, cost-plus basis by "letter contracts" from Webb to the companies, the tasks and terms sketched in the broadest and vaguest language. Announcement of the awards prompted surprise and consternation in the aerospace industry, and many companies complained they had not been given an opportunity to submit competing proposals, that the two giants would obtain unfair advantages by their influence at the apex of the manned space flight program. The GE contract was widely regarded by the industry as "a consolation prize for the company's failing to win the competition to build the Apollo capsule," a contract that had been awarded to North American Aviation.[36]

The contract with AT&T gave birth to a new wholly owned subsidiary: Bellcomm, Inc., whose initial capital stock ($500,000) was provided in equal amounts by AT&T and its major production subsidiary, Western Electric Company. An AT&T vice president was named as chairman of the new company's board, whose other eight seats were divided equally among officials of AT&T, Western Electric, Bell Telephone Laboratories, and Bellcomm itself. John A. Hornbeck, a vice president of Bell Labs, was named president.[37]

Bellcomm and GE were to divide "staff" and "line" duties in the management of the Apollo moon program. Bellcomm would act as the technical staff to NASA Manned Space Flight officials, collating technical information, preparing documentation, and giving advice, all based on actual R&D performed by others. Hornbeck: "We work with papers, pencils, slide rules, and we have no hardware work or experimental laboratories on which to spend money." [38] GE, on the other hand, was to serve in a "line" capacity, providing a large number of man-hours for integrating hardware design and construction by many different companies, supervising the wedding of these into complete systems encompassing launch vehicle, spacecraft, launch site, and on-board and earth-based electronic control and communications. GE would have final responsibility for the technical "checkout" of these assemblies. The requirement for integration and checkout, the Senate Space Committee explained, "arises from the unique characteristics of the Apollo project," including "the serious consequences of failure as measured in terms of national prestige, human life, and cost; the complexity of the total project in terms of the technology required and the numbers and types of industrial contractors involved; and the long period of operations in the space environment with no significant maintenance or repair possible." [39]

Both contracts bore a faint odor of the Ramo-Wooldridge role in missile programs. Defending them, Webb stressed the unique scientific and technological resources and management experience of the companies which, he said, "the government needs . . . and cannot acquire any other way." The Bellcomm contract would make available to NASA's Washington headquarters—on a full-time basis, in a single package, and without delay—some two hundred Bell employees constituting a high-level, established, fully proven team. In addition, all the technical knowledge and skills of the parent company would be available for Bellcomm to draw upon. "We do not want to change the program without the best analysis by the best people in America," Webb said, and on a regular basis so that the Bell system would follow through and "take responsibility for the advice they gave us." [40]

Webb's pitch, the details subsequently disclosed, and the cries of protest from other contractors raised sharp questions in Congress. But NASA and company officials smoothed the surface, emphasizing the uniqueness of Bell resources and the anticipated "low cost" of the "paper studies" to be performed. They dwelt at length on the distinction between "management/technical direction" and "technical advice," with only the latter to be provided. Bellcomm's president was asked: "Do you make any decisions which might be called quasi-management decisions?" Hornbeck replied: "I am not sure exactly what you mean. I think we are a management tool in the sense that, say, Mr. Holmes can use us for information. Or as an input to the decision-making pool." [41] D. Brainerd Holmes confused the issue somewhat: "GE and Bellcomm are not personal advisers to me in any sense of the word. . . . The Manned Space Flight Advisory Committee is the only advisory group that I have. Bellcomm and GE are line organizations." [42] His difficulty may have reflected a semantic difference, or it may have reflected something deeper—perhaps a contest over authority in the Apollo project which figured just a few months later in his angry (but publicly reticent) resignation. The contracts did not lack congressional defenders, and most legislators appeared to accept the propriety and wisdom of the arrangement.

The GE and Bellcomm contracts raise three kinds of issues: the delegation, either deliberately or by default, of public authority and responsibility to profit-making private corporations; possible conflicts of interest; and possible sheltering of costly, wasteful, or improper practices in the service of NASA and company interests.

Public control and responsibility. Things have not turned out exactly as forecast. The "unique resources" of AT&T somehow did not prove adequate to service NASA's requirements, and Bellcomm, like GE, went into the field to recruit substantial numbers of new scientists-engineers, pirating them from other contractors, university faculties, civilian industry, new Ph.D. recipients, and even from NASA itself. One congressman in exasperation asked: "There is one point I don't understand. Why is Bellcomm hiring people from the outside in competition with NASA? At least one-third of the personnel of Bellcomm are from the outside. Why don't you have people of that nature hired by NASA?" [43] The legitimate bench scientists of the Bell laboratory who were supposed to be in Washington offices adjacent to the director of the Manned Space Flight Program instead remained at the Bell lab and were only indirectly on Bellcomm's payroll. Hornbeck: "The laboratories give us a bill that we pay and pass on to the government." [44] The "unique resources" of AT&T proved inadequate in other respects as well, and Bellcomm moved into the business of subcontracting parts of its work. Hornbeck: "We have contracts with other contractors, yes." [45] If there is "a special contribution of competence some other organization can make, we go to them immediately." [46]

The two contractors, in spite of earlier assurances, became deeply involved in program management, supervising and coordinating not only the army of other contractors but to some extent NASA "in-house centers" as well. On paper the centers were charged with subsystem and component integration for programs and contractors under their jurisdiction. But in actuality General Electric came to act as the instrument of NASA headquarters in coordinating and integrating the functioning of the centers. The lords of NASA officialdom also claimed a role in "coordinating" contractors, described by Dr. George E. Mueller (who succeeded Holmes in 1963): "In order to provide the proper coordination, we have set up executive groups for both the Apollo and Gemini programs to provide liaison at the very top level. These groups consist of the presidents or chairmen of corporations and our own management council, to be sure everybody knows what their companies are doing, and how they are interacting with NASA." [47] These infrequent ceremonial occasions were endowed with the cushiest tone. They may have been personally flattering to the participants, but they could hardly be considered a genuine day-to-day management or technical coordinating device. It would be interesting to examine the minutes.

Bellcomm's primary work is digesting and collating scientific information and R&D reports in order to project detailed requirements of future missions and to prepare specifications for R&D or hardware procurement. Hornbeck: "Considerable Bellcomm effort has been devoted to assisting NASA in the preparation of Apollo systems specifications . . . to define the objectives for the Apollo systems and its major subsystems; to define the technical approach to be used to accomplish these objectives; to establish system and critical subsystem performance requirements . . . to specify performance parameters which affect crew safety, design margins, capability for growth and program schedules, and to establish a uniform set of system design data." [48] To implement the specifications and integrate the systems, GE developed large technical teams at all the NASA centers. As the center of gravity shifts from concept and design to check-out, GE designs and builds the equipment. It is certainly a misnomer to call this "engineering support." The writing of specifications, the definition of requirements for reliability assessment, and the technical monitoring of the program add up to overall technical directions and management, rendering NASA claims to "in-house" management rather transparent.

The real situation was hinted at in NASA's description of GE's role as helping "to insure that the overall systems engineering requirements are being met in detail"; [49] or in description of Bellcomm's function as helping "to continually re-evaluate not only the environment but also the ability of the contractors to deliver guidance equipment, propulsion equipment, and all the other elements of the system to meet the specifications." [50]

In the preparation of documents, Bellcomm's research is done in NASA's own files, those of the field centers, the contractors, and open scientific literature. Where the necessary information cannot be found in these sources, Bellcomm's job is aimed at formulating the problem rather than finding the solution. [51] Collating material and information already in the pipeline and "'defining problems" are related processes, but the latter is not necessarily a dependent variable of the former. An element of subjective values, of judgment, of preference is involved in spelling out the nature of a problem and in weighing alternative approaches to its solution. Defining a problem involves an implicit or explicit hypothesis which is in fact an a priori commitment (though sometimes tentative) to a line of action. This commitment is an essential factor in selecting and interpreting raw data. The so-called "supporting role" of Bellcomm as a decision-making input may rather be a determining influence in the policy choices of NASA officials. Perhaps the Bellcomm input only rationalizes choices already made, or reflects interests of the community of major contractors in which AT&T (and GE as well) has assumed a leadership role, or reflects other kinds of diverse motives and interests. That this process is at work was revealed in the most important NASA technical decision: selection of the mode of flight for the manned lunar landing.

According to Dr. Joseph E. Shea, a group of contractors (under the direction of NASA's Office of Systems) reviewed all the data available and defined additional studies required: "By June 1962, the results of all the previous study activities had been analyzed against the criteria of mission success, safety, schedule, cost, complexity, and growth potential. The analysis of more than a million man-hours of technical work showed that lunar-orbit rendezvous was the most desirable mode of flight." [52] The subjective component was identified when Presidential Science Adviser Wiesner (with strong PSAC support) challenged this decision, both for the manner in which it was made and for its substance, raising questions which led President Kennedy to order that all the documents be submitted to PSAC for review. While PSAC deliberated, Webb went ahead to place contracts for development of the system. The President asked that in the future NASA submit crucial policy questions to PSAC *before* making decisions, but the relations between the Space Agency and White House science advisers grew increasingly distant, the Space Council under Vice President Johnson supervening—Bellcomm retaining in effect the primary role of technical (*qua* political) decision-making.

Without considering NASA's argument that an agency so young and in such a hurry could not possibly command the personnel resources necessary to conduct this function in-house, there is cause for concern at this gravitation of public authority into private hands. Cardinal choices are made obscurely and inaccessibly by men who lack a legitimate and broad public mandate and are unaccountable to independent or pluralistic technical or political judgment.

Conflicts of interest. Announcement of the two awards greatly alarmed the contractor community, which views AT&T and GE with considerable anxiety. Both companies enjoy unequaled security, AT&T's built upon its size and firm monopoly position, GE's arising from its solid foundation and quasi-monopolistic position in the civilian economy. But NASA and the two giants quickly reassured everyone, and the initial protests sputtered into silence. Thereafter the very size and stability of the companies seemed to quiet fears, enabling them to act as elder statesmen and honest brokers in maintaining some degree of fairness and consideration over the cartelized aerospace contract market. This appeared to make it possible for NASA to win acceptance for the appointments.

But it did not prevent the initial rash of protests which focused on concern that AT&T and GE would hog a disproportionate share, that they would use their surveillance role over other contractors to acquire industrial secrets that protected other companies, and that their already preponderant role in the corporate hierarchy would be enhanced to a degree incompatible with the delicate balance maintained among financial and managerial elites.

NASA officials acknowledged the problem and made clear that safeguards against conflicts of interest would be negotiated in the final contracts to protect other contractors, government, and public. NASA expected GE to manufacture "some of the equipment" for the check-out system, but some would be bought on a competitive basis from other companies. The contract would specify NASA's continuing control over each "make or buy" decision, and GE would be prohibited from production contracts in phases of the space program outside its check-out responsibilities. In regard to AT&T, NASA officials declared that the Bellcomm contract would contain a rigid and absolute hardware ban.[53]

NASA demonstrated little sympathy with protests about proprietary information. Holmes declared his opinion that in "over 90 percent of the activities" the companies would not gain "any additional knowledge which would be to their advantage . . ." He then put his finger on a viewpoint that the agency in other moods did not assert: "There is very little proprietary information in the program. Companies claim proprietary but to my mind if all the work done has been used by the government and it is paid for by the government there is nothing proprietary and we'll use it as we see fit." [54]

As the Space Agency moved to implement these assurances, it

managed to contain but not entirely dispel anxiety. As described by NASA's Shea:

During the early phases there was indeed on the part of contractors and our center people a type of resentment. . . . The working relationship is now finally easing itself out, but again it can only come about from contacts on the working level, technical contacts between GE people, the Bellcomm people, and both centers and their contractors, so that you finally establish a technical respect to make these contracts go.[55]

Webb laid down the general principle that the Bell System would not seek work from NASA in areas related to the program, "except where, in the opinion of the Administrator or the Deputy Administrator, an exception to this general rule should be made." Such exceptions "would be rare" and would arise only where the Bell System was in a unique position to perform a valuable service, or where the national interest would be uniquely served. In addition, specific exceptions to this limitation were stated explicitly: communications, tracking, and guidance. Webb stated that this should "not affect the dealings of Bell System companies with NASA," which "would remain free to deal with NASA" in the furnishing of communications for the man-in-space program "on the same basis as if the new arrangements had not been made." [56] Representative Chet Holifield characterized this as a "a pious statement that there shall be no conflict of interest involved, and then there is a complete exception . . . this is a very unique situation." [57]

The hardware restraint in substance meant that neither GE nor AT&T would suffer any rigid exclusion whatsoever, and whenever either might enjoy a competitive advantage because of its position, the Administrator would decide whether to permit bidding. It soon was clear that the constraints built into the contracts were permissive and trusting. As expressed by Bellcomm's president: "In general, I would say . . . we are constraining ourselves . . ." The counsel of the House Committee on Government Operations asked, ". . . on a self-disciplinary basis?" Answered affirmatively, the counsel concluded: "But basically there is not very much of on-going business that you are shut out of. There is practically nothing that you are shut out of." [58]

Bellcomm's Hornbeck asserted that Bellcomm was "a separate and independent" organization and "as a company is not in the hardware business at all," [59] a view rendered laughable when he described how he became Bellcomm president: "The president of the Bell Telephone Laboratories called me into his office one morning and said, 'We have a new job for you.' And I said, 'Yes, sir.' That is all that I said." [60]

The whole conflict-of-interest issue, according to Hornbeck, is not to be decided simply between government and contractors; "it really gets decided on whether your industrial competitors are going to let you deal with them . . . They are the ones who really police this thing in the final analysis. . . ." [61] From these policemen there has been only sniping criticism since 1962. AT&T and GE have been responsible to the interests of their brethren in the contracting community and have kept their fences well mended.

GE is involved in a vast production of equipment for the conduct of its check-out work and produces large quantities of hardware for NASA—intended as government-furnished equipment for other companies in the lunar program. The company has vigorously bid on a number of contracts in the space program with little evidence of restraint imposed either by government or by self-discipline. The same might be said for the Bell System, whose Western Electric Company has maintained an important hardware role.

NASA's conflict-of-interest restraints have also been rendered farcical because they do not apply at all to GE and AT&T contracting for the Department of Defense. As a result of the 1962 Bell Report, President Kennedy instructed all the agencies (in consultation with the Attorney General) to coordinate their relations with contractors in order to avoid this discrepancy,[62] but this was never effectively enforced. In 1965 GE was awarded prime-contractor status for the Air Force's multi-billion-dollar manned orbiting laboratory, adding to its already substantial Defense Department space role. It also won a half-billion-dollar contract for jet engines for the C-5A seven-hundred-passenger military transport, adding to its variety of hardware undertakings for all military services. These awards exemplified the trend toward fewer and bigger contracts and the collapse of McNamara's efforts to control interdepartmental conflicts of interest. The issue will remain muted so long as the major contractors continue to work out among themselves a cartelization of the government market under which the "competitive" position of each is maintained during periods of relative stability, broken only occasionally by the crisis of sudden political change when the wolf pack abruptly turns and devours one of its own.

Cost and waste. In estimating costs and establishing fees under the Bellcomm and GE contracts, NASA declares that it uses a "worth of service" standard (unlike the "financial need" standard imposed by the Defense Department on non-profits). NASA is "purchasing experience, technical judgment, and advice. In such a contract, capital investment requirements are extremely low and the fee bears no meaningful relation to the dollar investment, but it is rather related to the value of the service performed." NASA thereby denies the relevancy of any objective basis upon which to judge comparability and fairness of fee, since "worth of service" is a tenuous and subjective standard lacking any yardstick.

Under a Bellcomm contract in force in 1963, the fee was 7.87 percent of estimated costs on a contract of approximately $8 million for the year. Actual capitalization was $700,000, roughly equal to its fee for the year. Such an early profit, equal to 100 percent of capital investment, embarrasses NASA, whose officials prefer to relate the fee to the overall investment of the Bell System in the Bell Telephone Laboratories, well over $100 million. In view of the risk-free nature of the investment and the advantages gained for future profit, this is something of a conceit which symbolizes, if nothing else, the ambiguity of the Contract State in terms of the conventional slogans of "free enterprise" and "public-spirited" dedication to the public interest.

For fiscal 1964 the Bellcomm contract was increased by roughly a third and was expected to more than double thereafter. Repeating the words of its president, "It is a very simple budget picture. . . . We work with paper, pencils, slide rules, and we have no hardware work or experimental laboratories on which to spend money."[63] Considering that Bellcomm employed ninety-six technical staff during the first eighteen months of the contract, the cost of this pad-and-pencil work averaged about $85,000 per man. The increase in total value after 1964 was partly due to further augmentation of personnel, plus the leasing of a "computer in residence" at a monthly rent of $900,000![64] The leap from slide rule to computer came as a shock to some congressmen: "If you will look through the testimony previously given, and if you find any reference to . . . a computer in Bellcomm, I would like to see it."[65]

GE was to have another "computer in residence," just as a vast variety of government contractors were acquiring similar equipment, largely duplicating equipment already leased or owned in the federal establishment. This is a problem which continues to perturb the General Accounting Office, the Department of Defense, and the Joint Economic Committee—if not NASA. The federal government invested in 1965 $3 billion for purchase and rental of data-processing equipment, about 30 percent of total computer sales.[66] The GAO contends that a large part of this cost represents duplication and under-utilization springing from status-seeking on the part of federal agencies and contractors, as well as contractors' desire to acquire at public expense a capability that can be profitably extended to other business operations.

The GE contract had a 1963 value of $30 million, swelling in the following year to $153 million. NASA estimates that through 1970 the total value of the contract will approximate $625 million.[67]

Brains, Profits, and the Public Interest

The emergence of the braintrust industry accompanies the growth of government R&D and raises difficult and persistent questions. Is it possible to do the legitimate job of developing and procuring the new weapons and technology of national purpose while maintaining public control and economic and political pluralism? Is it possible to make full use of the resources of private industry for both government and consumer purposes and achieve a balanced investment in national values?

In spite of bland claims that the new braintrusters demonstrate a high degree of pluralism, independence, and objectivity, it is clear that they have to a considerable extent failed to protect public interests and public control, serving instead powerful coalitions of special interests on whose behalf they often invade the policy-making realms of state. There is a residual consensus in the U.S. that profit-making institutions cannot provide independence and objectivity in assuming government management and decision-making functions. There is increasing recognition that neither can the non-profits. Congressman Charles S. Gubser nailed the notion when he re-phrased claims made on behalf of the Aerospace Corporation: its "independence" was real enough, but "it ought to be called independence—of the Appropriations Committee." [68]

The institutions of contractor and braintruster have begun to manifest some of the worst aspects of inefficient, incestuous, and self-perpetuating government bureaucracy. There are multiple layers of contractors on top of contractors, with systems engineers and technical directors supervising systems engineers and technical directors, resulting in wasteful and unnecessary fractionization of contractor functions. This quasi-governmental bureaucracy is shielded from Hoover Commissions and Bell Reports which might check its unhealthy proliferation. There seems to be no present way to subject this hierarchy to the broad provisions of constitutional process; rather, it is a bureaucracy vested with anonymity, its empire-building inaccessible to traditional correctives. Like all powerful bureaucracies, it finds a way to blunt every instrument of administrative control.

There are legitimate jobs to be done by private industry, and because these cannot be deferred, government is forced to accept private stewardship of the tasks. The system is rationalized in terms of preserving free enterprise and private initiative, while the non-profit route, adopted originally as a corrective, is also subverted.

Just as Nageeb Halaby believes that people will work better "for capital gains or for bonuses" than for government salaries, so the corporate hierarchy, pirating the scientific and engineering talent of government and the universities, insists that "profit and the prospect of profit are the only energizers" that will mobilize national resources for conducting the government's business.[69] Defending the Bellcomm contract, Holmes told Congress the Bell Systems ". . . are not interested in doing business on a task basis, since it is not to their advantage."[70] Ivan Getting, president of Aerospace Corporation, provided a delightful bit of persiflage in a statement attacking Department of Defense efforts to reduce Aerospace fees by negotiating on the basis of "financial need" rather than the ambiguous "worth of service." Such a concept "is in fundamental conflict with traditional basic American national economic ideas. It is far more suggestive of the familiar Marxian doctrine to the opposition of which our nation is dedicated."[71] The point was bluntly stated by GE executive George L. Heller: "Over the long pull, defense must be a self-sustaining, competitive, profitable business, and the sooner it is accepted as such by all the vital elements of our economy, the sooner we will reap full benefits in military preparedness and in rising levels of living." Government must provide "the encouragement and incentives for superior performance and risk-taking." Together, the combination of industry and government "will place the responsibility for our national security in the hands of the best managerial, technical, and military talent in the world."[72]

These claims are specious. They seek to rationalize a situation—that already exists—in which profiteering at the expense of grave and urgent national needs has been normalized as the result of a series of emergency improvisations. Reiteration of this view in the public forum is a sign that the system is already under attack; it is a symptom of unease and guilt, and therefore perhaps of hope.

CHAPTER XIV

"THE SPORTY COURSE": WASTE AND PROFIT

Milton A. Fogelman, former Air Force contract officer, was pressed hard by the investigating subcommittee of the House Armed Services Committee in the 1965 Aerospace Corporation hearings. Responsible for negotiating the contract and setting the fees, Fogelman was handicapped by the corporation's refusal to document its financial requirement, a course encouraged by higher Air Force officials. Congressman Porter Hardy: "You were actually operating in the dark." Fogelman agreed. "It was a sporty course," he said.[1]

The unknowing and involuntary sport of the Contract State has been the American taxpayer, and those who have made sport with him are those who have been and are the beneficiaries of the system. The catalog of atrocities against the public purse includes "duplication, confusion, and do-nothing exercises,"[2] overstated costs, unconscionable hidden profits, instances of deliberate fraud, collusive price-fixing, overruns of cost, overcomplication of products, egregious technical errors, and company—as well as government—mismanagement. The recent public record of these atrocities represents only a small sampling but could fill a volume the size of the Manhattan telephone directory.[3]

The U.S. General Accounting Office has been an effective source of auditing data apprising Congress, the administration, and the public of the abuses of the Contract State, even though the full impact of GAO reports has been ignored. GAO work has been an invaluable tool of McNamara's cost-reduction efforts in the Department of Defense. While the limited auditing staff must confine its attention to about 5 percent of contract awards (in dollar value), the results have eloquently testified to the accuracy of its findings, leading to voluntary refunds of excessive charges in many cases, Justice Department action in others, and programs to prevent recurrences. GAO work suggests the horrendous scale of waste and profit hidden in the other 95 percent of contract awards, as well as the difficulty of administrative reforms, *post facto* recoveries, or court actions.[4]

Joseph Campbell, U.S. Comptroller General, summarized during the 1965 investigation of GAO (in the course of which he resigned "for reasons of health") the findings of his staff:

> (1) excessive prices in relation to available pricing information, (2) acceptance and payment by the government for defective equipment, (3) charges to the government for costs applicable to contractors' commercial work, (4) contractors' use of government-owned facilities for commercial work for extended periods without payment of rent to the government, (5) duplicate billings to the government, (6) unreasonable or excessive costs, and (7) excessive progress payments held by contractors without payment of interest thereon. The majority of our contract audit reports deal with excessive prices resulting from the failure of the agencies to request, or the contractors to furnish, current, accurate, and complete pricing data or from failure to adequately evaluate such data when negotiating prices.[5]

The overall dimensions of defalcation may be judged by the fact that the 5 percent sample reported in 1961 indicated ascertainable overcharges of $60 million; in regard to the other 95 percent there was "every reason to feel" the same pattern existed.[6] A projection of the sample would indicate contract waste for the decade of the 1950's well over $1 billion. And the margin appears to be increasing: a summary of GAO studies (over the period May 1, 1963, to May 1, 1964) disclosed ascertainable waste of about half a billion dollars in a 5 percent sample of procurements.[7] An audit sample comprising 237 awards (over an eight-year period) revealed "excessive or erroneous payments" of $124 million, half of which was later recovered.[8]

GAO's authority extends only over negotiated and sole-source, cost-plus contracts, on the assumption that open, competitively bid contracts contain a built-in cost yardstick and should not give government access to contractor records. (However, where evidence of collusion may be discovered it becomes a matter for the Justice Department.) With the emergence of the Contract State, negotiated contracting has tended to pre-empt the field, comprising almost 90 percent of dollar volume as late as 1965, when McNamara had already begun to reverse the trend, at least in the Defense Department.[9] R&D contracting has become the key to the magic circle and a guarantee of open-ended future contracts for R&D, systems engineering, and hardware production. According to former Assistant Secretary of Defense Eugene Fubini, the government charges R&D funds "for everything—I mean everything—right through the building of the first group of airplanes. Only then do we go into production costs." [10] Of total R&D awards, more than 85 percent are "pre-selected," that is, made without any formal solicitations for proposals from sources other than the company that receives the contract. Of these, almost 60 percent are "sole-source" procurements. Virtually all R&D awards are negotiated, even where selection follows solicitation of multiple-source proposals.[11]

Experience shows that government realizes at least a 25 percent saving in procurement by competitive fixed-price bidding as compared to negotiation: the price of the fluid for hydraulic equipment dropped from twenty-five to fifteen cents a gallon when switched to competitive bidding; costs of electric motors fell from $614 to $280, a reduction of 54 percent; eight-inch howitzers dropped from $8,000 to $4,000, and so on.[12] With the crippling of the Jet Propulsion Lab and Redstone Arsenal, the government lost its only cost yardstick in space systems development. Under negotiated contracts, therefore, there is no longer any objective cost yardstick available to protect public interests; nor is there a means to discourage collusion in multiple-source negotiated procurements, or to safeguard competitive bidding from corruption.

An interesting yardstick is found in two instances involving amateurs who for pennies performed the same work that contractors do for hundreds of thousands or millions. An enlisted man and a civilian at Wright-Patterson Air Force Base used $100 worth of spare parts to construct an electronic tracking component—for whose development a contractor had sought $100,000. They did the job in two months, while the contractor had estimated six.[13] In a similar case, a group of California engineers, amateur radio enthusiasts, in their spare time designed and built a series of communications satellites operating on ham frequency bands. In 1961 and 1962, two of their "Oscar" satellites (launched by the Air Force) broadcast the word "hi" in Morse code to radio amateurs around the world. After this much-acclaimed success they undertook a more ambitious project, Oscar III, an active translator satellite capable of relaying twenty-five conversations at 1,000-mile ranges, launched for them in mid-1964. Made from parts donated by aerospace firms (surplus, paid for by the taxpayer, who else?), Oscars I and II were valued at several hundred dollars apiece in hardware, plus a couple of thousand man-hours of design, construction, and testing. Oscar III contained about $1,000 in parts.[14] NASA, the Defense Department, and the new Comsat Corporation spent tens of millions for similar spacecraft.

Quality of Management

Negotiated awards give virtually complete discretionary authority to agencies and individual contract officers, and the scope of this practice has become so enlarged in the last fifteen years that it defies administrative control. The result is that the bulk of contracting has been subject to practically no uniform standards, permitting commitment of large sums of money on the basis of grossly inadequate specifications, cost estimates, and review. The agency and the contract officer are almost free agents, able to apply any kind of subjective standards or none at all, dependent on the contractors themselves for negotiating data, and unarmed to protect the public interest if indeed there were a will to do so.

In a typical case the Air Force awarded a series of electronics R&D contracts, each larger than the one before, to a firm that had no previous experience, was organized in anticipation of the contracts, and based its whole financial credit on the collateral of the contract officer's verbal commitment. He granted the company sole-source status on the basis of the claim (later proved false) that the techniques to be used in performing the contract involved "proprietary" data.[15] This particular case was investigated by Congress and, though of minor magnitude, laid bare "routine" practices. Beginning with an initial award of less than $50,000, the contract was expanded within three years to over $1 million on a sole-source (and subsequently on a "phony" multiple-source) basis. The contract officer readily admitted that the "guide for evaluators" permitted him to swing the award to anyone he wished. In his defense he declared: "I believe the sole-source basis for extension of research contracts now is quite a normal and expected thing." [16] The company used contract money even for its original capitalization and managed, through direct-cost reimbursement, overhead, and fees, to acquire very substantial equities (buildings, lab equipment, etc.) which constituted concealed profits.

This syndrome becomes really vicious in awards to the major contracting empires; it destroys any notion of cost control and public accountability, converting government contracting into a purely political process. The pattern was illustrated by Congressman Earl Wilson (Indiana) who in 1963 made daily speeches of protest in the House. He cited, among others, a case involving the Collins Radio Company (Cedar Rapids, Iowa), with whom the Navy negotiated a contract for the development of an advanced walkie-talkie radio. After it was developed at a cost of over a million dollars), Collins was granted sole-source status for construction of 670 sets. The company, 134 of its employees former Navy officers (including a rear admiral, a captain, and several commanders and contract officers), was highly successful in winning negotiated and sole-source contracts although it had a dismal record in competitive procurements.

Wilson cited a series of similar cases involving respected major contractors, their corporate ranks heavily larded with retired admirals and generals whose salaries ranged to $100,000 a year.[17] As of 1960, according to a congressional report, the employment of retired military officers by the ten top defense contractors was: General Dynamics, 186; Lockheed, 171; North American, 92; Martin Company, 63; Boeing, 61; Douglas, 40; General Electric, 26; United Aircraft, 24; Hughes Aircraft, 22; AT&T, 6.[18] These figures illustrate the tendency but do not tell the whole story, including movement into company ranks of military men who resigned before retirement eligibility, and many civilian employees whom the military often prefers to put in the "patsy" role of contract negotiation and supervision.

Top officials of NASA and the Air Force, foremost exponents of the Contract State, demonstrate the persistent habit, like poet Walt Whitman, of "celebrating themselves." Webb and General Schriever, in all their public utterances, aggressively declare the uniqueness of management techniques and capabilities embodied in their own operations, apparently attributable to their wisdom and innate nobility. Schriever frequently complains "that the military performance of a vital function for the nation perhaps is not given as much attention and as much coverage, you might say, from the public media standpoint as I think would be desirable . . . in the area of management of scientific and technical programs . . ." [19] A psychologist might find in this habit some insecurity feelings, the need for constant explanation reflecting a basic anxiety which arises from actual insulation from the blood and guts of real management responsibility by multiple layers of prime and systems contractors. Thus Webb's and Schriever's protestations might be viewed as arising from the opposite of "management magic," a need to reassure themselves and the public of their usefulness.

In addition to possible self-doubt, Schriever and Webb face recurrent tremors of skepticism from other sources. The GAO has not been an ardent admirer of their management practices, ascribing "erroneous cost allocations and overhead distributions" by contractors to poor management, and citing "the need for improved contract administration and auditing by the responsible government officials." The GAO finds "a great deal of laxity on the part of all concerned with respect to protecting the government's interests . . ." [20] The House Armed Services Subcommittee frequently disturbs the self-esteem of NASA and its managers, citing instances of contract officers unable to obtain guidance from higher authority in enforcing legal and rational standards, noting the case of Colonel Hamby who "worked diligently to bring order out of the chaos" and was transferred "after he refused to bend before the willful pressures" of contractors. This "calls into question the ability of the Air Force to cope with the management problems attending its organization of space programs and will require continued congressional surveillance." [21]

Admiral Rickover does not hesitate to ascribe blame for the *Thresher* submarine tragedy to inept and transient Navy management. He declared in 1965 that "not much has been done to correct that situation." [22]

A contractor-employed engineer takes sharp exception to the "technical illiterates" who manage the contractors, finding their "only true effectiveness" is "engineer baiting, engineer downgrading, obstructing engineer effectiveness, and abuse of authority." [23] A government engineer charges that Air Force management is really "a sort of high-priced hardware-sitting service." Any effort by "the conscientious engineer" to enforce the government's own "badly mimeographed" specifications and directives "is frustrated in a most vicious manner. If he is the insistent type, he is given unsatisfactory reports, denied promotions, training slots, and intimidated with threats." Successful government project officers, he alleges, have "absorbed the principle that their work is to shuffle papers, put out glowing reports, attend meetings, go on TDY's [temporary duty away—with increased pay]." [24]

The key to competent and honest management of contractor performance rests in the hands of the lowly project officer and the staff resident in the contractor's plant. He is charged with detailed responsibility for directing and evaluating the work and providing objective advice to higher decision-makers and to contract negotiators. Not only must he be technically qualified, dedicated to professional standards and to the public interest, but he must feel confident that his superiors will recognize, value, and require such motivation, endowing his role with their own authority and assuring him support in his inevitable adversary relationship to the company he supervises. It is at the workbench level that the quality of governmental monitoring is tested, not at the top where exalted officials perform ceremonial and political functions, dealing like ventriloquists' dummies with such issues as the contractors are willing to lay before them. There is serious evidence (demonstrated for example in the relation of Colonel Hamby and Mr. Fogelman to the Aerospace Corporation) that high officials use contract and project officers as fall guys for implicit headquarters decisions that will not bear public scrutiny and that do not accord with official regulations. The "lack of official guidance" from headquarters is a means of realizing a variety of values which are not shared by the public, including the desire of parochial and rival agencies to circumvent congressional and departmental budget controls, to cover up mistakes, and to build financial security and economic power for themselves and for their corporate accomplices. The result is a *quid pro quo* guarantee of future contracts and fiscal charity in the government's relations with favorite contractors.

Facilitating this system is the practice of rotating personnel between diverse assignments, maintaining a lack of continuity and assuring that project and contract officers will be dependent upon the contractor for the information required to negotiate with and supervise him. Congressman Holifield:

> The military officer . . . is unacquainted and unskilled and untrained in the art of procurement, and yet there have been occasions where his line of command caused him to overrule the civilian employee who had many years of experience in procurement. . . . This committee has recommended many times that where skilled professional ability is required officers be given proper advancement in their profession without the necessity of switching them every two years from highly complicated work into field work which is completely different.[25]

An example of the quality of management came to the surface in 1962 as a result of the efforts of Wernher von Braun to achieve real government authority over the Centaur (liquid hydrogen upper-stage booster) project. The Air Force succeeded in winning control of the project in 1958, but responsibility was transferred to NASA one year later. NASA assigned technical responsibility to the Marshall Space Flight Center. A comic opera ensued in which von Braun's efforts to assert authority over the program were completely blocked by his NASA superiors who refused to molest the Air Force management team at the contractor's plant. With a new administration in 1961 and a change in NASA leadership, von Braun intensified his efforts, protesting the veto power the Air Force seemed to have over NASA. A congressional inquiry followed in which he euphemized the situation yet succeeded in conveying a sharp indictment of Air Force management:

> I think what we felt was lack of depth of penetration of the program on the part of government personnel, in very general terms. We believed that this staff of eight people—they were all very competent people, no question about it, but eight people is just an inadequate coverage on the part of the government, no matter whether it is NASA or the Air Force, to stay on top of a program of this magnitude, involving engine development going on in Florida and vehicle development in Southern California and static testing in . . . I think we could have provided more protection . . . by simply appraising potential problem areas better.

Von Braun agreed that quality control had been inadequate and had contributed to the slippages and technical errors that afflicted Centaur.[26] The subcommittee wondered how the contractor (General Dynamics/Astronautics)

> could have made such a fundamental mistake in cryogenic engineering in failing to discover the leak problem in the intermediate bulkhead until after full-scale tanks and vehicles had been built and transported to Cape Canaveral. It was difficult . . . to understand how such fundamental difficulties could have escaped the intervention of an alert technical management.[27]

Upon at last winning his point, von Braun felt it necessary to assign 140 technical people to supervise the contractor where the Air Force before had only eight, most of them clerical.[28]

Sample Catalog of Atrocities

The voluminous documentation of systematic waste and profiteering is buried in GAO and congressional reports. Neither Congress nor the public has manifested much interest or concern over the implications of this situation; the self-serving rationalizations of bureaucrats and company officials have received most of the press's emphasis, hiding the subject under a smiling and complacent mask.

The cases cited here illustrate only the kinds of abuse, not the scope or magnitude. A list must include:

Faulty (and in some cases fraudulent) cost estimates provided by contractors as the basis of contract negotiations;

Improper awards of sole-source status;

The use of subcontracting by prime contractors to conceal excess profits, to levy an extra fee on services rendered by others, to deny patentable or "proprietary" information paid for by the government in order to enable the contractor to use the data to maintain sole-source status and/or to charge the government several times for its use;

Questionable charges to overhead on contracts;

Calculation of fees to cover costs that cannot legally be assigned to other categories, to cover contractor mistakes, and to evade standards of cost accounting;

Duplication and overlapping costs;

Overruns and unnecessary gold-plating motivated by a desire to extend or inflate contract size or win additional contracts.

Negotiating costs. The pattern of negotiated contracting (whether multiple- or sole-source) was described by Senator Stuart Symington as involving some "pretty fancy finagling when it came to profit." Everybody knows how "some of these companies have deliberately quoted tens if not hundreds of millions of dollars below cost in order to get contracts, knowing that after they do that there will probably be . . . engineering changes . . . and they could really charge just about what they want . . ." Once a contract is won, they charge well above reasonable costs and "it is almost impossible to take it away from them, unless you want to wreck the program, where time is so important, and therefore to a certain extent the government is in their hands." [29]

Lacking a yardstick, unable to write adequate specifications or properly to evaluate proposals, the government is forced to depend upon the contractor's own cost targets. NASA's budget is largely made by contractors. Brainerd Holmes (referring to the Gemini spacecraft): "The fiscal 1963 congressional budget request was made at the suggestion of the contractor. The increase reflects McDonnell's six months of actual experience . . . It appears the contractor's optimism was not warranted, at least in terms of the dollars involved." Asked (by Congressman Emilio Daddario) to reconcile this inability to estimate costs with the fact that the Gemini capsule contract was awarded to McDonnell without competition on the grounds of its "great experience" in the Mercury program, Holmes responded limply: ". . . there was only one company that could give us the fastest possible application of experience . . ." [30] According to Holmes, the great bulk of NASA's budget request was founded on contractor figures without real in-house appraisal; NASA merely adds a percentage for "the things we think the contractor has not included." [31]

The result of this system (according to GAO) is five types of overcharges: negotiations based on erroneous cost information, negotiations based on cost information inconsistent with experienced costs, negotiations based on price proposals containing unwarranted provisions for contingencies, negotiations of final prices not consistent with contract terms, and negotiation of an inappropriate type of contract.[32]

Typical examples:

In the Bomarc (interceptor) missile program, the Boeing Company overstated costs, resulting in overcharges of $23 million.

In 1961 the Navy incurred $3.3 billion of "unnecessary" cost by procuring under negotiated contracts $14 million worth of spare parts available under competitive procurement at a lower price.

Westinghouse added almost $1.5 million to the costs of propulsion machinery for the aircraft carrier *USS Enterprise* in order to provide for contingencies which according to the GAO were "unwarranted" and did not in fact materialize. The money was not restituted until after exposure.

Bethlehem Steel charged the government an extra $5 million for construction of the nuclear frigate *USS Bainbridge* to cover unwarranted contingencies and "cost overstatements in its statement of actual and estimated costs." [33]

The Sperry Gyroscope Company overstated target costs by more than $3 million in producing a gyroscope, resulting in increased cost of over $1 million for target and incentive fees.[34]

In most of these cases the companies tried to square accounts with the government *after they had been caught*. Many such cases are not settled voluntarily and are referred to the Justice Department under the Truth in Negotiation Law, on suspicion that fraudulent certification of costs has been made.[35] But the record indicates little further action under criminal statutes. The gallery of suspected rogues includes the most respectable corporations in the country—Boeing, Fairchild, Burroughs, Cessna, Clevite, Foster-Wheeler, General Dynamics, Lockheed, Magnavox, Philco, RCA, Thompson-Ramo-Wooldridge, Western Electric, Westinghouse, General Electric, and Bendix. Most of these cases were dropped or allowed to die in abeyance by the Justice Department on grounds of "insufficient evidence" or "low probability of success in litigation." [36]

There is recurring evidence of price-fixing collusion among major government suppliers, which suggests that the practice is considered a normal part of business operations. Admiral Rickover notes that when corporate officials certify cost-estimating data, they are not held personally responsible for its accuracy. As a "legal person" the corporation becomes "the man who was not there" in the face of issues of legal responsibility. Rickover: "The corporation has now won the rights of a citizen under the law. Why should not the corporation and its officials likewise have the corollary obligations of a citizen?" [37]

Subcontracting. The contractor community carefully juggles books to show moderate percentages of total profit, usually about 3 percent of contract value. Total contract value is preferred as the basis for stating percentage profits, for calculation based on invested capital would tend in many cases (especially where government provides the plant and equipment) to show considerably larger margins. The low rate of declared profit is used to prove the contractor's spirit of public service and sacrifice, his efficient management, the oft-repeated contention that performance of contracts is a "bargain" compared with the 6 to 7 percent profit margins of manufacture for the civilian market; and to allege that too-great concern for excess profits is insulting to the corporations and might persuade them to abandon this public service.

When profits cannot be concealed by stuffing overhead or inflating direct costs, the common method is through subcontracting. The subcontractor escapes direct responsibility to laws and regulations that aim to protect the public interest. The prime contractor charges the entire subcontract cost, including profit fee for the sub, to his own direct costs. As a result, as Admiral Hyman Rickover points out, profits may be hidden in cost figures all but impossible to check. He cites "Company X" which reported a profit of 15 percent but was subcontracting work for parts to other divisions of its own empire at unchecked, non-competitive prices, thereby further augmenting real profits.[38] This practice, transparent in the case of a wholly owned subsidiary, is equally suspect when it occurs between nominally independent companies whose financial structure and directorships are interlocked. It is suspect when, after a series of subcontracts, the prime contractor suddenly merges with or acquires the sub through paper manipulation. It is suspect when a trade-off between the major contractors occurs under different contracts, each prime subcontracting a share of the work to the other.

Some instances:

In 1962 the Permanent Investigation Subcommittee (under Senator John L. McClellan) of the Government Operations Committee found instances of "real" profits more than 40 percent of contract value arising through this device. McClellan cited a three-tier pyramid which he called typical: at the bottom, Western Steel Company produced components for the Nike system under a subcontract from Douglas Aircraft, charging a profit fee of $1.5 million on top of $14 million; in turn, Douglas (subcontracting to Western Electric) added another $1.5 million fee to its $15.5 million cost; at the top, prime contractor Western Electric added an additional million dollars and charged the Army a total of over $18 million on the basic $14 million procurement.[39] The total prime contract was a $1.5 billion award to Western Electric which subcontracted about 75 percent of the work, mostly to Douglas, which in turn subcontracted 80 percent of the work. At the apex of the whole contract, Western Electric charged the government a profit fee of more than $100 million, amounting to 31.3 percent profit on that portion of the work done by itself; yet on the total contract value, the profit falls to an apparently reasonable 7 percent.[40] A true perspective on the unconscionable profits in such a procurement must include not only the multi-tier fees on fees, but also the fat hidden in overstated direct costs and in overhead.

GAO reported a 1963 case involving the Melpar Corporation, in which the government was charged 41 percent more than Melpar paid to the subcontractor who actually did the work in question. Melpar claimed the difference as due to its own efficiency and cost savings which it should be allowed to retain as an incentive fee. GAO commented: "It appeared that a significant portion of this amount was attributable more to the negotiated price based on overstated cost estimates. . . ."[41]

In procuring spare parts for aircraft during 1959-1961, the Air Force permitted the prime manufacturers to follow the subcontract route with the result that the government paid $51 million for parts that could have been bought directly from the prime's suppliers for $19 million less.[42]

The instances audited by GAO often lead to contract re-negotiation and cost reductions but are a small segment of total contracting. At the same time the primes argue: their profit added to subcontractor profit is justifiable since it costs money for them to oversee the work and to apply quality controls that meet government specifications; they cannot and should not control prices set by "independent" companies with whom they do business; and subcontracting forces the prime to assume the risks of subcontractor overruns and costly errors, and he should therefore enjoy any windfalls just as he must absorb any losses.

On examination these arguments break down. It is true that the government deals with the prime and that the prime is legally responsible for the performance of his subs. But it is also true (in the words of a GAO official) that "the prime has no criminal responsibility as far as the sub is concerned," [43] that quality control by the prime is often mythical. When subcontractor products prove defective, as has happened, the contracting agency has granted a brand-new contract for remedying the defect—with additional multi-tiered fees. There is virtually no risk involved for anyone, except possibly for the nation and the public purse; the federal agencies, committed to the prime contractor's financial stability (as a "national defense resource"), re-negotiate to pick up overrun costs in one way or another. It is the sheerest fabrication that major primes lack interest and influence over subcontractor prices: in fact, the two characteristic patterns are either sharp competitive pricing by subs or regular dependency of subs upon primes. If anything, primes deal with independent subs with the hard-eyed shrewdness of Yankee horse traders, often requiring the subcontractor to absorb costs of tooling and beginning production which in their own activity is generously allocated to government. Subcontractors "hope to get business for several years with the prime contractor. And in the hope to get that, and to meet competition with other prospective vendors, he absorbs this tooling cost." [44] In short, prime contractors do not suffer the handicaps that government has all too willingly endured in supinely accepting contractor cost estimates and prices. In the words of Admiral Rickover: "The government is the only customer I know of who is always wrong." [45]

Overhead. In the average industry working for the government, according to Jerome Wiesner, "overhead will run somewhere between 100 to 200 percent. It is a rare industry that is as low as 100 percent. . . ." [46] Although the Bureau of the Budget has established rules for controlling overhead, the real burden rests upon the agencies and their contract officers. The record shows them to be overly permissive and strangely lacking in curiosity. For example, in 1961 the Defense

Appropriations Act made illegal the use of overhead charges for company advertising and promotion. As a result, contract officers permitted an undiminished volume of such advertising, subsidizing politically potent industry trade journals and buying favorable agency and industry treatment in the news, while, in deference to law, including a few help-wanted lines of type.

Company profit is taxed about fifty cents on the dollar; therefore a profit dollar is worth only half as much as an overhead dollar which is not taxed at all. Admiral Rickover: ". . . much ingenuity is used by business in making expenditures for plant repairs and rearrangements, tools, manufacturing control techniques, computer programs, and the like which may be charged to overhead but which are actually intended to improve the company's overall commercial capability." [47] Contractors enjoy almost complete freedom in the decisions to incur overhead expenses and to allocate them to contracts, various companies in their empires, and between commercial and government operations. Rickover: "The government agency may never know how much the equipment actually cost to produce or how much profit the contractor makes in producing it. . . . It seems clear to me that contractors have a far greater incentive to increase these tax-free overhead costs than they have to cut costs and pay taxes on the profits." [48] He estimates that contractor true profits are at least 15 percent rather than the 3 percent level which their bookkeeping shows.

Typical overhead stuffing:

The 1962 McClellan investigations revealed that the Douglas Corporation was still charging as overhead the developmental costs of the DC-3 transport aircraft (undertaken in the 1930's and largely unimproved since that time) on its Nike missile contract.

The 1959 Hebert investigations disclosed that General Motors had pocketed excess profits of more than $17 million on a $375 million contract, most of the excess assigned to "unnecessary" overhead.

Chrysler Corporation, which produced T-43 Army tanks, was reimbursed in overhead for rehabilitating an automobile plant which (when tank production was completed) was available for commercial production.

A syndicate of American firms engaged in building a military installation in Spain received excess administrative overhead of almost $7 million, according to a 1963 GAO report.

The question of overhead allowance for data-processing equipment has become a permanent problem. GAO reported in 1964 that Boeing was charging to overhead a leased computer facility—the leasing cost exceeded the purchase price by almost $4 million over five years and by $17 million over ten. Similarly, the Martin-Marietta Corporation incurred "unnecessary" overhead of almost $8 million over a five-year period through a similar lease arrangement, with anticipated "unnecessary" costs estimated to rise to $40 million if the practice continued another five years.[49]

Rickover cited an example of a company which maintains a large product engineering group and charges this group as overhead proportioned between its commercial and government work. But its government work involves little product engineering, practically all of which is applied to commercial products; government is actually subsidizing and financing the company's civilian operations.

The House Armed Services Committee investigating unit conducted hearings in 1964 on overhead allowances made for "morale and recreation benefits" of contractor employees,[50] based on a GAO audit of thirty-six plants which claimed $6 million for such programs. It was found: In October 1963 Aerojet/General leased Disneyland for one day to play host to its Los Angeles employees—cost, $101,000. During 1963 a Martin Company plant in Colorado held its annual crab feast for some supervisory employees, flying the crabs in from Baltimore—cost of the crabs FOB Baltimore, $1,800; cost of the crabs delivered, $5,600. Twenty-three plants charged government about $1.5 million for losses in providing food services at plant sites—only seven had properly offset such losses against surplus revenues from commissions on plant vending machines, the rest (representing about $1 million) charging the entire amount as overhead.

"In several instances," the GAO reported, "we noted that the losses were caused by the furnishing of food to the contractor's executives without charge or at abnormally low prices, whereas little loss was incurred and sometimes small profits were made on food furnished to the other employees."[51] In some firms, vending machine revenues were high, accumulating funds of over $1 million in some cases, more than enough to cover food service losses. What happened, the House committee noted, was that "gross losses are passed on to the government while gross profits are usually turned over to employee organizations" which are controlled by the corporation, are sometimes invested in corporation activities, and are rarely disbursed for employee welfare.

In the category of party-throwing, the largest expenditure was found at the Hughes Aircraft Company, where during 1962 $53,000 was claimed as overhead for activities that included parties held in private homes, retirement and farewell dinners, Christmas parties, dances, catered parties for employees and wives, and cocktail and theater parties for middle management.[52] At six plants, contractors spent about a half-million for Christmas turkey gifts charged to overhead. IBM charged government another half-million largely representing the cost of operating three country clubs located at various places in New York State, in spite of the fact that "overall operations of the corporations are primarily concerned with commercial sales," not with direct work for government.[53] GAO commented: "We believe it is apparent that some of the expenditures for employee recreation and morale are rather elaborate. . . . To the extent that a company is engaged almost exclusively in government contract work . . . and thus is in a position to pass such costs on to the government, there is little incentive for the company to limit such expenditures." [54]

The study of Aerospace Corporation indicated a salary structure (charged as direct cost) that greatly exceeded "comparability with salaries paid in private industry." The so-called non-profit company perfected a simple device for evading Defense Department efforts to hold reimbursable salaries to such a standard: for one year that part of salaries which might be questioned as direct cost was charged to overhead; in following years it was submitted (and accepted) as direct cost.[55] As overhead, government paid almost half a million dollars to Aerospace for relocation expenses involved in moving plant facilities and personnel from El Segundo to San Bernadino, California, a distance of seventy miles. Of the 170 employees moved, the Air Force allowed for each an amount equal to the sum of one month's salary, plus the cost of moving household goods, plus the cost of mileage by private car—an average of $2,000 per employee (Aerospace had asked $4,000). The average, however, is not indicative of the distribution of these funds: one executive received $7,000, including $3,000 to cover loss sustained on sale of his old residence and $3,000 brokerage fees and closing costs. (GAO noted: "Clearly against Aerospace's own relocation policy." [56]) Many executives had collected thousands of dollars but were not significantly closer to the new site, and one was farther away.

A similar case involved Lockheed. It was a general practice to pay relocated employees thirty days' *per diem* allowance (plus regular salary) regardless of circumstances in each case or number of days of work missed. This, GAO noted, did not conform to the company's own policy, and 70 percent of relocated employees received unnecessary payments up to $1,000 per move.[57]

Technical errors, overruns, etc. The cost-type contract not only fails to provide incentives for efficiency, timely performance, responsible engineering, and adequate quality control, but in fact it positively discourages all these values: "In many cases there have been incentives not to finish," Jerome Wiesner declared.[58] The House Select Committee on R&D Contracting reported in 1964 that "the practice known as 'gold-plating'—that is, stretching out a production or development contract by introducing extensive design modification—is another usage wasteful of manpower as well as money that must be checked." [59]

Technical errors, costly mistakes, continuation of spending for technically obsolete or infeasible programs—all such errors spring from the character of the contract system, the contract form, and the quality of contract management. Inadequate specifications, the launching of large production programs on the basis of inadequate engineering knowledge, failure to evaluate and intervene in a timely fashion in supervising the contractor's work are traceable to the lack of R&D in-house capability in the government and the absence of objective engineering advice and technical direction. In 1964 GAO officials pleaded with Congress to insure that careful "experimentation, development, testing, be accomplished before volume production is undertaken to prevent the dissipation of resources for material that is unusable, requires expensive modification, or is no better than less costly material already available." [60]

A sample of recent findings:

The Army spent about $300 million for development and production of a missile system which did not meet performance standards, a fact that was "known at the point in time when the Army ordered successively increasing quantities. . . ." Because of these deficiencies, existing older weapons had to be deployed in lieu of the new system.

The Navy spent (over a ten-year period) half a billion for the P-6M seaplane and "did not receive a single serviceable aircraft," having ordered quantity production before "a reasonably satisfactory plane had been developed." Engineering errors and contractor (Martin Company) failure to adhere to testing procedures were known to the Navy, but they only provided occasion for additional contracts and fees. GAO estimated waste due to design and engineering errors at more than $200 million.

The Army awarded $3 million in contracts for radiation measuring instruments "even though it was aware . . . that the instruments were not suitable. . . ." Thereafter, over $663,000 was expended to modify the instruments, which eventually had to be scrapped anyhow.

The Navy incurred "unnecessary" costs of $376,000 when it ordered a new type of radar which would operate in a frequency band as yet unapproved by the Director of Naval Communications. After production was completed the frequency band of the sets had to be changed.

The Army incurred unnecessary costs of over $7 million for production of a trailer to transport the Honest John rocket, "when knowledge was available that it had design limitations and did not represent an improvement over existing equipment." Trailers equally adequate, some of which were already in use, would have cost about $3 million.

The case of the Navy F-3H jet fighter was a dismal story of technical errors involving eleven crashes of test aircraft and the deaths of several pilots. In the words of Congressman Holifield, "It turned out that it was a very poor procurement." [61] The engine was not correctly mated to the airframe and was underpowered. In spite of this, money continued to be spent (up to $302 million), virtually all wasted. A few years later the admiral in charge of this procurement retired to reappear as vice president of the contractor (McDonnell Aircraft).

A Thompson-Ramo-Wooldridge case involved fuel pumps procured under subcontract for B-47 aircraft at a 110 percent mark-up. When they proved defective, a new contract was written with TRW to correct the mistake, leading one congressman to exclaim: "Now, that in itself is not understandable to me. . . . Somewhere along the line it looks like we are operating a rather peculiar system when we are allowed to at least—not penalize, but add another profit of 110 percent to correct their own error." [62]

In 1965 the GAO charged NASA with mismanagement of the Nimbus Meteorological Satellite project, resulting in "unnecessary" costs of over $1 million. Soon after award of the contract to General Electric, tests showed that estimated satellite weight would exceed design specifications, which in turn had already exceeded the launch capability of the vehicle to be used. In spite of this, the project manager allowed GE to continue working for six months toward the original design goal "even though it was clear" the effort "would be futile. . . ." [63]

Rickover cited a construction contract awarded in 1961. After the job was finished the contractor presented claims for a 200 percent increase in contract price. Although denied by the contract officer, the company appealed and its claims were endorsed by the contract review board.

Contract firms maintain large legal staffs which specialize in the business of escalating the price of contracts, either (in Rickover's words) "by pressing contracting officers for unwarranted and unreasonable contract changes or by dragging unfounded claims before various administrative review boards and courts. . . . A great many contracting officers over the years have authorized changes that were not really changes, simply because they had available to them neither the time nor the talents to successfully fight off the claims lawyers." [64]

Not only is it incredibly difficult to cancel a program that has gone wrong or grown obsolete, but when such programs are canceled the government reimburses the contractors for losses claimed to result from the cancellation. This is often a substantial part of the money required for completion of the contract, leading agencies to argue in Congress that such programs be continued for whatever good may come out of them, even if the end product is plainly unworkable or no longer needed.[65] This tends to confirm a suspicion that contractor cancellation costs are related to agency ambitions to keep programs limping along for another year in spite of administration attempts to cancel them. There is proof of this in subsequent "reactivation" of several such programs (for example, large solid-fueled rockets) which NASA and the Air Force, with the connivance of contractors and Congress, managed to save year after year from executive budget deletion.

The "sporty course" represented by this sample catalog of waste and profit is an intrinsic part of the Contract State. It creates the new economic federalism which transfers to "private" centers of power a quasi-public authority and all of the characteristics of an overweight bureaucracy. "One of the biggest problems of a bureaucracy is that you don't have a cost sheet. You can't go broke. . . . It is hard to establish procedures that show you exactly what every item costs. And . . . you don't have a memory." [66]

What does "every item cost" in terms of the erosion of democratic processes and public decision-making? In terms of the corruption of business and public life? In terms of the concentration of economic power, the stagnation of the civilian economy, and growing inequities of income, employment, and social justice? What does "every item cost" in terms of hardening socio-economic cleavages? The growing barriers to class mobility? What are the costs in terms of overall allocation of national resources—human, technological, and natural? The quality of American life?

What does "every item cost"? Even the richest society cannot afford to evade, postpone, or ignore the question for long.

CHAPTER

XV

PATENTS AND POWER

Uncertainty and controversy surround the question of private versus public rights in patents that result from the government's massive R&D investment. The issue, muddled by abstruse legalisms or simple Populist slogans, must be understood in terms of the contemporary role of innovation and the structure of economic power in highly capitalized societies.

Originally the patent was an incentive and protective device to spur the efforts of individual inventors, but its role has been fundamentally altered by an economy in which further innovation requires expensive facilities and years of toil by teams of highly trained specialists. The day of the basement inventor who was able to make significant contributions with little more than string and sealing wax is virtually dead, like that of the neighborhood grocer who has been swallowed up by chain supermarkets.

The patent system arose simultaneously with the printing art in the fifteenth century and was closely associated with the institution of copyrights. Both grant a time-limited monopoly to reward creativity. The system is now accepted throughout the world. Even the Soviet Union has a system of cash rewards and recognition, and in the last decade (as it achieved technological parity with the West) has shown a willingness to conclude patent treaties agreeing to pay for foreign patents previously pirated, being motivated by possession of valuable pioneering technology of its own.

Under U.S. law, patents on invention vest exclusive rights for seventeen years and are not renewable, although new patents may be issued for improved or related processes and devices. To be patentable, a thing must be "new" (that is, not previously patented, in use, or described in any printed publication) and "useful" (capable of reduction to practice, or already operative). It may be a process or method, a machine, a fabricated article, or a composition of matter. Its originality must be verified by a "patent search," and when granted it is subject to challenge in the courts.

The traditional theory of the system is to serve the public's interest in innovation by vesting a property right in the innovator. Indicative of the patent's changing role has been a stream of court actions against monopoly corporations which use extensive patent pools to keep the benefits of invention out of the hands of competitors. (The tendency of the federal courts in recent years has been to instigate a system of compulsory licensing, forcing patent-based monopolies to permit use by others.) The fact that patents are declining in importance while copyrights maintain full vigor suggests the changing social processes which have undermined the role of the individual in technological innovation, but not in literary utterance.

Invention and technological innovation today are products of organized and regular channels. In the words of one expert: "Any outsider or rebel, whether in a corporate laboratory, the government, or some basement, who would like to invent otherwise might almost as well think of starting his own great newspaper or his own telephone system or navy." [1] Almost three-fourths (64 percent) [2] of currently patented inventions are made by professional scientists and engineers and preassigned by them to the corporations where they are employed.[3] The basement inventor still exists, but his contribution is small and limited to areas which do not need elaborate laboratory facilities, such as novelty toys and household gadgets. Even here industrial R&D submerges the individual: when the old-fashioned innovator produces an item of commercial promise, he is often forced to sell the rights to large companies to get production and marketing, and he faces the danger that corporate engineers will "invent around" and freeze him out entirely.

The individual inventor has faded from significance and the large industrial lab has taken over. Such institutions as the Bell Telephone Laboratory have hundreds of scientists-engineers assigned over a period of years to a wide range of problems. These professionals are willing to alienate their stake in possible patents in exchange for the opportunity to use the research facilities and to enjoy the security of long-term employment. Property rights in a patent represent a small chance at a big reward: this risk the salaried technologist prefers to pass along to his employer. In large, patents no longer provide direct incentives for the individual, and corporation incentives, as we shall see, are considerably more complex than those subsumed by the traditional system.

Since the turn of the century the industrial lab has replaced the individual inventor, and since 1940 government contract funds have overwhelmed commercial support of these labs. According to a 1964 congressional study, small companies and unorganized inventors support only 2 percent of patentable invention. Large private industries perform over 70 percent, but more than half of this (about 60 percent) is supported by the federal government through contracting or tax benefits. A few very large industrial corporations do the lion's share: of the total R&D expended from company funds, only 10 percent came from companies having fewer than five hundred employees (although such firms provided 35 percent of total employment).[4] Most of the R&D which private industry finances is contract-related, paid for by government through overhead on contracts and/or as a component of price in competitive procurement.

The fact that a handful of aerospace industries accounts for the bulk of so-called private funds invested in research suggests the real extent of the public commitment to work embodied in new patent applications.[5] Left to itself, commercial invention would not produce the technology required by military and other public needs. This fact validates the increasing federal R&D investment; but the mode of government contracting and the changing structure of economic power raise a variety of questions about the patent as part of the process of industrial concentration and cartelization.

During the postwar decades, in which government R&D contracting has grown exponentially, patent policy has been diverse, confused, and fluctuating. Two clear-cut patterns have emerged: the "contractor-title" (often called "license") doctrine under which private contractors retain full ownership of patents arising from government work, but the government receives a free and non-exclusive license for patent use which can be extended to other companies performing government contracts; and the "public-title" doctrine under which government acquires ownership of such patents, making them freely available for use by all American industry.

The "contractor-title" pattern was official policy for military procurement until 1961. It had an honorable origin in the traditional system under which virtually all R&D was done in-house or in university laboratories, and hardware was purchased from companies whose business was mostly civilian. (This also became the policy of the National Science Foundation, the Bureau of Standards, and the Weather Bureau, which largely deal with universities and independent laboratories.) This system was upset by the accelerating technology of war. "Public-title" represents a departure initiated during the atomic bomb project when there was general consensus that atomic energy should remain under the exclusive control of government for national security reasons. The Atomic Energy Act of 1946 gave statutory force to the doctrine and, in like fashion, the 1958 founding legislation for NASA laid down the same principle for space. The Departments of Agriculture and Interior traditionally adhered to this practice to give maximum aid to farm, mining, and manufacturing production.[6]

Both NASA and AEC statutes provide for waiver of government patent title when such action is determined to be "clearly in the public interest." NASA has introduced a waiver policy so liberal that it undermines the essence of its legal charter, claiming this move a practical necessity in dealing with Air Force contractors who preferred the pattern to which they were accustomed, and might refuse to accept R&D work on any other terms.

A Smoldering Controversy

During the late Eisenhower years, a vocal congressional minority persistently attacked "contractor-title" and liberal waivers as a "give-away." When in 1961 the new administration took its bearings on this issue, a curious collision occurred: McNamara began to tighten Defense Department practice by asserting government title to patents in new R&D contracts, while NASA's Webb sought even greater dilution of the statutory limits upon the Space Agency's waiver authority. This generated congressional interest, and in early 1962 the House Space Committee offered a bill to delete NASA's statutory "public-title" regulation. The Justice Department opposed the change, arguing that "contractor-title" would increase the costs of government R&D and promote monopoly among aerospace companies. The legislative proposal was put aside when President Kennedy promised that the issue would be reviewed and a uniform administration policy enunciated.

A presidential directive was issued in October 1963, but it soon proved to be an invitation to increased confusion rather than a guideline for clarity and uniformity. The memorandum provided: (1) the government would normally acquire rights to inventions made in the course of work on a government contract, except when the head of the agency concerned certified that the public interest would best be served by some other arrangement; (2) in cases where the government contract called for R&D directly related to the field in which the contractor had an established commercial position, the contractor would normally acquire rights to inventions. Where the contractor retained title, and the patent related entirely to government work, the government would receive a free, non-exclusive license. On the other hand, where the contractor could claim that the major input of the patent came from the company's commercial know-how, the Kennedy directive required the contractor to take effective steps within three years to bring the patent into practical use or make it available for licensing on reasonable terms. In addition, the government would have the right to insist that licenses be freely granted to others to the extent that the invention was relevant to public needs.[7]

NASA interpreted the directive as endorsement of its own liberal waiver policy. On the assumption that space technology was rich in commercial promise and spin-off whose exploitation would be fostered by patent monopolies, Webb announced a broadening of his agency's waiver system, citing the President's directive as justification. In the future NASA would waive virtually all government title rights *in advance* at the time a contract was signed, even without examining specific patent applications. The National Association of Manufacturers, the U.S. Chamber of Commerce, and the Machinery and Allied Products Institute indicated satisfaction with the new regulations, hailing them as a true expression of the President's intentions which they hoped would be extended to all agencies.[8]

What was specifically on their minds was the Defense Department, where McNamara was energetically going in exactly the opposite direction, also using the White House directive as his authority. The Defense Secretary attempted to tighten up the automatic "contractor-title" practice of his predecessors in order to protect the government against double charges, gain better control of "sole-source" contracting based on patents, and combat the tendency toward concentration in the contractor community. His efforts kept the patent issue alive in Congress, where most legislators sought ways of confounding his aims while a hardy and vocal few leaped to his defense.

A three-year running debate ensued. Each year Louisiana Senator Russell B. Long led an effort to amend NASA authorization bills to narrow Webb's waiver authority, while other legislators fought to protect traditional Defense Department practices against the rigors of McNamara. By 1965 the legislative docket was crowded with a variety of invention bills of contradictory purpose, and President Johnson sought a compromise by appointing a "Commission on the Patent System" to conduct "an independent and complete examination" of the entire institution.

Three schools of thought have crystallized in the controversy. Two argue extreme positions in favor of an absolute and uniform policy, one for private and the other for government patent ownership; in the middle are those who favor a flexible set of guidelines such as those attempted in the 1963 Kennedy directive. In the course of the three-year impasse, the third group has been growing in size but is unable to agree on concrete standards for drawing the line. The result has been that the flexible position has increasingly become a way of defending the status quo, of supporting Webb and undermining McNamara.

The critics of "contractor-title" argue:

The taxpayer bears the cost of R&D and therefore the government should take title, otherwise the public may be forced to pay the cost many times over in future government contracts or in purchasing a commercial product that utilizes the patent. As stated by Lee Loevinger, head of the Justice Department's anti-trust section: "It would be paradoxical to tax the public to raise funds for scientific research, on the premise that this research advances the general good, and then give the results of this research to a private company for exclusive commercial exploitation." [9] By denying any private property right in what is essentially a public asset, the government insures the widest possible access, making the invention available for commercial use to all comers without additional cost and thereby maximizing the benefits.

The vesting of private property rights in certain areas of contemporary technology may also inhibit and obstruct international cooperation. This argument was used with great effect during the early years of atomic energy development, when complete government monopoly was favored as a means of not only protecting the U.S. monopoly but keeping the problem of international control unencumbered by private interests. In 1962 the Department of State argued that the surrender of government patent rights on weather control and forecasting inventions would frustrate the nation's foreign policy by making it "more difficult and, it could be, impossible" to carry out a commitment to share the new technology with the world and to conduct a joint effort with the Soviet Union.[10] Similar questions were raised concerning a communications satellite.

The most impassioned argument against "contractor-title" is that it contributes to the concentration of economic power, buttressing a few large companies' quasi-monopoly position in the bidding for government contracts, excluding small business and increasing contract costs. Senator Wayne Morse estimates that the NASA waiver policy has the effect of "transferring the results of 1965's $4.4 billion taxpayer-financed R&D to the sole benefit of a limited number of large corporations, since only about 11 percent of NASA's R&D funds go to small business." [11] Admiral Rickover declared: firms receiving Pentagon money "are a relatively few huge corporate entities already possessing great concentrated economic power. They are not ailing segments of the economy in need of public aid or subsidy. Nor are there any real reasons to offer patent giveaways in order to induce them to accept Defense Department research grants or contracts. . . . To claim that agencies cannot get firms to sign such contracts unless patent rights are given away strikes me as fanciful nonsense." [12]

The House Select Committee on R&D declared in a 1964 study that

> . . . a small number of giant firms and a few defense and space-related areas, with their facilities located principally in three states . . . receive the overwhelming preponderance of the government's multi-billion-dollar research award. Clearly, if the resulting technical discoveries are permitted to remain within these narrow confines . . . a disproportionate amount of the benefits will be channeled into the hands of the few and a further economic concentration will take place.[13]

The Senate Select Committee on Small Business in 1963 saw "contractor-title" practices as "seriously hampering" rather than aiding the efforts of small business to bid on government work, thereby perpetuating "the prevailing distribution malapportionment. . . ." [14]

Senator Russell Long has collected a long list of indictments against government patent policy during his crusade. He has attacked General Electric for what he calls the "bluff" of their refusing to do space work unless they retain all patents as corporation property, noting that the company stood charged with collusion to make light bulbs last a shorter period of time in order to increase sales. A pharmaceutical company was cited which held the exclusive right to market a test kit to detect the onset of a form of mental retardation in infants. The kits had been developed under government grants and could have been produced for about six dollars. The company sold them commercially for $262. Long asked, "Why should we spend our money to achieve this type of result?" He noted that the Kefauver investigation of the drug industry revealed that the public "was being charged somewhere around one hundred or two hundred times the cost" of producing tranquilizing drugs. "If we are going to pay for this development, why should we not get the benefits . . . ? Why let some private corporation hold these advantages for its own good . . . ?" [15]

On the other hand, the defenders of "contractor-title" argue:

This practice is basic to maintaining the values of a "free enterprise system." Private industry needs this incentive to accept government R&D work and to undertake early commercial development of new inventions. "Contractor-title" helps rather than hurts small business in that it provides a means of competing with the giants through acquiring ownership rights to originality and pioneering.

No new patent is worth anything, the defenders declare, unless large additional sums are invested for its commercial development; firms will be reluctant to make such investments unless their priority is protected by exclusive patent rights.

Charles L. Shelton, a spokesman for the aerospace industry, asserts that "government-title" causes industry to shy away from public contracts and "dries up" cross-feeding between government and commercial research.[16] As stated in a 1965 report of the House Select Committee on Government Research: "A business firm will be reluctant to pioneer in production and marketing of something new which it does not own and which it could, therefore, not protect against competition that would simply cash in on its ice-breaking efforts." [17]

The contractors argue that since the bulk of all R&D is sponsored by the government—characterized as "unprecedented control over technological advances"—a "government-title" policy would further strengthen the hand of bureaucracy, contradicting the traditional role of the free market, subjecting "a large proportion of U.S. technology to the limitations inherent in any complex activity controlled by central planning." [18]

Industry raises a question concerning the difficulty of distinguishing between the contribution made to new patents by federal money and that made by existing competence and previously acquired contractor know-how. It argues that all patentable items contain a large amount of proprietary information which, if title be assigned to the government, deprives the company of its property without due process of law or just compensation, jeopardizing its competitive position and undermining economic initiative.

In favor of the present "flexible" policy:

In 1965 Donald B. Hornig, Presidential Science Adviser, threw his support in favor of flexibility, declaring: "The differing objectives and circumstances under which federal research is conducted rules out the possibility that a blanket [policy] could take them properly into account." There are circumstances, he said, where the government "would like to take advantage of the fact that the prospective contractor has made a substantial private investment in the field of interest. The granting of some commercial rights may be necessary to attract private investment in developing and commercializing the investment." On the other hand, there "may be opportunities through a licensing program to exploit the inventions abroad that could be of economic benefit to the United States." [19]

Clarifying the Issues

All the arguments contain valid propositions and describe legitimate problems, but they suffer equally from a certain artificiality and legal formalism, failing to take cognizance of the declining importance of patents as an institution. All attempts to impose a "government-title" policy have been substantially undermined by industry's willingness to abandon patents entirely in favor of such devices as secrecy, non-reporting of patentable innovations, and claims of "proprietary information" (that is, non-patentable know-how in which the company has an investment). In addition, contractors have asserted (and government has accepted) the proposition that company R&D funded as overhead on government work creates no public right whatever.

The advocates of an absolute "giveaway" policy have at least recognized that no proffered alternative holds promise for arresting this kind of private aggrandizement, and may have other untoward effects. All efforts at a more restrictive policy have failed: the prime contractors simply do not report processes or inventions that they can use for maintaining "sole-source" status or for commercial exploitation. Rather, they hold such information as proprietary, finding adequate protection by this means. They may patent such information later to codify its proprietary status, waiting for a time when the extent of government funding cannot be proven. By secrecy and proprietary claims, they achieve the bonus opportunity of recharging the government for use of the process or product in other contracts, on occasion reselling it as a separate package to government or to other companies. The quality of government's management of contractors is such that genuine discovery and evaluation of this practice is impossible. Even if adequate public control existed, the problems would be formidable.

The trend toward a broad waiver policy recognizes that government is helpless to change the real situation by a theoretically stronger assertion of title rights. The strictures of patent laws are part of a totally transformed structure of economic power in which the simple laws of Adam Smith cannot be usefully applied, in which neither plain nor fancy enforcement can break the circle of concentration or reinvigorate the patent as an instrument of economic growth.

Patents have declined in significance not only in the government contract but also in the commercial market, where their new role still retains considerable value for corporations. The value lies in the use of patents and "patent pools" as part of the panoply of disposable resources available to large corporations in enhancing and maintaining their place in an increasingly concentrated and cartelized society. But even this is of declining importance. A complete monopoly does not need patents to deny competitors an entrance to the civilian market; but it may find it useful to acquire patents on substitutes which may at some future time offer indirect competition. Such patents are used to postpone the introduction of price advantages in order to enable the patent holder to depreciate his existing capital investment most advantageously in maintaining a stable profit margin. A full monopoly may also find patents useful, as we have seen, in maintaining its favored position in contract negotiations with government, as a means of denying government the opportunity to create a capability elsewhere in the economy for multiple-source or competitive procurement.

No monopoly position is permanent and complete. It is subject to a variety of threats arising from the vicissitudes of government and politics, as well as the unpredictable nature of technological development and the changing structure of demand. The most valuable patent is one that offers a strong price advantage to a product which has no good substitutes. A pool of such patents during the early stages of a new technology can be used to create an unassailable monopoly whose power may be perpetuated indefinitely by virtue of assured profits, know-how, and capital equipment. The pool endows the holder with a unique ability to exploit follow-on innovations so as to maintain the advantage of lead time over would-be competitors. The individual patent (which expires after seventeen years) comes to be of minor and expendable importance, subordinate to the continuous building of a fresh patent pool and the rising threshold of know-how, capital equipment, financial resources, and an established share of the market.

Until 1955 about 21 percent of all patents were issued to less than 1 percent of U.S. corporations,[20] and the discrepancy has increased with the growth of government R&D. Self-perpetuating patent pools tend "to create industrial monopolies of unlimited duration . . . forming a frightening barrier to all competition, a monopoly against consumers of the goods, and a monopsony to the free-lance and foreign inventors who find no market for their patents but this . . . pool." [21] The tendency of the Justice Department to bring successful actions against such pools as were held by RCA and General Electric (forcing licensing to other producers) is symptomatic of the turnabout. Patents may still encourage invention, but in a way which is socially quite undesirable and has effects opposite to early commercial exploitation and greater productivity at cheaper prices.

Today patents afford small business or individual inventors little or no protection. A patent suit is one of the most expensive forms of litigation, and big corporations with relatively unlimited funds, attorneys, and time, holding a pool of thousands of patents and maintaining large R&D establishments, can if they wish find a legal basis on which to hound any competitor. "Patent litigation and the threat of it and the necessity of being ready for it . . . favor the great corporation and force the small opponent into bankruptcy or to selling his patent on the big fellow's terms." [22] For example, the DeForest patent on television was eventually upheld in the courts, but in the meantime DeForest had sold it, gone through several bankruptcies, and left the field. While entangling the small opponent in litigation, the large corporation offers him a way out through sale of his rights, backed by the threat of indefinite litigation and the possibility that the corporate labs will succeed in inventing around him (they usually do).

The 1964 Joint Economic Committee study notes that this leads to several inevitable consequences: the decline of patenting itself, the decline of litigation in favor of out-of-court settlements, and "the strengthening of the giant and the ruthless over the small and the conscientious." [23] In sum, the record of forty years blasts the corporation argument that government R&D contracts must preserve private patent rights in order to sustain small business and the free-lance inventor; neither has been able to maintain its place in the economy by patent ownership.

The large corporations also use patents to strengthen their bargaining position with each other in the politics of corporate control. They often seek exclusive patent rights to innovations which they have no intention of using, to prevent their use by others or to force the concession of licenses by other patentees. Patent-pooling and cross-licensing, heavily employed by the chemical, petroleum, and auto industries, symbolize what might be called "the balance of power" existing at any given moment among the great quasi-monopolistic corporations. Some hundred industries regulate their competitive relationships in these ways (as well as with other devices, of course).[24] Litigation among large corporations erupts from time to time (the Sperry Gyroscope Company won a suit against the National Semi-Conductor Corporation in 1964; Technical Measurement Corporation of New Haven, Connecticut, has a suit against Nuclear-Chicago Corporation of Des Plaines, Illinois, for patent infringement; etc.). These represent outcroppings of deeper political engagements which often end in the rejuggling of corporate power, mergers, acquisitions, or the disappearance of companies. The process is most severe in its continuous gobbling up of small and new companies by the giants who tend to maintain by cartelization a nice balance among themselves. But even the giants exist in the tension of the elitist politics of the corporate and banking community, which is not immune from periodic upheavals. The ups and downs of politics, banking, changing government and commercial market requirements, and so on, occasionally expose one of the major corporations. This may happen because of possible overcommitment in an area of de-emphasized technology, or because a company's managerial clique suffers relative eclipse in the circle of the economically powerful, or a social fall from grace in the country club. Whatever the cause, the process of corporation politics and the struggle for survival among corporate elites continues relentlessly, all of the participants ready to turn upon and carve up the empire of a momentarily weakened member. Whether involving the Goliaths or the Davids, patent litigation may accompany the combat, but as a very minor element of harassment or defense.

Patents today derive their most immediate cash value in the pursuit of government contracts, where they can mean unchallengeable "sole-source" status and an inside track to public decision-making and future awards. But "proprietary data" can achieve the same role and, given the inside track to start with, some pretext of proprietary data is easily found. Where sole-source status genuinely depends upon exclusive patent-holding (frequently the case in certain chemical and drug industries), the holder achieves real benefits, at least until competitors succeed in "inventing around" the patent. For a time, the owning company can refuse to grant licenses to anyone, charging whatever price the government and commercial market will bear.

After initial windfall profits it will usually soften its grip, granting a limited number of licenses to competitors (who are close to inventing around anyway, and/or have patents of their own to trade) in order to avoid cutthroat price-cutting and to keep the number of suppliers small. In effect it thus cartelizes the market and maintains, by collusion or by what is called "price leadership," artificially high price levels while avoiding the appearance of monopoly and the danger of anti-trust action.[25]

The Defense Department sought in 1965 to head off new patent legislation which would have denied government any protection at all against this vicious system. John M. Malloy, Deputy Assistant Secretary for Procurement, told a Senate committee that the government's right to use patented invention without a license, while paying fair compensation to the patent owner, was "essential if government procurement was not to be held up by unsubstantial but time-consuming allegations or by exorbitant royalty demands and by refusals to license second sources." [26] Congressional attempts to make existing "Buy America" provisions more binding upon Defense Department procurement would also protect the system from competition by foreign firms which hold (under their own country's laws) similar patents.[27]

Industrial secrets and technological know-how have become more valuable than patent-holding, conferring stronger advantages in maintaining or acquiring monopoly positions in both the consumer and the government contract markets. In the latter category, the proprietary claim is superior to patent-holding because it represents an unexamined basis for sole-source contract awards and an unauditable basis for cost estimates in contract negotiations.

Space Administrator Webb uses the trend to industrial secrecy to justify NASA's waiver practices: "Incentives for contractors to conceal and protect new technological developments as trade secrets rather than to disclose them as patentable inventions are not desirable in government contracting for research and development." [28] This is true, but it is doubtful that even the most liberal waiver policy, such as that presently pursued by NASA, will discourage or reverse the trend from patents to secrecy. In 1964 the General Accounting Office told a congressional subcommittee: "Defense Department contractors have not only failed to submit the patent and technical data on a timely basis as required, but in some cases they have treated such material as 'proprietary,' thereby increasing the cost to the government for procurement of items developed largely at government expense." The weakness of government in-house engineering, technical management, and evaluation was demonstrated, a GAO spokesman declared, by the "laxity on the part of all concerned with respect to protecting the government's interest and obtaining the full benefits paid for by the government in the areas of patents, patent rights, and technical data, drawings, and specifications. . . . And, as a result, it has not been possible to increase competitive procurement to the extent possible. . . ." [29]

Lockheed failed to disclose fifty-eight inventions that had been made entirely under defense contracts during a two-year period; Thompson-Ramo-Wooldridge subsidiaries failed to disclose some eighteen inventions over a three-and-a-half-year period. Not only had public patent rights been jeopardized (that is, royalty-free use or title), the GAO reported, but the government "may have paid royalties unnecessarily for the use of such inventions and may have lost its rights to such inventions because of intervening patents by third parties." [30]

NASA's D. Brainerd Holmes testified candidly to the quality of company claims of proprietary information a few weeks before his forced resignation: "There is very little proprietary information in the program. Companies claim proprietary, but to my mind if all the work done has been used by the government and is paid for by the government there is nothing proprietary . . . I have been on the other side of the fence and claimed proprietary. . . ." [31]

The proprietary claim and secrecy now far exceed patents as a useful contractor device, frequently denying government the technical data required to evaluate performance or to write specifications for new contracts, thus further placing government in the control of private corporations. Contractors have not hesitated to multiple-charge for work already funded by government, sometimes selling the technical data as a separate procurement, or selling it to other contractors who pass on the cost to the government.[32] For example, the Navy contracted to pay Westinghouse more than $1 million for a technical data package which the government had already acquired unlimited rights to use under other contracts.[33]

The congressional Joint Economic Committee noted: "Perhaps no point was stressed so much during the subcommittee's 1964 hearings as the need to have an adequate package of engineering drawing and detailed specifications in order that genuine advertised competitive bidding might be achieved with all the benefits flowing therefrom." The committee noted that under McNamara the Defense Department was at last beginning to enforce the government's rights in this area by formulating detailed directives for contract officers and by establishing an "Office of Technical Data and Standardization" [34] to monitor the problem more adequately.

Two other aspects of the normalized procedures of the Contract State have recently been exposed: the legal disability of the government to acquire either patent rights, royalty-free licenses, or proprietary information from subcontractors (this fact gives the primes a new technique of evasion),[35] and the fact that a vast amount of government R&D is entirely beyond public reach because it is allocated to overhead. The latter category includes a substantial amount that "represents participation in contractor's independent research and development programs" rather than direct contracts or grants. Department of Defense estimates show as much as $900 million annually being spent in this way, charged to "procurement and production funds" rather than R&D. The effect of this enables certain agencies (through contractors) to engage in programs disapproved at the departmental level; it enables contractors to support commercial operations with public funds.[36]

Longer View of the Imbroglio

The trend from patents to secrecy is not, as has been argued by Webb and others, a sufficient reason for government to renounce its patent rights completely and drop efforts to afford the public some kind of protection against the misuse of federal R&D for private purposes. It is true the issue is a thorny one which, in the words of a congressional report, "continues to be an economic and policy question of great importance and one in need of an early resolution." [37] The problem will, however, not be solved on the limited basis of legalistic formulas, although the situation would be greatly improved by effective implementation by the whole federal government of the reforms initiated in the Defense Department by McNamara.

Congressional debate on patent law and R&D contract policy has taken very slim cognizance of the real implications. The small band of Liberals that opposes "giveaways" fails to grasp the inefficacy of the solution proposed (strict public-title) to solve the problems which it vividly describes.

Under classical theory, the patent is supposed to eliminate secrecy by making it unnecessary, facilitating early application of new invention to production which will benefit the whole society. But in a mature, highly concentrated economy, industrial secrecy plays a vastly more significant role than patent-holding. Important technological breakthroughs, not in themselves patentable, may be kept secret and reserved from production until further development leads to a patentable derivative. The hope of a patent forbids all publication for more than a year before application and (in order to stay on the safe side) often discourages publication even longer until "all the developments, bearings, useful hints that the inventor of its rivals might derive . . . have been worked out and either found of small value or patent applied for. . . ." [38]

As a result the patent process tends to defeat its social purpose even where it is still applicable; it facilitates non-use and secrecy for long periods of time and fosters incentives that do not redound to any general and positive value. This is the underlying cause for the increasing tendency of federal courts to impose compulsory licensing upon major corporate patent-holders. But both compulsory licensing and "government-title" practices in R&D contracting fail to redeem the social utility of patents. In fact, both remedies worsen the problem, accelerating the trend toward corporate abandonment of patents in favor of such underground devices as industrial secrecy and claims (whether proper or not) of "proprietary information." Reform of the patent system is a legitimate need, but at best it is a shallow approach which will inevitably be confounded by the real and untouched problems of structural reform. The fact that the public interest may be equally abused and disregarded under any proposed patent reform raises far more difficult questions about American society and government.

CHAPTER XVI

THE ECONOMICS OF AMBIGUITY: COMSAT AND SST

The economics of ambiguity are not unique to postwar America but have during the last two decades extended their embrace to ever larger and more fundamental segments of national life. Their impact is only partially seen in the issues surrounding R&D contracting, the status of the new braintrusts, patents, and such public science programs as nuclear energy to generate electricity and power merchant vessels, communications space satellites, and the development of the supersonic commercial aircraft. The ambiguity now encompasses all categories of private activity, including business, education, banking—the list could be extended indefinitely. This transformation of American society (by means of social planning through fragmented and distorted special-interest subsidies) has been wrought with little outward pain. In fact, the pump-priming effects have served to conceal and postpone the agonies of incipient stagnation and the growing cleavage between affluence and despair. But these problems have been aggravated, and there are signs that the day of reckoning approaches.

The Bell Report of 1962 penetrated to the basic ambiguity of economic developments which, it said, "have inevitably blurred the traditional dividing lines between the public and private sectors of our nation. A number of profound questions affecting the structure of our society are raised by our inability to apply the classical distinctions between what is public and what is private." [1] The patents imbroglio affords a glimpse into the process, raising a few of the larger questions. Similarly, the controversies surrounding the Communications Satellite Corporation (Comsat) and the program to develop a supersonic air transport (SST) afford further insights.

Comsat: Communication and Cooperation

The Communications Satellite Corporation has been hailed as the first step of private initiative and risk capital into space, the newest realm of economic growth and commercial potential: "The greatest success . . . will not be achieved without an independent industrial base which is both unafraid to risk company funds and largely uninhibited by government . . ." declared the House Space Committee. There is a growing feeling in the business community that "space constitutes a new frontier for American industry. Direct and indirect payoffs seem certain to follow in terms of new technical knowledge, and new products and services." This, the committee allowed, "to be consistent with American traditions is a job for private enterprise . . ." By leading the world in the commercial utilization of space, the nation will demonstrate its desire "to develop peaceful applications of space which will benefit all mankind." Since dramatic advances in technology "are being equated . . . to qualities of competing political systems," a communications satellite "would contribute heavily to American technological prestige." [2]

Congress in 1962 moved to implement this vision because, in the words of one supporter, "private enterprise would have more flexibility, a greater speed, more initiative, greater risk-taking" than would the government. Rather than spending public funds, another legislator explained, "we would be getting the benefits of these tax monies . . . no government subsidy is proposed . . . nor is it needed." The chairman of the Committee on Interstate and Foreign Commerce told the House: "So far as the research is concerned, that has been completed. We know that the research and development has by this time reached the stage that it can become operational." [3]

A year later congressmen had changed their tone, questioning the continued growth of the NASA communications satellite budget. In the House: "About 90 percent of the $51 million proposed for fiscal 1964 would be placed on contract with industry. Cannot the corporation place the same contracts and fund the work?" [4] In the Senate:

"Somebody has got to have an answer as to whether . . . we are going to continue to spend hundreds of millions of dollars in order to make this a profitable venture for the people who do not have to invest much of their own money." [5]

This juxtaposition of moods within a single year symbolizes the ambiguity underlying the formation of a creature which is neither wholly private nor public, but a hybrid like the new braintrusters. It is privately owned and controlled in the interest of returning an early profit to its shareholders, but it is subsidized by the taxpayer whose representatives have no direct authority over it. Yet they have committed American prestige to its success and cannot allow it to fail either financially or technologically.

To this private monopoly have been delegated important foreign policy objectives and diplomatic powers needed to establish American world leadership and develop the mechanisms of international cooperation for space communications. The national commitment guarantees that (regardless of economics) Comsat's future earnings will be assured as needed through a variety of indirect subsidies, including the funding of research and development, minimal charge for NASA launchings, and an advance commitment that government will subscribe to approximately 20 percent of Comsat's bulk traffic at commercial rates. Further and more perplexing, Comsat would be permitted to bid on government contracts itself (most of which it will subcontract to the same companies with which government already contracts), creating a troublesome circularity and converting the corporation, like so many other hybrids of the Contract State, into a public broker operating for private profit. By vesting Comsat with a variety of public and international responsibilities, the U.S. government in effect has granted the private corporation a species of diplomatic immunity from the laws of economics.

Comsat's origin, charter, and mode of operation provide a perfect instance of the novel questions raised by the economics of ambiguity. The record of its brief history demonstrates the social and political costs of the tendency to cling to the myths of private enterprise in an environment where they are not only inappropriate but a cloak for gross improprieties which corrupt the public forum and befuddle the public purpose.

Comsat has not only made international cooperation in space more difficult but has generated a number of domestic conflicts. More and more it relies on public subsidy and gives priority to its search for government contracts. In place of U.S. leadership and international collaboration, the world faces the prospect of three or four competing space communications systems which could transform the promise of cheap and universal dialogue into a new test of political loyalties and bloc cleavages.

The background. The military in the mid-1950's was first in the communications satellite field (with Project Advent, ultimately canceled), well before explicit authorization had been granted by the administration. In 1958, on the advice of PSAC, President Eisenhower formally approved a modest civilian developmental program under NASA and granted official status to a military space communications requirement. The Army moved to upstage the Air Force, orbiting an active repeater satellite during Christmas week of the same year. Known as Project Score (scoring against the Air Force?), the satellite was small, primitive, and powered for short life, but it succeeded in transmitting the President's Christmas message around the world.

During the next three years a variety of developments and motives came to a head in the heated debates that were to lead to Comsat's charter in late 1962. Apart from the issue of prestige (intensified by a succession of Soviet space achievements), a military incentive existed for space communications. From 1961 on, the Air Force threw its weight behind a proposed medium-altitude random-orbit system which it estimated could be deployed before 1965. The military requirement was based on the need for reliable world-wide communications able to survive a nuclear attack and maintain control of the Strategic Air Command, the nation's primary nuclear deterrent.

Motivation came also from the private communications carriers, especially AT&T, which saw the need for inevitable expansion of international communication channels and were convinced that satellites would be economically competitive and offer operational advantages besides. Cable traffic had been growing about 20 percent per year, and AT&T anticipated that the existing 350 two-way voice circuits between North America and Europe would be saturated by 1965; 1970 needs were projected at 1,400 circuits, and would quadruple again by 1980. Satellite systems were expected to provide the additional capacity more cheaply and more flexibly than underwater lines. Cables are commercially attractive only as point-to-point links and only over heavy traffic routes, are expensive to install and maintain, and are limited to carrying a narrow band of frequencies. Satellites promise coverage of all parts of the globe, cheaper maintenance and replacement costs, and almost unlimited coverage of the frequency bands, offering large numbers of voice or data transmission channels and the vastly broader bands required for television transmission. In spite of many technical unknowns (the number and lifetime of satellites, the actual rates of traffic, etc.), enough was known to create industry confidence that economic feasibility lay just around the corner.

These calculations made it appear obvious that the government monopoly in space research would soon require modification. Otherwise government would find itself involved in a commercial business, serving private users and common carriers in competition with privately owned cables. Pressed by the military, NASA, and the communications industry, the Kennedy administration resolved to divest government of this unwanted role by creating some kind of regulated public utility.

The issue was precipitated for JFK in an urgent form by a maneuver arranged by AT&T, NASA, and the Air Force in the last month of the Eisenhower administration (December 1960), as if the parties sought to confront the incoming administration with a *fait accompli*. The maneuver, if unchallenged, would have permitted the Air Force to circumvent both Congress and the Department of Defense while AT&T absorbed space communications into its already vast monopolistic empire. Pushing for a low-altitude random-orbit satellite system by 1964-1965 (when a reduced level of sunspot activity was expected to cut down normal communication channels some 50 to 60 percent), the Air Force solicited AT&T's aid. The latter turned to NASA and proposed launching forty to fifty such satellites on the basis of existing technology. As forerunners of the system, NASA and AT&T signed an agreement in July 1961 for development and testing of two to four active repeater satellites which AT&T proposed to design and build at its own expense, reimbursing NASA for the launching. The first launching (Telstar) was scheduled for April 1962.[6]

Word of the early stages of the maneuver immediately stirred the Defense Department, the Army, and the whole communications industry (General Electric, RCA, Western Union, etc.). The issue was joined, forcing Kennedy action and leading to the compromise which created the new corporation, a Defense Department decision to create a government-owned military system, and a persistence of controversy surrounding both programs.

AT&T's rivals were quick to point out the inferiority of the low-altitude random-orbit proposal, both technically and economically, as a long-range system. AT&T defended the Air Force plan on the ground that a high-orbit synchronous system would be less applicable for telephone service because of the six-tenths of a second delay in round-trip transmission. The *démarche* was blocked by the clamor of the communications industry, but it was obvious to all that radio frequencies had to be managed as a scarce public resource and that there existed a valid public interest in an operational system. Therefore the contestants realized that some kind of compromise arrangement was unavoidable. In the face of this recognition, efforts were bent toward shaping the new organization so as to maximize individual roles: the weaker companies concerted their influence to prevent a dominant AT&T role while they asserted for all a common interest in profits.

In early 1961 the clamor was out in the open, and the new President instructed the Federal Communications Commission to make a formal study of the questions of ownership and operation of the venture. Twelve firms expressed views in the FCC inquiry, all accepting the inevitability of a joint plan but disagreeing vigorously on its composition: AT&T recommended restricting participation to the international carriers, a suggestion which the national carriers (led by GE, Lockheed, and Western Union) rejected. The equipment manufacturers demanded the right to participate as well, but the carriers joined ranks to resist this notion.[7]

To strengthen its hand, GE announced in April 1961 formation of an independent company (assets $1 million) to be known as "Communications Satellites, Inc." This availed them little. One month later the FCC announced a tentative decision to limit the undertaking to international common carriers alone, calling upon them to establish an ad hoc committee to work out a plan for organizing and operating the venture.[8] The situation grew confused when the Justice Department objected to the FCC's ruling on the ground that the international carriers would shift the cost of the space system to domestic telephone users whether they used the satellites or not. The government's attorneys charged that the composition of the ad hoc committee raised anti-trust issues and might complicate the international aspects of space cooperation. A meeting between the ad hoc group, NASA, the FCC, the Justice Department, and the State Department failed to establish consensus and left the administration's position in a shambles, shifting considerable initiative to Congress.

The ad hoc committee then drafted a plan for a "non-profit" corporation which would be the carriers' common carrier, servicing them all at cost, permitting each of the international carriers to charge fees that would enable them to retain profits from Comsat operations in proportion to the volume of their use of the joint facilities. Under this proposal the government would be represented on the board of directors. The corporation itself would not own or operate the ground stations, which would be owned by existing carriers, but would merely plan and build the satellites and coordinate their use. The plan was embodied in a draft bill submitted in the Senate in January 1962 by Senator Robert S. Kerr, chairman of the Space Committee. President Kennedy sent down an administration bill embodying much of the same proposal but strengthening the provisions for government regulation of international problems, guaranteeing the FCC regulatory powers over the company, and building-in devices to insure competition and participation by all industry members in all Comsat activities.

Senator Estes Kefauver led a bloc of ten Democratic senators against the bill which they charged was a giveaway of public resources to private monopoly. The effort managed to mount a four-month filibuster but failed to modify the bill's provisions and, in the hot days of August, finally collapsed.

Under the new law Comsat was created as a private, profit-making corporation with a monopoly in the U.S. use of space for international communications. The communications industry was guaranteed (and limited to) 50 percent of the voting stock, while the rest would be offered to the public in small lots "to insure the widest distribution." [9] The sale of stock was to be arranged by a board of incorporators appointed by the President, after which a fifteen-man board would be chosen, six elected by the carriers, six elected by public stockholders, and three appointed by the President. Congress struck from the administration bill language authorizing government officials to inspect books and records of the corporation on the ground that the government's interests were adequately protected by its three seats on the board and by the normal regulatory powers of the FCC.[10] In order to prevent undue AT&T influence, individual companies, regardless of stock holdings, were forbidden to vote (either directly or indirectly) for more than three candidates of the six to be elected by the industry.

The presidentially appointed incorporators opened the sale of stock ($200 million) in July 1964, and in a hectic first meeting of shareholders in September a regular board was constituted, half elected by the carriers (who owned half the stock) and half by the public who subscribed to the balance. For the carriers—AT&T, owning 58 percent of the carriers' stock, elected three of its representatives to the board; IT&T, with 21 percent, elected two; the smaller companies combined to elect a nominee of the Hawaiian Telephone Company. Of the remaining carriers' share, General Telephone bought 7 percent, RCA 5 percent, while others purchased the remaining 9 percent. The four largest carriers thus emerged with 90.9 percent of the industry holdings and 45.5 percent of the total stock issue. For the public—a group of original incorporators led by Leo D. Welch (former head of Standard Oil of New Jersey and interim board chairman) collected enough proxies to elect themselves to the six public seats. The public share was distributed among an estimated 190,000 holders in small lots (less than a hundred shares each), potentially assuring the carriers a dominant role. Welch was named chairman, and Joseph V. Charyk (also a board member) company president. This election precipitated great disorder on the floor of the first stockholder meeting, as small holders charged "an inside job," forced a recount, and later filed suit to depose the "public members" of the new board. President Johnson filled the government's three seats with Frederick Donner, General Motors chairman, George Meany, AFL-CIO president, and Clarke Kerr, president of the University of California.

Congress continued to be perturbed by the ambiguous nature of its handiwork, unable to comprehend why budget requests for NASA and Defense Department space communications continued an upward trend, representing a hidden subsidy to a company which had been expected to put this activity on a strictly commercial basis. The vigor with which Comsat sought a wide variety of government contracts suggested that the government would continue to pump-prime the operation, having by its creation merely awarded unearned private equities and dividends, increasing rather than diminishing the burden on the taxpayer. Legislators were disturbed by the high salaries Comsat paid its officers: the board chairman received an annual salary ($125,000) greater than the President of the United States, while the second ranking officer was paid ($80,000) more than twice as much as the U.S. Vice President. Senator Symington queried a Comsat official: "Could you give an example of a similar job in the communications industry, a company starting out with uncertainty as to its future profits, where anything comparable to these salaries has ever been paid before . . . ?" The official replied: "I hope this corporation will prove to be a gold mine, a real gold mine." [11]

A child of the economics of ambiguity, Comsat was born and raised in controversy. The extraordinary market surges of its stock brought charges of "manipulation" and undue "insider" profit-taking. The company became deeply involved in the fight within government over plans for a military communications system, and disturbed some members of Congress by plans to hasten profits by contracting to provide services for NASA and the Defense Department on a large scale immediately, while phasing in commercial activity gradually. It early showed a tendency to strike out on its own as a corporate body, bypassing and antagonizing the large common carriers by seeking and obtaining an FCC ruling to permit it to build and own its own ground terminals, clashing head-on with the major television networks in its efforts to extend its international monopoly into the area of domestic video transmission. Its rate schedule was attacked by all potential users as inordinately high, and pressure was brought on Congress and the FCC for a reduction. In international "cooperation," Comsat raised hackles in European countries whose government-owned systems objected to the domination of space communications by a private, profit-making American monopoly. France, Britain, and the Soviet bloc launched plans for government-to-government systems of their own. The NATO pair were perhaps bargaining for better terms from the U.S. company, but the Soviets pressed ahead with a full and ambitious system of their own, offering generous terms to the nations of Europe, Japan, and the underdeveloped world. Largely as a result of the ground-station dispute, Welch was forced to resign as chairman in 1965, but the shape of things to come remained in a state of turbulence.

The shape of domestic controversy. The glamor of space, the government commitment to company success, and manipulation by professional speculators accounted for the astounding success of Comsat's 1964 initial stock offering.

In spite of the fact that 1966 was set as the earliest target date for full commercial operations (with the possibility that the experimental "Early Bird" in late 1964 might offer limited commercial service), and the prediction that there would be no profits for at least eight years, the initial offering (at $20 per share) was quickly subscribed and began a breath-taking ascent in price, doubling and then tripling within months and continuing to climb. The carrier's share, which was oversubscribed, thus netted the buyers more than a 300 percent paper profit while the corporation had hardly begun to operate.

In September 1964 Comsat trading moved from over-the-counter to the New York Stock Exchange, where speculation and short-selling by brokers and radically fluctuating prices might be better controlled. But the volatile rise continued anyway, reaching more than three times par in early December. The exchange was forced on several occasions to halt trading, finally raising the margin requirement from 70 to 100 percent in an attempt to stabilize the situation. This move succeeded in leveling the rate of trading at prices between 42 and 66 a share (par 20), leading in mid-1965 to a reduction of the margin requirement once again.

The second shareholders' meeting (May 1965) once more was lacking in decorum, degenerating into a raucous shouting match during which at least two stockholders were ejected bodily after the board handily re-elected itself and all the officers.

We have seen that the politics of service ambitions were deeply involved in the controversies from which Comsat emerged. Defense Department in-fighting continued to play an important role, with both McNamara and the service branches, primarily the Air Force, seeking to use Comsat to promote their own ends. With the corporation launched, AT&T pushed the low-altitude random-orbit satellite system in which the Air Force continued to show interest on the assumption that it might achieve operational status several years in advance of the cheaper, more efficient, high-altitude synchronous orbit system. AT&T and the Air Force both intimated that a Comsat commitment to the low-orbit plan would help win a major Defense Department contract. However, in the meantime, a synchronous orbit (23,500-mile) launch vehicle was being perfected. Comsat officers resisted AT&T and Air Force pressure, reluctant to commit the company's entire operation to an expensive system which might be on the verge of obsolescence. The situation thereby worked for McNamara and against the Air Force.

McNamara's role was expanded when President Johnson designated him to head a National Communications System Committee which sought a merging of all the nation's security communications, including those of State, NASA, and Defense. Originally cool to the Air Force–AT&T plan as premature, and seeking to buy time for other technological options to become available, McNamara agreed to urgings that the Defense Department explore the use of Comsat services for its own global communications system. The Air Force, on the other hand (aware of the declining fortunes of AT&T's proposal), switched in the opposite direction, insisting that the military required an entirely separate network for various technical and security reasons. It now demanded immediate funding for twenty-four to thirty satellites in low random orbits. McNamara hedged by agreeing to include the item in the fiscal 1965 budget, declaring, however, that no contracts would be awarded until discussions with Comsat were complete.[12] McNamara could depend on the support of influential congressmen personally committed to Comsat success and to the conviction that government should avoid competitive undertakings. In addition, it was becoming clearer that the synchronous system would soon achieve technical feasibility.

By negotiating with Comsat McNamara gained time during which new technology was proven and a wiser long-range military commitment funded. By mid-1964 the Defense Department/Comsat discussions were at an impasse. It was becoming clear that Comsat's commercial requirements were divergent from a security system that required national control, broader geographical coverage, anti-jamming frequencies, and other features which if developed by the corporation would double government's cost. Such security use of Comsat equipment would pose serious obstacles to the international arrangements Comsat was seeking for commercial cooperation, and would restrict the kind of military control essential to strategic requirements.

While McNamara was buying time he had not eliminated the possibility of a low-orbit random system (but under Defense Department control), or a mix of this with intermediate-altitude and synchronous satellites in order to achieve maximum coverage and reliability. What he wanted was a full evaluation of all the options in terms of cost/effectiveness. He never accepted the impetuous Air Force demand for an operational system, any operational system, by 1965. During this period, the Titan IIIC booster reached a stage of development which encouraged the Defense Department to consider multiple launchings of its own communications satellites into intermediate-altitude random orbits which required fewer satellites for global coverage and were expected to cost considerably less than the system advocated by the Air Force.[13] Accordingly, the Air Force was at last directed in 1964 to proceed with the development of such a system. The House Military Operations Subcommittee under Representative Holifield took exception to this plan, urging instead a mix of synchronous equatorial and intermediate-altitude polar orbits, rapping the Defense Department's procrastination.[14]

As the Defense Department moved ahead, Comsat energetically looked for ways to win government business, proposing to build a completely separate satellite system under contract, or to act as service manager for a system built for the Defense Department by other companies. Impelled by congressional urging, discussions were resumed between the Defense Department and Comsat. The latter went ahead, awarding a $100,000 contract to Hughes Aircraft for preliminary design and engineering, looking toward procurement of twenty-four synchronous-orbit satellites which it proposed to furnish government agencies. The Defense Department demurred, but a ready customer appeared in the person of James Webb.

NASA was under pressure from Congress to reduce its own space communication programs. Its Syncom satellites (synchronous orbit), including telemetry and command stations, were transferred to the Defense Department in 1965. Now it appeared NASA had a requirement for a space communications system to support the Apollo moon program. This was a blessing to Comsat which thereby stood to gain a substantial subsidy for creating its commercial system. NASA indicated interest in contracting with Comsat for five synchronous satellites, as well as earth terminals, for a communications system for space exploration, particularly for Apollo and follow-on programs. Comsat sought bids on permanent ground stations in Hawaii and Washington state, which would become not only links in its commercial system but would provide spacecraft tracking and telemetry capabilities for NASA.

The Defense Department assured Congress that any excess capacity of its space communications satellites would not be used for normal military traffic currently handled by the commercial carriers and constituting a substantial portion of their business (20-30 percent): "I wish to emphasize," Secretary Vance told the Senate Space Committee, "that . . . it is and will continue to be our policy to make use of commercial communications facilities for the transmission of the bulk of our traffic." [15] As though AT&T needs a government subsidy of its already bloated profits. This curious leitmotif reveals something about Congress and the American economy that should excite somebody's interest.

Controversy has embroiled the communications industry over Comsat's desire to own its ground stations, to control domestic satellite transmissions, and to deal directly with all users instead of leasing all its facilities to the international carriers.

All planned systems require sensitive and powerful terminals to send and receive via the satellite relay. Terminal ownership holds the key to the future role of Comsat and the carriers in international and domestic competition. If each carrier owns terminals, Comsat would be denied an independent role and limited to designing and building satellites and coordinating their use. It would be denied opportunity to acquire equity in the system and would not be permitted growth, diversification, and profits as an independent center of corporate power. Such an arrangement would be in keeping with the original concept of the entity as the carriers' carrier, nothing more.

Comsat's demand for ownership and control of ground stations is defended on technical grounds as necessary for efficiency, as the only means by which the corporation can conserve and control demands made by users upon its satellites. RCA vigorously objects, charging that a heavy Comsat financial commitment in ground stations would tend to freeze technology, create a monopoly where none is needed, raise the cost of the service, and reduce American carriers to "second-class citizenship," since terminals in foreign countries will have diverse ownership: "We believe American carriers are entitled to treatment no less favorable." Comsat would cease to be the "carriers' carrier" and pre-empt the role of present carriers in offering communications services directly to the ultimate user, according to RCA spokesmen.

The issue erupted into the open when in August 1964 Comsat asked the FCC for permission to acquire exclusive ownership and control of ground stations. Five common carriers, a union representing communications employees, and an association of independent telephone companies filed statements asking a negative ruling. AT&T supported the Comsat position on temporary terminal operation but stipulated that ownership should be on a fifty-fifty basis with the carriers, and that Comsat operating rights should not extend beyond the initial three or four stations.

It was charged that Comsat ownership would not only by-pass established carriers but create a serious conflict of interest—Comsat would be in a position to give priority to its own business over that of other carriers. The objectors derided Comsat's use of the term "initial" with respect to the stations, arguing that the arrangement once established would become permanent. But it appeared in mid-1965 that the bumptious young corporation had won its point. An FCC vote (5 to 2) approved its request, but the decision was qualified as an "interim policy limited to the critical early years" and had as its purpose to expedite construction of three stations. Inclined to minimize the qualification, Comsat opened negotiations with AT&T to purchase the latter's operational station at Andover, Maine. All the common carriers registered objections to the ruling, which IT&T described as "a devastating blow." [16]

At this writing, the ultimate outcome of the terminals issue cannot be gauged, but the 1965 resignation of Comsat Chairman Welch (the corporation's driving architect) suggests forces at work behind the scenes to reverse designs for independence and to reassure the carriers.

In the founding law, Comsat's role in domestic transmission was left vague because of the prevailing thought that no major cost or technical advantages were to be gained in this area by abandoning existing long-lines and microwave relays. In 1965 the cost experience of Early Bird astounded the experts. David Sarnoff declared that even direct transmission of television from satellite to home sets was moving into the realm of economic feasibility. The American Broadcasting Company notified the FCC of plans for a domestic system to relay network television to affiliated stations all over the country. NBC and CBS indicated interest in the plan which, by eliminating dependence on telephone lines, would mean a substantial savings on the $50 million annual leasing costs which the networks paid AT&T. ABC spokesmen said they were convinced of the economic practicality for each network to own a satellite for domestic distribution, but they agreed to consider a joint plan with the other networks. United Press International joined the trend, announcing that it would seek its own system for transmission of telephotos, facsimile, and Teletype messages.

In its filing with the Federal Communications Commission, ABC left no doubt as to why it wanted to become a private carrier, owning both the satellite and ground stations:

> In this space age there is no reason that ABC and its affiliates should be restricted to mountain peak and tall-tower techniques of yore; they should be permitted to take advantage of space-age developments, such as synchronous stationary satellites, the utility and dependability of which have now been amply demonstrated. Neither AT&T nor Comsat has been, or should be, permitted a monopoly on the use of satellites for domestic purposes.[17]

Comsat parried by issuing a statement that "establishment of a communications satellite system for commercial purposes was entrusted to the corporation by law," and that no dividing line existed between international and domestic transmissions. It warned that domestic frequencies were heavily congested and interference problems would be greatly worsened by separate satellite systems. Comsat spokesmen questioned the economic feasibility of such use, pointing out that AT&T provided coast-to-coast TV transmission at about $300 per hour compared with the rate from New York to Europe via Early

Bird set at about twenty times as much. But Comsat also promised to pursue the creation of a domestic system for television and high-speed aircraft communications.

At the same time, the FCC was the center of a new controversy over the rate schedule proposed for transatlantic use of Early Bird: $6,700 an hour for TV and $4,200 a month for each telephone voice channel, available only to common carriers who in turn would retail the time to their customers at an even higher price. NBC announced it could make no use of Early Bird at these rates; CBS and ABC said they would rely on air shipments of film and tape, except for possible special programs and urgent news. The CBS statement declared that the cost structure proposed for the use of Early Bird "militates against its use on a routine basis." AT&T accepted the telephone rate and indicated its intention to lease one hundred voice channels.

The FCC granted permission for the start of commercial operations but denied the proposed rates, ordering Comsat to hold all revenues in a special fund pending the outcome of further deliberations. A month later (July 1965) the FCC ruled that Comsat could not deal directly with the television networks but, pending final resolution, that the traditional international carriers take turns in acting as middlemen (to neutralize any competitive advantage) until the matter be ultimately resolved. At the same time Comsat issued a new rate schedule about 15 percent lower, allowing users greater flexibility for transmissions of less than thirty minutes.

Eruption of international disputes. The extravagant claims for the contribution of space communications to international understanding have been confounded by the fact that Comsat has been a new source of festering division, suspicion, and conflict among Western nations, and has estranged the Eastern bloc.

The American undertaking began as a unilateral one which deliberately demurred at international planning on the ground that the Soviets would use such delay to establish priority in the field. "The strategic value of controlling such a potentially powerful mass communications medium," the Senate report declared, "could not have escaped the scientific and technological planners in the USSR. . . . Extended discussion of our efforts . . . may deter us from moving fast enough to make an outstanding achievement in communications satellites. . . . Agreement on certain international aspects . . . is secondary and gives the advantage to the Soviets." [18]

As a result, Comsat's plans ran into determined opposition throughout the world, and the Russians used the satellite issue as a symbol of U.S. "imperialism." Western European governments refused to acquiesce in the American proposal and declined to negotiate a string of bilateral agreements; instead they formed a union of telecommunications authorities and insisted on bargaining as a united front. They sharply rejected an American proposal that their participation be restricted to ownership and operation of ground stations, and insisted upon a share of ownership and control of the whole system, based on an agreement between governments rather than between private business entities. The negotiations climaxed in August 1964 with the adoption of provisional arrangements satisfactory neither to Comsat nor to anyone else.

Europeans viewed the formation of the private corporation with misgivings from the start. In late 1962 a movement was initiated to launch an independent European system; at the same time, however, European governments sought clarification of U.S. intentions, going ahead with their own plans as a means of enhancing their bargaining posture. In March 1963 the British government announced agreement among twelve British companies to pursue an independent plan in which other countries were invited to participate. The official "dismay" of the State Department and its cable requesting further information signaled the beginning of diplomatic bargaining and the end of the period of U.S. unilateral action.

Julian Amery, British Minister of Aviation, indicated that his government had ordered design studies for satellites to be placed in orbit by the European launcher development organization, declaring he was hopeful that the European system would be established independently of America's: "No one country should have a monopoly of space." [19] The United Nations Committee on the Peaceful Uses of Outer Space (meeting in early 1963) brought home to the State Department the fact that Comsat was causing America to lose rather than win friends. Both neutrals and allied delegates were moving behind Soviet proposals to exclude private companies from space activities, threatening to isolate the U.S. even from its friends. The Soviets argued that private ownership of spacecraft compromised the principle of the liability of states for space activity; to combat this charge, the U.S. government agreed to accept liability for Comsat (thereby upsetting one of the arguments that had been used in promoting Comsat in Congress). But even this act failed to assuage the

Russians, who charged the time would come when the U.S. would deliver a lecture on free enterprise and claim it had no control over the activities of a "United Fruit Company in outer space." [20]

Serious negotiations commenced with NATO allies, and the U.S. sought to extricate itself from embarrassment by working out a formula to resolve European unease. The State Department found itself in a curious situation in which it acted in behalf of the private corporation as well as the American people, and in which the corporation held important veto powers. The officers of Comsat were also placed in the highly delicate position of negotiating to protect the interests of their company as well as those of their nation, interests not necessarily congruent. The Europeans made no bones about their willingness to build a competing system unless granted a substantial role in the management and control of Comsat. Forced to abandon its insistence on bilateral arrangements, Board Chairman Welch and President Charyk opened negotiations with the All-European Conference of Telecommunications in late 1963, trying to persuade it to join the U.S. program; without such support Comsat plainly faced a dismal future. American emphasis on prestige had in fact backfired; now all the developed nations were sensitive about their political prestige as well as their economic self-interest. They could not afford to boost America's image at the expense of their own. Herein lies the self-defeating nature of a search for the empty abstraction of prestige.

There appeared to be obvious friction between Comsat and the State Department as to who should negotiate what, leading perhaps to administration second thoughts about the wisdom of the 1962 law. In fact, the State Department had not shown much enthusiasm for Comsat from the beginning. At the start of discussions the State Department dissociated itself, taking the position that it would not interfere with Comsat so long as no moves were made in violent conflict with U.S. foreign policy. But as talks proceeded the department was forced to bring pressures to bear in behalf of compromise with the nation's allies. Early in the negotiations the U.S. had flatly refused any official exchange with European countries of technical information regarding communications satellites and launch vehicles (in order to avoid contributing to the strategic missile and space ambitions of some participants). This policy contained the implication, repugnant to Europeans, that they might have to depend on American producers for hardware if they joined the arrangement.

The search for a formula proved to be tortuous, but the American corporation held some strong cards: its priority in the field, the incomparable space commitments and capability of the United States, the difficulty of maintaining a European united front, the fact that all the European nations were already adjusted to private American-owned monopolies in conventional telephonic and telegraphic communications.

Early in 1964 Britain reversed position, declaring itself in favor of joining the American plan, offering to purchase a 10 percent interest in Comsat and urging similar action by all European nations. (The offer was later raised to 20 percent or $40 million.) In subsequent weeks most of the governments of Europe imitated this move, insisting that Comsat permit them to buy in and thereby gain a share in policy decisions and management. The American corporation was taken aback. In spite of having made this suggestion at an earlier negotiating stage, Comsat now resisted the move, announcing that further investments were not needed. Spokesmen warned against exaggerating the profit potential, claiming that European zeal to invest was out of proportion to anticipated traffic and revenues. European demands for a share of the company's space hardware contracts were turned aside with statements that competitive bids would be welcome but that a "political" allocation could not be entertained because of the "commercial nature of the enterprise." (Comsat was to collide with the FCC because of its practice of awarding contracts to American producers without competitive bidding.) The Europeans adopted a new tack, proposing that each nation own (either publicly or privately) the ground terminals on its own territory, but that control of the satellite system itself be internationalized "in the least necessary time." [21] This was accepted "in principle" by U.S. negotiators who stipulated that the issue be deferred.

On this note the preliminary talks closed—inconclusively, without signs that a real basis for agreement could be found. The sixteen European and Commonwealth nations renewed threats to launch an independent system while the U.S. government publicly painted a rosy picture of progress. Some European spokesmen let it be known they would not support "another NATO—nominally international but in fact . . . run mostly as the U.S. sees fit." [22]

Behind the scenes, however, real gut negotiations were intensified on an informal basis. The U.S. was determined that the resumed talks scheduled for Washington in mid-summer (1964) not be allowed to fail. The State Department moved decisively into the picture, making it clear that some kind of consortium of governments *and* private agencies would be established to resolve the impasse, with interim apportionment of ownership and control in accordance with anticipated use by each of the participating nations. The U.S. proposed that an interim plan be effective at least until 1969, and that an interim international committee be established with weighted voting according to anticipated use. As the major user, the United States would have 61 percent of the votes, but, to overcome European fears, an "effective majority" requiring the concurrence of at least two other nations would be acceptable. The main lines of this proposal were agreeable to the Europeans, but they insisted that the "effective majority" require more than two concurring votes.

In August the voting issue was finally resolved and an agreement signed. Two governing groups were recognized. Comsat would act as the official U.S. representative and as general manager for design, development, construction, operation, and maintenance of the space portion of the system. An Interim Communication Satellite Committee was established, composed of all member nations whose estimated usage exceeded 1.5 percent. On the international committee each nation would have a vote equal to its quota. Comsat was granted a preponderant role, initially set at 61 percent but dropping, as other nations invested, to a fixed minimum of at least 50.6 percent. To prevent complete American domination, the pact defined a quorum as the voting power of the largest member plus 8.5 percent of the rest (in practice this meant that Comsat, the British, and one other representative could hold meetings); and fourteen areas were enumerated in which the body could not act without an "effective majority," that is, equal to the votes of the largest member plus 12.5 percent. The U.S. thereby was subject to a veto (if all other members joined together to withhold the necessary 12.5 percent voting power) in such areas as setting rates, access or withdrawal of members, and choice of technical designs. The pact provided that the international committee would contract its work among the member nations in proportion to each member's voting quota whenever "proposals are determined to be comparable in terms of quality and price." All contracts were to be made on the basis of open and competitive bids. Ground stations would be nationally owned but subject to jurisdiction and certification of the International Committee in order to control their locations and technical characteristics, and to insure "equitable and non-discriminatory" access for all signatories.

Everyone was unhappy with the agreement. The Russians denounced it as containing doctrines directly opposed to the U.N. Declaration of

Principles (governing activities in outer space). The Soviet delegate to the International Telecommunications Union declared that the AT&T monopoly really controlled Comsat, and Comsat controlled the International Committee which was therefore "simply a private company with participation of foreign capital." [23] Another Soviet delegate at an international space law conference called upon the United Nations to organize a truly international communications system "on the basis of equality."

Both Russia and France went ahead with plans for independent space networks. De Gaulle sought to continue the project proposed earlier by the All-Europe group, but now found France's efforts unaided—in spite of which he continued a long-term development program which orbited its first relay in November 1965. The Soviet program is close behind U.S. efforts. In addition to a network of fifteen or more low-level active repeater satellites for domestic transmissions, the Russians orbited a second multi-channel high-altitude device in late 1965, intended, according to Tass, to "verify the possibility of organizing a communications system." [24] A Soviet-French accord was reached under which the Russians in December 1965 beamed a seventy-five-minute color television program from Moscow to Paris.

Among the signatories, meanwhile, there was no relief from clashes. The successful launching of Early Bird (May 1965) was marked by transmission to Europe of a presidential address on U.S. NATO policy, aimed at countering separatist views on the continent and undercutting French dissidence. No advance arrangements were made at the European end and several ministries protested the breach in protocol. The publication of Comsat's rate schedule caused a reaction similar to that of U.S. television networks. The European Broadcasting Union termed the charges "prohibitive," and the British Broadcasting Corporation filed a protest with the FCC, urging that the fees be cut in half.

Space communications service via Early Bird was formally inaugurated in June 1965 with ceremonial speeches by six chiefs of state. There was unintentional irony in President Johnson's declaration: "In these times, the choice of mankind is a clear choice between cooperation and catastrophe. Cooperation begins in the better understanding that better communications bring." [25]

The resignation in mid-1965 of Leo D. Welch as chief Comsat officer appeared to mark the end of the company's drive for an independent corporate role and acceptance of the passive role marked out for it by the international carriers. To his place was named James McCormack, highly respected MIT vice president and retired (in 1955) two-star Air Force general, whose experience and credentials make him an integral part of the Contract State. In the last years of military service he was the Air Force R&D director. Upon moving to MIT he took charge of the school's Lincoln and Instrumentation laboratories, two creatures of government contracting closely associated with the Air Force and the work of its industrial contractors. He was a key organizer of the Institute of Defense Analysis, serving as its president for three years (1955-1958). He was also a director of two Air Force-sponsored non-profit braintrusts, MITRE and Aerospace, until his assignment to Comsat. In addition he sat on the boards of Eastern Airlines, Bulova Watch Company, Boston's Federal Reserve Bank, and the New England Telephone and Telegraph Company, an AT&T subsidiary. Though certainly an honest and accomplished man (he had to hold the high esteem of many people to win these positions), his profile beautifully reveals the interlocking nature of the Air Force, its major contractors, key braintrusters of both his parent service and the Department of Defense, and the banking community. As a man closely associated with both AT&T and the Air Force, it will be surprising indeed if Comsat under his leadership incurs their displeasure.

NASA announced in late 1965 that it would provide $6 million over a three-month period for "initial" work on the synchronous satellite system which Comsat would contract to Hughes Aircraft. In September 1966 Comsat expected to have in operation for NASA a tracking and data acquisition link between ships around the world and NASA's control center in Texas for the Apollo moon landing and subsequent space explorations. Comsat was also expected to supply portable ground stations for the vessels. As this initial NASA contract grows, Comsat can contemplate a secure future, early earnings, and a substantial subsidy for its commercial operations.

B-70 and SST

In the six decades since Kitty Hawk, technology has moved to ever larger and faster aircraft: since World War II it has reached to the speed of sound (for commercial carriers) and substantially beyond (for military aircraft). In many respects military flight reached its apogee in World War II and, in spite of technical advances, has fallen into decline as a defensive or offensive weapon. It has preserved a role only in tactical and limited warfare and as a shuttle service for troops and equipment. This is a paradox: as technical capabilities for range and speed become almost infinite, the major military investment shifts to aircraft with pre-World War II characteristics and to slow-moving, short-range helicopters. The decline of strategic aircraft is the source of current pressures to boost commercial air transport to the full limits of technology, raising the complex issues surrounding the supersonic transport.

Traditionally, commercial flight technology was the beneficiary of the national investment in military R&D. Both the military and the National Advisory Committee on Aeronautics (NASA's predecessor) had aircraft development programs, most of them conducted in-house (Langley Research Center, and others), some of them by the airframe manufacturers under contracts that were primarily for design and serial production rather than R&D. This system represented a sizable subsidy to the private airlines (which enjoyed other help as well through mail contracts, public funding for airport construction and maintenance, etc.) but was a legitimate spin-off from military requirements. The low funding levels of prewar years, substantial government in-house facilities, the general use of real competitive bidding by airframe companies, and technical management by government agencies tended to impose a reality principle upon the technological development of commercial transport, protecting the public purse and keeping the national investment in balance with other public needs.

The phasing-out of strategic aircraft accompanied the rise of missilery (trends primarily due to the increasing efficacy of radar-controlled ground-to-air defense missiles) and has been responsible for the crisis that now faces private airlines and airframe manufacturers. Private R&D has become prohibitively expensive. And in any case, the airframe fabricators are unaccustomed to funding advanced developments for the commercial market. The phase-out of military production forces them to spread developmental costs over fewer commercial units, raising costs to the already subsidy-dependent airlines and tending to freeze technological "progress." Regardless of the question as to what constitutes "progress," the problem is real. As military air power has been phased out, the federal government has sought substitute means for maintaining the financial health, competitiveness, and international standing of the private carriers, augmenting R&D funding (under NASA), grants-in-aid for airport modernization and construction, tax benefits, and other indirect subsidies. Rejecting the pattern of public ownership that prevails throughout the world, and reducing direct subsidies which are politically exposed and vulnerable, the U.S. has allowed the scope of hidden public aid to swell.

The movement of the industry toward supersonic flight represents a leap into enormously high costs ($8 to $10 billion, not including additional costs for modified ground facilities, traffic control, etc.), an unproven technology, and unknown parameters of public acceptance, safety, and operating efficiency. In the early 1960's, as the administration denied Air Force ambitions for a large-scale crash program for supersonic bombers (such as the B-70), President Kennedy committed the nation to the goal of developing a commercial SST, offering the industry a cost-sharing formula under which the government would fund 75 percent of the first $1 billion in R&D costs. This was done in the name of maintaining American leadership in air transportation, aiding our effort in the prestige struggle with the Russians, and overtaking independent British-French efforts in this area which might enable them to dominate the world commercial airline market, aggravate the U.S. balance-of-payments problem, and injure American companies and jobs.

Other motives were present as well, including widespread faith in the beneficence and inevitability of technological advance for its own sake, the demand of airframe manufacturers for new work and employment in existing production facilities, and (perhaps most important) the Air Force's fight to preserve a strategic role for manned aircraft, to reverse or circumvent the administration's decision against massive expenditures for development, production, and deployment of supersonic bombers.

The Air Force and the XB-70 (later RB-70, then RS-70) played a central role in the political economy of the commercial SST program, almost as if the traditional spin-off relationship were reversed. Instead of the commercial airlines benefiting from military R&D, the Air Force sought to continue large-scale funding of supersonic bombers as a spin-off from the commercial program, arguing with great conviction the importance of competing with the Russians in the arena of commercial air while in fact enlisting contractors, private airlines, and a considerable segment of Congress in what was essentially an effort to surmount the B-70 ban.

The Kennedy shared-cost proposal to the aircraft industry embodied a firm test of financial feasibility and social utility by imposing upon airline participation the harsh laws of economics. Unaccustomed to funding the development of new technology from their own revenues, the private airlines were forced to re-examine their blithe claims. So they neither put up nor shut up, rejecting the proposal and continuing a strenuous campaign to impose upon the public purse the obligation to assume *all* the costs and risks of SST, not for the sake of the airlines but for reasons of national prestige. But Kennedy was not easily deflected from the cost-sharing test, reflecting the administration's lack of conviction in the diplomatic and economic utility of a crash program. As a result the airlines threw their whole weight behind the Air Force and the manufacturers to persuade members of Congress to override attempts to bury the B-70. In the legislative halls this coalition prevailed, keeping the bomber (or its variant forms) alive with an annual ritual by which more funds were appropriated for the project than had been requested. When McNamara refused to spend the money, Congress turned to funding a larger B-70 program in the NASA budget.

By 1965 both SST and the B-70 enjoyed increasing budget allocations. Government forged ahead with the former even without a firm agreement on cost-sharing from the industry, which more than ever resisted anything more than nominal investment of its own funds. However, both programs represented a compromise. Having named the Defense Secretary (in 1964) chairman of an SST policy committee, President Johnson endorsed a phased program which allowed the target date to slip from 1967 to the mid-1970's. The Mach-3 Air Force program was held to gradual developmental steps, partly as a means of indirect support to the civilian SST, partly as an exploratory program in the development and working of rare metals, and partly to create new technical capabilities which might prove useful for space and other future military needs.

The background of decision. The mid-1963 decision to commit the nation to a stepped-up SST program climaxed a debate that had raged since 1959, when President Eisenhower first cut the B-70 budget by more than 90 percent. In the battle that ensued, Air Force General Thomas White sought to invent new missions to justify the program to Congress, joined by FAA head (and former Air Force general) E. R. Quesada and industry spokesmen, who told Congress that the B-70 was necessary for its own sake and would give the nation the civilian SST as a bonus.

After his inauguration President Kennedy was pressed hard by Air Force advocates, but he determined to build up strategic-missile and limited-war systems, ordering a 50 percent cut in the B-70 program (on which over $793 million had already been spent) and a five-year stretch-out in producing a prototype. Congress restored the funds but

McNamara refused to spend them. The Air Force was nettled and set off on a series of B-70 promotional campaigns throughout the country—increasingly couched in exhortations about the miracles that supersonic flight would bring to the private traveler.

As pressure switched to the SST, the administration leaned toward a Mach-2 (about 1,500 miles per hour at sea level) airliner favored by some of the airlines. The president of National Airlines considered the Mach-3 "usable only in transoceanic flight because of altitude requirements." Because of the twelve-mile cruising altitude required to prevent sonic damage over land, a Mach-3 (2,000 miles per hour) plane might actually make transcontinental flight longer than do slower aircraft, he said.[26] The problems of Mach-3 technology were beyond "the state of the art." Supersonic military planes suffered from short range and lack of economy and efficiency. Engineering for a Mach-2 was on firmer ground, involving few radically new materials and techniques. In terms of costs, the B-70 experience so far had not been reassuring.

The step-up from Mach-2 to -3 involves a drastic discontinuity: propulsion power is not linearly related to speed (a power increase does not proportionately increase speed), so Mach-3 would require vastly more powerful engines which would burn fuel prodigiously. Conventional materials would not endure at Mach-3 speed, so new alloys stronger than steel and lighter than aluminum would have to be developed. More power and more fuel would mean high operating costs and reduced payloads. Sonic booms would mean higher-altitude flightpaths, making flights longer and fuel consumption less efficient. Airport runways would need to be lengthened, possibly causing some airports to move still further from urban centers, increasing time lost to and from airports. At Mach-3, pilot reaction times would be too slow and minor emergencies would become major ones; the whole flight plan might need to be programed by computer, converting the plane to a kind of winged missile, raising new problems for traffic management and safety.

Because of these factors the British-French (Concorde) and the Russian SST programs had settled for Mach-2. This fact prompted arguments that the U.S. leapfrog these competing programs by reaching for Mach-3, just as the U.S. space effort was reaching for the moon. These arguments finally persuaded the President; but he also insisted on a cost-sharing basis as a hedge against contractor and industry importunities and as a test of economic realism.

Mach-3 was defended as the best way to "maintain U.S. international leadership in air transportation." [27] By leapfrogging Concorde and the Russians, the American SST (planned to be operational a year or two after the Concorde) would cause airlines of the world to wait for the "better" aircraft, thereby undermining the Concorde's market. "We estimate it would contribute in the year 1975 about $800 million worth of exports," FAA head Halaby said, "a favorable balance of payments . . . ," and would maintain employment and pre-eminence in U.S. commercial aircraft manufacturing.[28] The administration was fully aware of the effect of the cost-sharing plan, depending upon it to deter a technically premature effort: "If we thought the development costs were going to be over a billion dollars, I think we would advise either termination or postponement of the program. . . . We aren't determined that this airplane will be built at any cost. This is a commercial . . . transport, and we can't think of it as we do the space or military program." [29]

Free Enterprise flunks the test. JFK was forced to formulate an administration position on SST in 1963 because of congressional impetuosity on behalf of SST advocates,. Senator Mike Monroney, chairman of the Aviation Subcommittee of the Senate Space Committee, moved to fund the program under NASA's budget and scheduled hearings to provide a pro-SST platform.

In June the President announced his decision, committing government to commence "immediately a new program in partnership with private industry to develop at the earliest practical date" an SST prototype "superior to that being built in any other country . . . neither the economics nor the politics of international competition permit us to stand still . . ." [30] He requested an initial $60 million and spelled out the cost-sharing plan. A $750 million ceiling was set on government support, part of which was to be recovered through royalty payments when the plane became operational. Private industry would be expected to provide an additional $250 million on a 1 to 3 cash basis. An earnest fee was built into the plan, requiring each airline to deposit $100,000 (with the FAA) for each SST it planned to buy; deliveries would be made in the order of such deposits. This feature forced SST advocates to put up or shut up—providing the administration with a basis for using the cost-sharing test to hold back public funding in proportion to industry's reticence.

The Monroney hearing shaped up as a vehement industry attack on these guidelines. The transoceanic lines and the airframe manufacturers insisted that government pay the entire cost and postpone a decision on how to recapture the funds. Administration spokesmen argued that anything less than 25 percent industry participation would threaten "the private enterprise character" of the undertaking.

The hearings and the public campaign which accompanied them led the President to seek ways of placating Congress. After frequent discussions with Monroney, he won agreement for a re-study of the issue, and a panel of notables was appointed under Eugene R. Black, former World Bank president. For a time Monroney leaned toward the administration's views. He told an Airline Association conference that the proposal "must be considered final . . . I am afraid that many of our finest businessmen still hope to consider this project in the nature of a defense contract. This effort must stand alone as a civil aviation need. . . ." [31]

The Black report was submitted to Kennedy's successor during the 1963 Christmas recess: "We conclude that one of the basic philosophies of the current program, namely that of tying the United States effort to the Concorde and . . . compressing time of development and construction is dangerous, technically and economically." [32] The principle of cost-sharing was endorsed, but the panel urged that a concession be made to industry by modifying the formula to 1 to 9 (that is, reducing the industry's share from 25 percent to 10 percent), a recommendation that did not satisfy the airlines and was rejected by the President, who adhered to the original formula.

The hornets of controversy buzzed through Washington as the FAA moved ahead with a design competition to establish firmer cost estimates. The administration characterized industry complaints as a "bargaining posture" which would change when the manufacturers (North American, Lockheed, and Boeing on the airframe; General Electric, United Aircraft, and Curtiss-Wright on the power plant) were forced to submit competitive bids. The result was devastating: none of the six proposals agreed to the 1 to 3 financing plan, and most placed development costs well above $1 billion. Halaby: ". . . none of the aircraft proposed met the established range, payload, or economic criteria for a satisfactory transport." [33] A final run-off to select one airframe and one engine manufacturer was set for 1965, after which the work would commence on the cost-sharing basis with the winners. The losing bidders would be reimbursed in full for the design competition costs.

At this time Air Force Chief of Staff Curtis LeMay unleashed a sensation on Capitol Hill by challenging the reliability of strategic missiles, directly contradicting Secretary McNamara and calling for immediate large-scale funding of the supersonic bomber. The House

Armed Services Committee overruled the Defense Department and put the B-70 back in the budget, an action confirmed by Congress. President Johnson decided to throw a curve ball of his own. This was election year and he could not permit LeMay to provide the Republicans with a dangerous issue. With ceremonial emphasis and theatrical staging, he unveiled a "top-secret" supersonic military aircraft (the A-11). He emphasized that the craft, already undergoing test flights, had solved the problem of tooling and alloying titanium to withstand Mach-3 flight temperatures, thereby moving the nation closer to an economical SST and maintaining preparedness for any strategic use that might be found for supersonic manned bombers. At the same time McNamara conducted a lengthy briefing to remove the cloud that had been cast over the reliability of missiles. Actually, LBJ's ploy was an elaborate deception carried out without outright falsehood by suggestive innuendo. Built by Lockheed under an Air Force contract awarded in 1959, the A-11 (later re-designated the YF12A) was already well known to key legislators, the contracting agency, and aircraft manufacturers. The President was deliberately evasive about the plane's mission (later disclosed as interception) and the reasons for secrecy; he explained that the timing of the disclosure permitted "the orderly exploitation of this advanced technology in our military and commercial programs." [34] The press speculated that the plane was a replacement for the slower and more vulnerable U-2 "secret" reconnaissance aircraft. The aircraft industry insisted that the disclosure in no way modified its SST position, the Air Force grumbled that it did not in any way obviate the argument for the B-70, and Congress criticized the administration for by-passing the legislative trustees of the public purse. In all, the episode was delightful.

At this point the President appointed a top-level administration policy board for SST under Secretary McNamara, while continuing to fund engineering studies at the rate of $2 million a month. It began to appear that SST would be an expensive monster which could be put in service only by means of unprecedented government subsidies which the administration did not favor. The British-French Concorde, a much simpler technical accomplishment, was well behind schedule and its development cost was already twice that predicted, leading the British Labor government to second thoughts and to bickering with the French. Official U.S. estimates were turning less confident, some predicting at least $2 billion needed for the job. Moves were underway in Congress to enable aircraft producers to escape the cost-sharing formula without evoking charges of "giveaway." Senator Warren G.

Magnuson, chairman of the Commerce Committee, proposed a government guarantee against company losses if too few planes were sold to meet development costs under the 1 to 3 formula; others proposed a long-term government guarantee of loans; a substantial bloc rallied around the Air Force B-70 as a backhanded method for underwriting SST.

President Johnson's response was a devious compromise, similar to the one employed a year later in containing pressure for an Air Force manned orbiting laboratory. He agreed to support a new military aircraft as "a long-range advance strategic reconnaissance plane, capable of world-wide reconnaissance for military operations." [35] At the same time, he assured himself that McNamara's management would keep the manned-bomber program under strict low-budget control, with careful technical and strategic evaluation at each step before larger funding or future cancellation. The new supersonic aircraft was designated the SR-71 (as if to suggest that it was a more advanced B-70 type); in fact, it was merely a modified version of the A-11. The editor of *Aviation Week* commented that this was "still another designation pasted on its titanium skin. But many congressmen were fooled, and unthinking daily newspapermen and wire service reporters failed to catch the deception and spread the news across the nation. . . ." [36]

Continued pressure from the industry, the Air Force, and the Congress led in late 1965 to a McNamara announcement of plans under consideration for modification of the controversial TFX movable wing, all-service fighter aircraft into a reconnaissance and tactical bomber.

These moves failed to assuage the aerospace industry or to isolate the Air Force from its advocates. Instead they shifted the emphasis and the pressure once more to the SST—for which the administration in the 1966 budget had requested no funds.

Senator Monroney joined the campaign to force an SST speed-up, emphasizing again the impact which the loss of commercial air leadership would have on the balance of payments: failure to build the plane would "choke off" 375,000 jobs, he declared, and mean a loss of $1 billion a year in passenger revenues. "If we capitulate, it would mean . . . a second-class airline industry," just as, he said, we now have second-class merchant marine, fishing, and railroad industries.[37] Once again Johnson was devious. With great formality he ordered a speed-up and asked Congress for $140 million to finance the first eleven months of an eighteen-month R&D program aimed at solving the toughest airframe and engine problems. Guided by the recommendations of McNamara's SST Policy Committee, the "speed-up" deferred until 1967 the far more expensive and critical decision as to whether test planes would be developed and how they would be financed. In effect, the administration sought to contain the SST coalition, seeking to pursue a reasonable exploratory program in a low key while avoiding an open clash with Congress. Forced to accept a larger scale of funding on both programs—commercial and military—the White House yet managed to prevent a politically dangerous stampede, satisfying no one and guaranteeing a continuation of the fight. But by spending millions the President thereby deferred the commitment to spend billions and managed to conserve the administration's view of the public interest as well as his own maneuverability.

Compared with Comsat, the SST episode provides a counter motif against the insulated economics of ambiguity. The flow of concealed public subsidies which brighten the financial prospects of the private corporation raises a number of interesting questions which one day, perhaps, may be properly examined in Congress or elsewhere. The answers could destroy the purblind myths of free enterprise and unmask the real ambiguities and dilemmas of the Contract State.

CHAPTER XVII

THE BELL REPORT: RATIONALITY AND REACTION

The abuses of federal contracting in the 1950's led to a reform effort by the Kennedy administration. For the most part, however, this was confounded by the united forces of the Contract State, whose influence and affluence continued to flourish and which achieved even greater legitimacy than before.

In his first year President Kennedy sought a means to discipline and control these forces. His most important move was to appoint a top-level committee under his Budget Director, David E. Bell (including PSAC's Wiesner, Defense Secretary McNamara, and Commerce Secretary Hodges, among others), to study federal contracting and procurement, with special attention to R&D, the new braintrusters, and government's own management capability. The group was charged with recommending policy revisions which might subject the military-industrial elite, about which Eisenhower had warned in his Farewell Address, to control by public authority. In a separate action the President enunciated a new code of ethics for contractors and part-time advisers, which attempted to impose more realistic (and therefore more effective) standards and guidelines.

JFK was prompted and aided in these efforts by the accumulating evidence of impropriety and the growing demand for remedial action.

In 1957 the Bar Association of New York City had sponsored a committee of former government attorneys to study conflict of interest in federal procurement and policy-making. Backed by a Ford Foundation grant, the group examined existing statutes and found them "ineffective to curb real risks, and crippling where enforced," concluding that a new integrated statute to replace the existing tangle should aim at three kinds of conduct: persons serving government should not be allowed to tamper with the wheels of government for personal advantage; they should not be allowed to help an individual private party or corporation to serve special interests; and they should not be allowed to use their offices for the purposes of personal economic gain.[1]

The Ramo-Wooldridge investigation of 1959 gave substance to this report, leading a number of congressmen to sponsor bills to write the recommendations into law. Eisenhower's Farewell Address nailed the issue on the door of the incoming administration and aroused congressional attention. In 1961 several vectors were in movement. Chief among them was Kennedy's appointment of the Bell Committee; he followed this with new conflict-of-interest directives. Two separate and somewhat competing investigations were initiated by the House of Representatives. A subcommittee of the Armed Services Committee (under the chairmanship of F. Edward Hebert, Democrat of Louisiana) sent questionnaires to directors and trustees of leading non-profit research corporations created and supported by the military; and the House Government Operations Committee scheduled hearings into the general problems of defense/space/R&D contracting and management.

The House Armed Services inquiry proved to be a tough and hard-nosed proceeding which managed to net a few instances of dubious propriety and legality. A typical item in its preliminary questionnaire:

> Do you now own or have you purchased or sold for your own account or for the account of others, any stocks or bonds or securities or shares or property of any nature, or any interest in any corporation, organization or entity having any contractual relationship with the military establishment while a trustee or director of this corporation?

On the other hand, the Government Operations Committee conducted a scholarly and ruminative inquiry (which filled a forbidding five volumes) addressed to more positive aspects of the problem but marked by an academic blandness which, in sum, seemed to favor the status quo. Using the Bell Report (which appeared in early 1962) as a springboard, the hearings emphasized those parts which seemed to endorse the existing system, blunting the cutting edge of the administration's position. The Bell Report's real emphasis was on the need to rebuild and preserve government's in-house competence for R&D, systems engineering and management, contract evaluation, and yardstick. The muting of these implications appeared to be a deliberate tactic to neutralize the report's explosive potential with a great lump of cotton batting, to muffle the alarum for fundamental reform with professorial kindliness and a studied detachment and calm.

In-House and the Economics of Ambiguity

Defense Secretary McNamara's management team provided the driving force behind the Bell Report. Dr. Harold Brown, then Defense Department chief scientist, later to be Air Force Secretary, emphasized the importance of government in-house in a talk to personnel of the Naval Research Laboratory. He sharply assailed the existing patterns of R&D activities and cited "four good and clear reasons" for preserving and enlarging government-operated labs:

1. They form a spearhead for continuous research peculiar to military needs and make it possible for the military to bring findings before the scientific community.
2. They permit objective advice and evaluation of outside scientific projects.
3. They provide organizations that can manage or help manage weapon-system programs, without distracting military officials who are assigned to operations.
4. They are an essential part of the system of technical education for military officers.[2]

The Brown statement marked a heady departure from the decade of silent erosion of the laboratories, raising hopes (at least among the few who understood these obscure issues) that the forthcoming Bell Report might reverse the trend. It also raised fears among the powerful constituencies of the Contract State, who began to organize to defend their system.

The presidential task force issued its report early in 1962, clearly defining the central issues: "Concern has been expressed," the Bell Report declared, "that the government's ability to perform essential management functions has diminished because of an increasing dependence on contractors to determine policies of a technical nature and to exercise the type of management functions which the government itself should perform." Questions were asked about the role of non-profit firms, the erosion of traditional university functions, and the improper use of cost-plus contracts which "have been criticized as providing insufficient incentives to keep costs down and to insure effective performance." The report summed up the basic philosophical issue in a paragraph that deserves to be carved in rock:

> . . . the developments of recent years have inevitably blurred the traditional dividing lines between the public and private sectors of our nation. The number of profound questions affecting the structure of our society are raised by our inability to apply the classical distinctions between what is public and what is private.[3]

The purposes to be served by federal R&D, the report declared, "are public purposes, considered by the President and the Congress to be of sufficient national importance to warrant the expenditure of public funds. The management and control of such programs must be firmly in the hands of full-time government officials clearly responsible to the President and the Congress." [4] Bell amplified the meaning:

> . . . the decisions which seem to us to be essential to be taken by government officials, rather than being contracted out to private bodies of any kind, are the decisions on what work is to be done, what objectives are to be set for the work, what time period and what costs are to be associated with the work, what the results expected are to be, and the evaluation, and the responsibilities for knowing whether the work has gone as it was supposed to go, and if it has not, what went wrong and why, and how it can be corrected on subsequent occasions.[5]

The report indicated a central factor of public control, that government "have on its staff exceptionally strong and able executives, scientists, and engineers fully qualified to weigh the views and advice of technical specialists. . . ." [6] There is a "serious trend toward eroding the competence of the government's research and development establishment—in part owing to the keen competition for scarce talent which has come from government contractors." It is "highly important to improve this situation . . . by sharply improving the working environment in the government, in order to attract and hold first-class scientists and technicians." [7] Jerome Wiesner testified that his job as Science Adviser to the President was to try "to bring good people into the agencies," and to make government "a place where scientists can have a most attractive career in science and engineering. . . ." But, he said, ". . . the situation is actually still deteriorating." The best-qualified people were still leaving and "once this begins . . . then the next echelon leaves . . . you may keep the buildings full, you may spend the money," but the trend continues implacably and the public interest is increasingly naked before hungry contractor cliques.[8]

The trends to open-ended, cost-plus contracting and wasteful and misleading brochuremanship were, the report noted, direct results of this erosion. Agencies were forced to depend upon solicited and unsolicited proposals from interested contractors "before usable and realistic specifications of the systems have been worked out in sufficient detail." This process "unnecessarily consumes large amounts of the best creative talent this country possesses . . . this practice can and should be substantially curtailed" by improving the government's ability "to accomplish feasibility studies or letting special contracts for that purpose before inviting proposals. In either event, it would require the acceptance of a greater degree of responsibility by government managers. . . ." [9] Adequate in-house could "provide alternate approaches" and "a way of judging what is done outside" which would go far in taking "the profit . . . out of brochuremanship." [10]

If government were able to write its own specifications and to conduct its own feasibility studies, it would no longer be captive to profit-making contractors and hard-sell techniques. To a large extent it could abandon the blind faith of cost-plus and bring hidden private profits and careless accounting under the discipline of open, competitive bidding for some of the research, most of the vastly more expensive development, and all of the associated and follow-on hardware production. By imposing this means of economic realism, the subsidized inflation of scarce resource costs might be subjected to downward pressure; this would check the erosion of technical competence in government, the universities, and civilian industry.[11]

Spokesmen of industry who testified on the report followed what appeared to be a consistent strategy of side-stepping its main impact. They concentrated on those parts of the document that emphasized the need to preserve the nation's broad technological capabilities and to continue to rely upon a mix between in-house, university, and industrial performance. Their testimony dwelt long and lovingly upon the positive contributions of industry, avoiding hard specifics and questions of public control and cost accounting. Nowhere in five volumes of hearings is the usual cantankerous and persistent search for gut issues characteristic of so many congressional inquiries. Instead, the hearings provided a lovefest between legislators and contractors, both of whom damned the report with faint praise.

Helge Holst of the Arthur D. Little Company (a management and

R&D firm of Cambridge, Massachusetts) declared that the report was too much concerned "with creating in-house capability . . . almost for its own sake. I oppose this because it seems to me that it is the route in which it will be the most difficult to get a two-way flow of advantages into the government and out of the government." He termed recommendations for enhancing government labs "working backward. Why not have the research and development take place in an environment which exists and which is created expressly for commercializing and propagating civilian benefits?" Building up the in-house will—"and this may sound a bit radical"—defeat the contractor's feelings of "civic obligation, if you will, to serve a tour of duty in the government or on a committee, or as a volunteer of some kind trying to give the government, the nation, our own nation, the benefit of whatever they can contribute." [12]

Industrial contractors, Holst continued,

> differ from the Bell Report in believing that in this day of specialization, and in an attempt to be efficient and economical, and to use scarce talent as effectively as possible, and not either to raid government agencies for the benefit of civilian or civilian agencies for the benefit of government, it is better for the government personnel to be concerned primarily with determining the needs of the nation . . . and then to bring in outsiders to try and see how you implement those needs.[13]

He reversed the argument of the Bell Report: "One of the dangers in in-house" is that "they are saved from competition. Both concept competition and cost competition. This is a real danger." [14] In the face of overwhelming evidence to the contrary (nowhere reflected in congressional questions), Holst blandly asserted the existence of real "concept and cost competition" among contractors, invoking the clichés of "free enterprise" and the "dead hand" of government, and the need to preserve the former against the latter.

Dr. Brown delivered a definitive and devastating judgment on these claims when he asked why "wouldn't it be possible to obtain the objective evaluation from a private contractor rather than an in-house capability?" It would have to be, he responded, "a private contractor of a different kind than now exists. I think it would have to be analogous to a professional firm, analogous to architect-engineering," and—sweeping aside all the assorted and persistent claims of the non-profits to qualify for this role—"there is no such organization that I know of." [15]

Specific Recommendations

The Bell Committee submitted a series of action recommendations which were adopted by the administration as official policy for all federal agencies. These fall into five general areas: improving the salary scale for federal management and technical personnel; enhancing the atmosphere in government laboratories by giving them a more flexible and creative role; experimenting with new kinds of government R&D institutions; revising and tightening the contract instrument in defining the role and terms of work of private industry; and finally, integrating and coordinating all executive policy-making and implementation in the conduct of the nation's scientific business.

Salaries. The report recommends that salaries for career government employees, particularly in professional grades, be raised "to provide greater comparability with salaries available in private activities." Salaries all down the line have lagged behind the rest of the economy. State governments pay hundreds of top-level people more than Uncle Sam pays the Secretary of State, and there is too little range at high levels. The sharpest discrepancy exists between contractor and government personnel doing comparable work, often side by side. Federal scientists can frequently increase their salaries by "$5,000, even more annually" by defecting to contractors.[16] The irony of the situation is that the higher salary is charged back against the contract, so the public pays the penalty both ways—paying the higher salary (with a company profit added) and losing the management capability.

The holder of a bachelor's degree could expect at least $1,000 per year more as starting salary with contractors than with government. After twenty years the average government employee received over $2,000 less than his contractor counterpart. At higher professional levels, the disparities increased. The government scientist or engineer with a Ph.D. could expect to receive $8,606, compared with the average $11,564 a year in industry. The director of a government laboratory, such as the Bureau of Standards, received $19,000 annually, his counterpart in a non-profit corporation upward of $40,000 plus many perquisites. The National Institutes of Health, before the 1964 congressional assaults, financed several university research professorships that paid at least $6,000 more than the maximum possible for NIH's chief officer and other scientists.

The RAND Corporation has the lowest scale among the non-profits, yet its salaries tower above the civil servant's. In 1962, 250 RAND staff members drew annual salaries above the then civil service maximum of $19,000, ranging upward to $35,000 annually. The non-profit Aerospace Corporation offered starting salaries to key personnel 30 to 40 percent higher than those of previous employment. A sample of forty-eight employees (studied by GAO) showed average annual increases of 18 percent, with some ranging to 100 percent. A physics professor earning $14,055 was employed at a starting salary of $28,000 with extravagant fringe benefits; a NASA employee earning $17,500 was hired at a salary of $24,000 plus fringe benefits and large annual increments; a Space Technology Laboratory employee at $22,500 was hired at a starting salary of $30,000, etc.[17] An Aerospace executive recruited from RAND went from $37,800 to $59,000 in just two years.[18] The president of Aerospace, Dr. Ivan Getting, received $75,000 a year plus 20 percent incentive compensation, plus paid-up life insurance (cost to Aerospace annually: $1,800), plus non-contributory retirement (cost to corporation: $9,000 annually), plus payment of country club fees, and so on.

Profit firms go even higher: while middle-management and professional employees enjoy salaries comparable to those of non-profits, brochure-writing technical people and top management enjoy salaries from $50,000 to $100,000, with a handful at the apex going well beyond, not to mention such added perquisites as stock options and bonuses. The author of this book has been offered immediate doubling of his university salary to join a "software" contract project with a California aerospace firm—on the basis of a twenty-minute telephone conversation.

The Bell Report called for "controls over the salaries and related benefits received by persons employed in the private sector on federally financed research and development work." It recognized that raising salaries may not be enough, for loose cost-plus contracting would enable contractors to bid the price up still higher in order to maintain a system which redounds to their profit advantage. Thus a tightening of all contracting practices and some kind of centralized review was considered necessary. But the task force stopped short of recommending direct public control of the salary structure of private firms, suggesting that indirect constraints, closer monitoring, and more competitive methods of contract award would do the job.

Government laboratories. The Bell Report calls for assignment to the in-house R&D establishment of "significant and challenging work" and a greater reliance by decision-makers upon laboratory personnel, rather than contractors or consultants, "in advising on and reaching program and management decisions." Government should "simplify and coordinate management controls" over in-house facilities, eliminating "unnecessary echelons of review and supervision," and giving laboratory directors "a discretionary allotment of funds for undirected work and more authority to command resources and make program and administrative decisions." Professional personnel at all levels must have opportunities for "periodic retraining and upgrading of their capacities" and be assured a stable prospect for career planning and advancement, maintaining professional mobility both within government and between government, the universities, and industry, while preserving standards of professionalism and objectivity. Government could no longer rely upon security of employment as a stabilizing incentive; but it continued to enjoy advantages (if it only would use them) in maintaining stable research programs which hold people's interest, and it could support the most pioneering kind of creativity which, under present usage, becomes for private contractors a cornucopia of unlimited and nonaccountable public funding.[19] "In-house laboratories should take chances. That is, they should work on things which have perhaps less than a 50-50 chance of working out," Brown said.

It is going to take a combination of actions on the part of the government and the laboratory managements. It is not enough to say that a laboratory must have a general research or a general development mission. There has to be something that is the most important thing that the laboratory does and that the laboratory must try to be better at than anyone else in the country is. That has been lost sight of, and it is not always easy to implement.[20]

New organization. Without being specific, the report canvassed the possibility for new types of institutions:

It might well take the form of a corporate body, perhaps with a board of trustees, with authority granted by the Congress to have some greater freedom in personnel selection, appointment, and compensation than the normal civil service system provides and with some considerable degree of independence in the use of funds which were made available to them. . . . This is about all there is to the idea at the present time. It is a bright idea and it needs a good deal more wringing out. . . .[21]

Such an institution should provide a means for reproducing within government some of the positive attributes of non-profit corporations, avoiding their limitations and pitfalls. Several institutes might be created under high-level supervision, independent of mission-oriented departments, each with authority to operate its own career merit system as does the Tennessee Valley Authority. This would follow the pattern employed by the British government in establishing an independent corporation with revolving research funds and almost complete freedom to support basic and applied projects. Commenting on this notion, Wiesner expressed the view that this should have been done in 1958-1960, instead of creating the numerous tribe of non-profits. Having been involved in setting up the notorious Aerospace Corporation, he observed: "It would be better if there had been some more formal mechanism which was available. . . ." A government-owned institute justified experimentation, and should be tried, he said, if the need that led to Aerospace "appears again." [22]

The contract instrument. With feasibility studies and real specifications by public servants, government could avoid having to evaluate contractor proposals by turning to agencies of dubious objectivity, unbound by official policy. This would make it possible, the report indicated, "to maximize the use of fixed-price contracts in the research and development effort," especially "in the late stages of development work, testing, and support services." Where cost-type contracts were retained, "greater attention should be paid to the development of incentive arrangements which would relate larger fees to lower costs, superior performance, and shorter delivery time," providing *real* penalties for inadequate performance. Realistic standards should be adopted and enforced uniformly by the whole government "with respect to costs which are eligible for reimbursement under cost-type contracts." The role of the non-profit corporations should support and implement, not replace, government control of its contractors: they should be denied "profit" fees above costs for the acquisition of major capital equipment, although some independent research might be allowed.

Integration and coordination. To spell out the detailed policy directives and budgetary changes required to adapt these new policies to the diverse requirements of many mission-directed agencies, the report directed the Office of Science and Technology to "study the means of exchanging information" to improve techniques for supervising and evaluating contractor performance and effectiveness. The National Science Foundation was instructed to improve the collation and dissemination of information about "current scientific and technical resources of the nation"; and the President's Science Advisory Committee was asked to study means of enhancing the exchange of scientific and technical information within the government, between government contractors, and, where applicable, with scientific and technical facilities serving the civilian economy.

The Salary Sequel

Symptomatic of the failure of the Johnson administration to implement the Bell Report were actions subsequently taken to improve the federal salary scale. President Kennedy sent a proposal to Congress in 1962 which called for two new top-level civil service grades, GS-19 and GS-20, and orderly increases in salaries of all grades over a three-year period with a healthier spread at the upper end. The bill provided that the Bureau of Labor Statistics regularly collect information on salaries paid by private industry and, based upon this data, make further upward revisions in future years to keep pace.

For two years a number of factors conspired to defeat the proposal in 1962, and when a salary bill was finally passed in 1964, its character had been changed. It became a general upward adjustment for all grades (as well as congressmen and top appointive positions in the judiciary and the executive), rather than a careful structural reform designed to achieve the purposes of the Bell Report. Congressmen were little interested in raising federal salaries above their own (then $22,500) and reluctant to raise their own for fear of voter reprisals. Furthermore, administration proposals granted only nominal increases in the numerous lower and middle grades, offering no political solace. Legislators (many indifferent or hostile to the direction pointed by the Bell Report) considered their power over federal salaries a form of political patronage to be spread so as to maximize the effects upon their districts where federal employees were voters. They were predisposed to support increases across the board at the price of defeating structural reform. In addition, this was a period of administration pressure against price and wage increases in the economy, generating pressures from labor and industry which linked their demands for price and wage increases to congressional action on the federal pay bill.

The Senate bill kept most of the structural reforms intact and introduced a general upward adjustment for the whole government, including Congress. The House defeated the legislation, and a series of cloakroom conferences followed during which the bill's character underwent revision. The proposed slope of the upper reaches was reduced and the curve at middle and lower grades flattened. The President was authorized to classify sixty "agency executives" ($19,000 to $26,000) at his own discretion (but within six months). The bill further guaranteed a minimum 3 percent increase for career employees in the lower and middle pay brackets; it permitted heads of department, in certain circumstances, to appoint exceptionally qualified individuals to steps above the minimum entrance step in grades GS-13 and up with the approval of the Civil Service Commission, and limited to 2,400 such super-grade positions.

As a result of the 1964 changes, the curve of professional federal employees was constructively raised and the slope slightly (but not adequately) modified. The median paid to highly qualified career employees rose from slightly more than $11,000 in 1962 to around $14,500 in 1965. Considering age and experience, median salaries approximate those of private industry with a slight falling off at the upper levels.[23] But the median is deceptive because it includes the salary levels of civilian industries where incipient stagnation and professional unemployment exist, and where professional salaries are low compared with those of government contractors. A median or average conceals the wide disparities in private industry in the upper reaches, the very disparities which—subsidized through government contracting—create the factor chiefly responsible for the drain of technical skills from the universities, the civilian economy, and from government. The legislation went in the right direction but not very far.

While taking this step forward, Congress found other ways of defeating its purpose. While authorizing more personnel slots at the higher grade levels, it clung to the habit of cutting administrative budgets more uncritically than facility and program funds, demonstrating the most liberality in supporting, even increasing, requests for funds for the army of private contractors. Positions shown on agency tables of organization remained unfilled not only because the salaries were still not competitive, but often because funds were not appropriated to pay the salaries.

Furthermore, industry remained free to outbid even the improved federal salary structure, pointing up a vexing problem which the Bell Report recognized as capable of undermining any salary reforms. The report noted that the standards of federal salaries should be comparable with industry; but in order that industry would not outbid these again and aggravate the whole situation, the document groped toward some kind of control over contractor salaries, at least in the area of negotiated, sole-source, and non-competitive awards. As stated by Dr. Brown: "We will study further the question of review of salaries accepted as direct charges under our contracts. This is a difficult question. . . . The Bell Report does not recommend a ceiling on salaries nor is there any intention on the part of the Department of Defense to do so. Such an attempt would be an unwarranted interference with free private enterprise." [24]

Saving the Laboratories

Under the leadership of Secretary McNamara, the Defense Department made the most vigorous and relatively successful effort to implement the Bell Report, meeting determined opposition at every step and forced to make many compromises. The department proceeded through the twin routes of revising the form and tightening the administration of contracts. It tried to broaden competitive bidding, review more rigorously decisions of contract officers, limit reimbursable overhead, create instruments of centralized auditing, and seek ways to rebuild in-house resources. In early 1962 McNamara ordered an analysis of the conditions which made such labs as the Naval Ordnance Test Station and the Naval Research Lab outstanding. The Defense Department's budget request for the following year reversed the trend of the previous decade by seeking funds to construct three new in-house facilities and to increase allocations to existing installations.[25]

The most positive response to these moves came from the Army and the Navy, whose traditionally strong laboratories had been sadly worn away by the Contract State. Delighted at the prospect of salvaging its depleted arsenal system, the Army adopted a five-year plan to "replace or upgrade existing facilities or to accommodate new missions." It moved to raise the salary classification of "bench-level scientists and engineers . . . actively engaged in research projects"; budgeted non-mission-oriented funds to laboratory directors for free-ranging exploratory work ($10 million for fiscal 1963); launched a new career program; and established R&D achievement awards to provide recognition for in-house scientists, engineers, and technicians.[26] The ever-resourceful Air Force tagged along, providing a bit of comic relief which served to illustrate the paucity of its in-house: to indicate compliance with the Bell Report, an announcement was made (in June 1962) that the Air Force Academy at Colorado Springs had established a "basic research laboratory" for the use of faculty and students. It would provide an "in-house research capability." [27]

After his initial incision, McNamara realized the need for more drastic surgery. In 1965 he moved to impose centralized control over all R&D activities of the separate services, seeking to loosen the parochial grips of multi-echelon command which by-passed departmental policy review. Government laboratories would henceforth report directly to a Defense Department command level, and lab directors would have greater authority to determine what problems to research and what facilities and personnel they would need. They were thereby to be freed from budgeting by self-serving and competitive services who, in the name of "mission-related research," frequently used them in the game of contract allocation. The services strongly resisted the moves with the argument that they "would divorce the labs from the mainstreams of weapon development and move them into an atmosphere of research for research's sake." [28] At this writing, the laboratory reorganization has not been consummated, and its critics have retreated to the position of proposing that a new pilot laboratory be set up first in order to test the soundness of McNamara concepts.

In NASA, whose in-house position was comparable to the Air Force, there was no McNamara, Brown, Rubel, or Fubini—there was only Webb, the accomplished salesman of space. Above him there was only President Johnson who, as he acquired a grip upon the governmental machinery and sought a formula for personal and national greatness, was converted from his long-time advocacy of the Contract State to the principles of his Defense Secretary. During 1964 he began to show a determination to enforce White House control over every part of the federal bureaucracy and to control the industrial-military elite. Sensitive to the long Texas shadow cast over him, Webb pushed for a major electronics in-house facility, all the while dismantling the essentially in-house character of the Jet Propulsion Laboratory and the Marshall Space Flight Center. Plans were announced in 1964 for an Electronics Research Center (ERC) in Boston. It was emphasized that there would be "a considerable amount of prototype hardware development" in addition to "paper research work." Many congressmen interpreted ERC as a pork-barrel payoff to the Kennedys and refused to authorize funds until "further study." The first substantial funding occurred in the 1965 budget (considerably less than NASA's request). Renting space here and there, ERC began operation and launched plans to begin construction of permanent facilities in the spring of 1966.

NASA officials provided *Missiles and Rockets* magazine the basis for some interesting prose: "Despite its comparative newness, ERC has already come up with two procedures unique to the Space Agency's research work. They are: laboratory chiefs will not simply monitor industry and university contracts that are awarded; they will manage. . . . Procurement officers have developed a new procedure called the Industry Registration Kit for R&D which will supplement the regular NASA registration forms." [29] A deputy chief of the Marshall Center fired back a sharp rejoinder, pointing out that in spite of "complaints from some of our contractors for our active participation in the research efforts," he and other Marshall people attempted actively to "manage" the contractors rather than, in the words of the original ERC article, "sitting back waiting for it to be completed and a final report submitted." [30] It would appear that in its anxiety to conform to administration policy, NASA headquarters overemphasized the uniqueness of the contracting practices of the new "in-house" facility.

There were other indications of the Space Agency's ambivalence on the in-house issue. First, practically all of ERC's 1965 funds were allocated for "out-house" contracts.[31] Second, ERC's 1966 budget allocated more money for administrative operations ($7.6 million) than for R&D ($5.2 million), and much of the latter was to be used for test equipment.[32] Third, and most authoritative, Webb testified in 1965 that ERC's function was to look beyond the current $2 billion of yearly contracting in electronics: ERC would be "a research center to plan outside research to be done by the most competent people in the country." [33] In appointing the director of ERC, Webb made clear whom he considered to be "the most competent people in the country." Instead of a career government science administrator, a university man, or a bench scientist, any of whom could be expected to exercise some independence of the contractor community, he typically chose a vice president of the Bendix Corporation, one of NASA's largest electronics contractors.

The failure to make real progress toward implementing the Bell Report was symbolized by the appointment in 1964 of another committee to study the same problems and to add to the mounting pile of such studies whose cumulative effect has been to defer rather than facilitate reform. The President's Science Advisory Committee called upon Dr. Emanuel R. Piore, a vice president of IBM, to head the study. Piore told reporters he hoped the study would produce "a fundamental statement of policy on why the government needs its own laboratories." Such a policy statement could prove useful, he said, since there were increasing complaints from private research industry about the competition from government laboratories.[34] His statement made it clear that all the Kennedy-Johnson battles to roll back the influence of the Contract State, to discipline the contractor cult in the name of larger national values, were still inconclusively joined. Partial victories had not secured a stable beachhead; government still faced a massive assault to sweep it back to the lush heyday that the Contract State achieved during the fifties and has enjoyed almost unmolested ever since.

With the erosion of in-house, government can hardly find scientific and engineering consultants, braintrusters, or qualified managers who at the same time are not present or future recipients of personal advantages arising from their governmental role, and therefore subject to the subtle corrupting influences of the epoch. At a time when government is sponsoring most scientific research both in universities and profit-making corporations, independence and objectivity have become preciously rare.

At this late time, what can be done to reform the institutions of the Contract State? Clearly, the implications and recommendations of the Bell Report provide a starting point. The salary differential between public and contractor personnel must be stabilized, both by raising the former and effectively controlling the latter. The in-house capability must be resurrected and government officials vested with genuine systems-engineering and technical direction responsibilities, somewhat on the pre-purge model of Redstone Arsenal or the Jet Propulsion Laboratory. This will eliminate dependence upon brochuremanship, facilitate cost accountancy, and make possible a turn to realistic incentive, cost-plus contracting for pioneering R&D, and competitive bidding for later development phases and for hardware production. The suggestion for experimental creation of a quasi-independent government corporation, a TVA for R&D, invites exploration. These recommendations provide a jumping-off place, but the ultimate test will be political.

The President will have to be bold and courageous in maintaining the integrity of his administration against the underground coalition of special interests which operates within Congress and the federal establishment. There must be a balanced approach to science and to technological innovation, subordinate to political judgment. There must be better understanding of the need to protect the independence of the universities as an irreplaceable resource. In David Bell's words, "Colleges and universities have a long tradition in basic research which must be utilized. They provide the unique intellectual environment combined with the process of graduate education, which has proved to be highly conducive to successful undirected and creative research by highly skilled specialists. . . ." [35]

The filter-down process of pump-priming the civilian economy by fostering ever-greater economic concentration and income inequality must be replaced by a frank acceptance of federal responsibility to control the tide of economic bigness, and to plan the conservation and growth of all sectors of the economy and the society. This is asking a great deal. There are many signs of promise, but the prospects are not reassuring. There are also strong forces endowed with all the momentum of custom, usage, and legitimacy which can insure continued distortion and patchwork as the Contract State goes marching on.

CHAPTER XVIII

THE BATTLES OF McNAMARA

Government contracting and procurement in the decades of the Cold War (now for the Defense Department alone over $25 billion annually, for NASA over $4 billion) have created a financial never-never land with far-reaching effects upon American life and destiny. The present form of the Contract State was shaped by a wide variety of domestic forces, legitimatized and propelled by the partly valid and partly hysterical national reaction to the inescapable fact of permanent diplomacy and a runaway technological arms race.

Complacent in the late 1940's in our unchallengeable military position (represented by atomic bombs and strategic bombers), the nation over-reacted to the evidence of the early fifties that the Soviet Union could and would successfully challenge us, stealing a march in missilery and then in space. The American response was a crash program to re-establish a strategic superiority that was becoming increasingly meaningless, self-vitiating, and diplomatically useless. In the pursuit of this chimera, all the institutions and standards painfully constructed in two great wars to safeguard the integrity of the state and the public interest were thrown to the winds.

By the late 1960's it was clear the U.S. was leading the world in the development and deployment of strategic weapons and had finally overtaken the Soviets in space. Just as evident was the fact that the expected diplomatic rewards were not to be realized by this route.

From the beginning of his administration, President Kennedy recognized the futility of an infinite strategic arms race which worked as a self-fulfilling prophecy to force both major adversaries to ever-higher levels of investment in strategic systems, each side forced to deny the other any theoretical advantages that might be achieved by this investment. The young President saw the logic in stabilizing this useless burden and sought to formalize the common interests of both contenders by a number of agreements (such as the 1963 Test-Ban Treaty), informal understandings (mutual cutbacks in plutonium production), and unilateral actions (U.S. declaration of adherence to the U.N. resolution against orbiting nuclear weapons in space).

This trend enabled both super-powers to conserve their resources for the more complex and diplomatically significant tasks of internal economic and political health and unrelenting global political change. In both areas their uncontrolled strategic arms race had not only been unavailing but was a complicating factor which distorted their domestic life while converting both into muscle-bound giants, increasingly unable to deal with their limited conflicts of interest with each other and the unruly quarrels and demands of the small nations, including their own blocs.

The crash missile programs of the 1950's may have been necessary, but both Kennedy and his successor realized that the powerful vested interests created domestically around government contracting could not be allowed to become a permanent institution. The moon race of the sixties extended the life of the system. It picked up the slack caused by the leveling of investment in strategic weapons, thereby making a strategic rapprochement with the Soviet Union more politically feasible by providing a substitute to feed the power structures of the Contract State. But once the diplomatic goal was realized, the administration addressed itself to disciplining and dismantling these forces. At stake was and is the integrity of presidential leadership in concerting the tremendous resources of the nation for improving the quality of American life, and for the pursuit of diplomatic objectives with the wide variety of military, political, economic, and social instruments required by an ambiguous, pluralistic world. The domestic vitality of the nation in the long run may prove to be the only ultimate weapon.

Robert A. McNamara, whose tour of duty as Secretary of Defense has set a longevity record for the office, has been the unflinching hero of the campaign to reform and control the Contract State. His efforts were initiated in the first months of the Kennedy administration;

but in spite of promising beginnings they were generally blunted and evaded through 1964 by the powerful coalition of interests in the military services, the Congress, and in the nation. Lyndon B. Johnson, as Senate Majority Leader in the 1950's, had become the key broker of this coalition, a role he shared with such colleagues as Senators Robert Kerr, Clinton Anderson, Stuart Symington, and others among whom he became pre-eminent, achieving for himself political power while others (such as his close assistant, Bobby Baker) were not above exploiting the opportunities of political brokerage for personal wealth. Johnson persisted in this central brokerage role as Vice President, using his influence over Kennedy's appointments and as head of NASA's Space Policy Board to perpetuate the system.

But with the mantle of the presidency comes the call to greatness and a perspective on the public and the national interest that is afforded by no other seat of responsibility. As President, LBJ quickly demonstrated an avid passion for freeing himself from the obligations incurred by a lifetime of political maneuver in the foothills of power, and for freeing the office of the presidency from the congressional shackles he himself had so effectively used to hobble and constrain a frustrated and sputtering Eisenhower. From turning off lights in empty White House rooms to protecting and supporting his Secretary of Defense, President Johnson mobilized his unique talents. He adopted the slogan of "declaring war" against all the immemorial ills of mankind, reflecting a wish to mobilize for positive domestic reform the same kind of national consensus so readily available for national security.

By mid-1964 McNamara's efforts in the Department of Defense were moving at last, and in the following year the President extended the McNamara budgeting and accounting method to the entire administration. The system, the President informed the press, would "bring the full promise of a finer life to every American at the lowest possible cost through the use of modern management tools." The Budget Director (Charles L. Schultze) provided further details: the system would enable the administration to "ask the right questions"; it would make the agencies "work all year long on future programs instead of crowding all the work into the last few hectic weeks of the year as the budget is being drawn up"; costs would be projected over several years rather than one year at a time. The President had ordered all agencies to be operating on the new system by 1967.[1]

The implication was clear: despite increased government spending to meet the needs of the war in Vietnam, the gang of government contractors grown fat and sassy at the taxpayers' expense would be in for leaner days. Future attempts to smother McNamara reforms by resourceful circumvention would confront a more united and disciplined federal establishment. McNamara's conduct of his office had "changed the name of the game. For the first time, the military services were dealing with a man capable of doing just exactly what Congress had instructed him to do—unifying the Pentagon into a single effective military establishment." [2] Backed by a team of tough and knowing scientific managers and managing scientists, McNamara by 1965 had done what no one believed possible. He had humbled the Air Force, unified procurement in selected areas, and initiated financial and budgetary reforms which made possible top-level control of the sprawling contractor and military empires. With grudging admiration, a persistent critic wrote: "For good or bad, the McNamara reorganization is here to stay. Therefore most agree the best thing to do is to play the game by the new rules."

Reviewing the impact of federal procurement in 1965, the congressional Joint Economic Committee concluded that "the prospect for an economical and efficient supply and general services systems . . . has never been so bright." [3]

But this appraisal, while justified, does not mean the battle has been conclusively won and the forces of the Contract State will quietly surrender their privileges and become loyal soldiers in the President's war against waste. Rather, it means that these forces, better organized than ever before, will absorb the shock like a feather pillow and puff out in new places, demonstrating determined, ingenious, agile durability and undiminished political and economic power. Administrative regulations, directives, laws, executive orders, and the zestful skill of the President may prove insufficient. Loopholes in rules and laws, inertial resistance in the lower echelons of bureaucracy, the continued intervention of influential congressional committees serving the concrete interests of constituents and the grubby facts of pork barrel, and the President's own participation in this process which, hopefully, he seeks to direct toward achieving creative goals in the name of the broader public good—all are factors which can obfuscate the best intentions of a chief executive. The battle against the Contract State will continue to be fought and refought before and after breakfast every day, just as it has been for many years. But the commitment of the President, the stack of unredeemed political debts acquired as chief broker of the old system, are substantial and positive assets he can exploit to sway the "great consensus" toward his vision of the future.

McNamara Begins

Upon appointment by President Kennedy, McNamara activated those traits of energy that had earned him a reputation for industrial management. Recruiting a team personally loyal to him and skilled in "scientific" management, he ordered the services to conduct hundreds of studies and reviews of on-going programs and future plans. Delegating great authority to the department's new comptroller general, the RAND Corporation's Charles J. Hitch, and to a staff of highly qualified scientists (headed by Dr. Harold Brown), he forced the military to do what no civilian had ever done before: to define and justify prevailing military strategy, and to relate it to weapons systems and budgetary requests. He allowed himself and his team to be educated by reports of the General Accounting Office, directed all personnel to take "prompt, clear, and positive action" upon each one, and assigned top Defense Department aides to assure that implementation was followed through. The Secretary regarded GAO reports as "a useful source of management assistance" and set a limit of sixty days for departmental response.[4]

Under his regime the Defense Department has shown itself quick to take remedial action upon the basis of these reports. For example, in 1965 action was initiated within weeks of receiving GAO information to see that income from vending machines and similar activities would be treated by contractors as a credit against overhead accounts, thus reducing government costs. Directives were issued to disallow planned losses on food services and careless "swindle-sheet" expense allowances.

His education completed, McNamara set about the difficult task of finding new contract yardsticks. He started with a clear and consistent reformulation of military strategy. A program of cost/effectiveness analysis was applied to all budget requests. Future budgeting was planned on a five-year "program package" basis. To correct the military propensity for rushing into large-scale contracts on the basis of inadequate information, he adopted a "step" method of moving from program definition and feasibility studies through gradually larger R&D and hardware contracts, attempting to weed out duplication and unpromising efforts at as early and inexpensive a stage as possible.

He played an important role in formulating the administration's new code of ethical practices and the recommendations of the Bell Report, forcefully implementing these in his own department. In 1962 he launched "Project 60," whch aimed at consolidating the Defense Department's control of all defense procurement, eliminating the unruly, fragmented, and refractory decentralization that had resulted in service duplication, competition, and evasion of departmental authority.

McNamara's example provided President Johnson with the method of his "war on waste," a government-wide cost reduction program instituted in October 1964.[5] In its first year the Defense Department claimed a saving of $4.1 billion, and a long-range goal of $6.1 billion per year was set for 1969.[6] In order to make the cost-reduction effort more than a public relations stunt, McNamara established rigid standards for evaluating such savings as the separate services might claim. When the General Accounting Office turned down his request for an independent audit, he established a central audit service within the department.

His reform campaign lacked admirers in segments of the military and Congress. The industrial contractors were virtually unanimous in attacking him. Fortunately his armor appeared to have few chinks. An aerospace journal expressed the typical sentiment: "Mr. McNamara has had his nose so close to the cost-reduction trees that he can't see the military forest. Waste can never be eliminated from the Pentagon, because the whole place is a waste . . . but waste or not, these military organizations and projects, and the industry which supports them, are an absolute necessity. . . ." A McNamara press conference statement ("We don't propose to turn defense industry into a WPA. We are going to buy what we need and only what we need") was interpreted as a "perhaps subconscious view of the industry which supplies these weapons" as an "unscrupulous giant feeding voraciously at the government trough." Three years, the editorial declared, is the limit of any Secretary's effectiveness, and "this is Mr. McNamara's fourth year." [7]

Congressional reaction has been ambivalent. "Economy" has always been one of the most positive slogans of the legislators, so long as it occurs in someone else's back yard. Senator Richard Russell symbolized the ambivalence, declaring his willingness to support any action to cut the staggering defense budget—and denouncing in the same breath the shut-down of an underutilized air base in Georgia. On the other hand, a number of congressmen have been strong admirers of the Secretary. For example, Senator Paul H. Douglas: "This has been greatly needed for years. You are the first man to have the courage to do it." As a whole, Congress has endured the McNamara regime and since 1964 has succeeded in sabotaging it only in isolated instances and in small detail, although this relative impotence did not reflect any lack of trying. McNamara's successful strides since 1964 can be understood as springing not only from the rightness of his cause but from the support of the Congress' better self and the powerful endorsement of President Johnson. In July 1965 a presidential memorandum asked all Cabinet officers and agency heads to take "home with you and read carefully" the fourteen-page McNamara Report on cost-cutting, adjuring them all to learn from the Pentagon's paragon.[8]

Cost/effectiveness and phased programing. In pre-McNamara days, military planning and budgeting were treated as independent activities, the first falling within the province of the Joint Chiefs of Staff, the second within that of the Comptroller and the civilian leadership. Planning for military forces and weapons systems was projected over a period of years, but budgeting was based on a single year and was put together in terms of lump categories of like things, such as military personnel, operations and maintenance, procurement, and so on. As described by Army General Maxwell Taylor in his book *The Uncertain Trumpet* (a bitter attack on strategic and military planning during the Eisenhower years), military plans were prepared without regard to resource limitations; actual capabilities were truncated by an arbitrary budget ceiling which imposed a meaningless political compromise between competing service ambitions, totally unrelated to strategic concepts.

As a result, in the words of former Comptroller General Hitch, "serious program imbalances developed. . . . Tactical air support, furnished by the Air Force, was far short of what was required by the Army ground forces. The amount of airlift available for the strategic deployment of the general purpose forces was never adequate in terms of the total deployment objectives. Combat stocks . . . were seriously unbalanced." [9] The lump categories had no relation at all to a consistent formulation of functional contingency plans for the use of military forces, and thus the budget process denied the Secretary and the President relevant information for decision-making.

McNamara changed this situation. Pentagon fighting strength was organized in "program packages" which forced functional cost analysis upon the services. The new method regarded as a single bundle all function-related budgetary items, from R&D to operational activities and procurement. For example, Polaris submarines were considered part of the same strategic package as the Air Force B-52's, Skybolts, and ICBM's. The overwhelming clarity of this packaging made it all too evident that many pet projects of the services—for which huge amounts had already been spent—were actually surplus to the nation's needs. In retrospect these reforms may appear simple and obvious, but they were and still are resisted because they tend to deny political maneuverability to the services and their constituents. The McNamara (Hitch) system was a common-sense approach to the government's business, but it raised fervent moans of "computerized" decision-making, of ignoring the advice of experienced military men, of letting a few callow civilians take over the nation's defense.

"The requirement for strategic retaliatory forces," McNamara sought to explain to Congress, "lends itself rather well to reasonably precise calculation." The number, types, and locations of aiming points determine the number and explosive yields of weapons required "to insure the destruction or substantial destruction of the target system." This in turn determines the size and character of the forces which can deliver these weapons in terms of accuracy, reliability, and ability to penetrate enemy defenses. The strategic requirements of the nation thus, he said, can be subjected to a useful comparative evaluation. Infinite power is not necessary or desirable, and impetuously proposed systems, whose deployment constitutes a heavy burden, could weaken rather than strengthen both the nuclear deterrent and other national requirements if military ambitions are permitted unevaluated sway.[10]

This procedure actually gives military expertise and judgment sharper focus and greater pertinence. It boils out real policy choices and exposes them to top-level decision. It avoids concealed, parochial, and *de facto* decisions in the lower echelons, and misdirected control by arbitrary budget ceilings.

The new system forces the military to define and justify its objectives, to point out alternate ways of accomplishing them, to calculate the cost and degree of effectiveness of every available alternative, or to lay out research and development undertakings to help establish such knowledge. Where unknowns exist or where new technology beckons, R&D programing becomes an intrinsic aspect of the calculation; each project or proposal is defined in steps which will "provide a much greater degree of certainty as to the scope, objectives, technical characteristics, management arrangements, and probable cost . . . prior to administrative and financial commitments to a full-scale development program." [11] By phasing the level of commitment from low-budget, widely diversified preliminary and program definition studies, the government can contain the blind tendencies of a runaway technology fueled by service rivalries and by the Contract State. It can retain flexibility and choice at every point of the process at lowest possible cost.

This is not to say that the system, once adopted, acquires an automatic effect. Far from it. It offers no respite from strenuous civilian leadership, without which it may be evaded or used against its sponsor to defeat the objectives of civilian political control and rational policy.

New code of ethics. In the Kennedy administration's early days, a new conflict-of-interest code was promulgated, representing an improvement over archaic and overly rigid requirements, but lacking real bite. The code was embodied in a Presidential Memorandum (February 1962) applied largely to part-time advisers and consultants; a strict hardware ban against firms holding contracts for technical advisory services was also imposed. A somewhat vague Defense Department directive was issued about the same time, stating that all department personnel, military and civilian, would be expected to exercise discretion where acceptance of a favor "might reasonably be interpreted" as affecting the objective performance of their duty. The soft impact of these directives served to alert the government hierarchy to the new leadership's intentions.

The Bell Report stated: "Still other standards are needed, and we recommend that you request the head of each department and agency which does a significant amount of contracting for research and development to develop . . . clear-cut codes of conduct, to provide standards and criteria to guide the public officials and private persons and organizations. . . ." [12] After three years it was found necessary to provide more precise and positive guidelines. In late 1964 a new Defense Department directive was issued, categorically forbidding acceptance of any favor, gratuity, or entertainment directly or indirectly "from any person, firm, corporation, or any other entity which has engaged, is engaged, or is endeavoring to engage in procurement activities or business transactions of any sort with any agency of the Department of Defense. . . ." Exceptions were made for circumstances where "the interests of the government may be served," such as when department people participate in meetings sponsored by industrial, technical, and professional associations, "where the host is the association and not an individual contractor." Acceptance of entertainment or hospitality from private companies at such affairs was specifically prohibited. Personnel might participate in contractor-sponsored activities on a strictly defined basis, such as the ceremonial launching of ships or the unveiling of new weapons systems.

The teeth of the new directive: any personnel accepting any favor, gratuity, or entertainment were required to report the incident and the circumstances within forty-eight hours! Discovery of instances not so reported would constitute prima-facie impropriety.[13] There was a clear implication that unreported exceptions turned up by Congress, the GAO, or the press would be subject to punishment.

Two major trade associations (which combined military men and contractors) promptly condemned the directive and accused McNamara of impugning their integrity. Retired Army General Arthur G. Trudeau, president of the American Ordnance Association, wrote to the Secretary his misgiving about "such stringent regulations. . . ." Air Force Major General Jess Larson, president of the Air Force Association, charged that the innocent had been tarred "with the brush that was meant for the guilty few." [14] Contractors' response was two-sided. They instructed employees to refrain from such acts as buying lunches for contract officials; at the same time, they began not to keep records of government personnel who were entertained on expense accounts (chargeable to contract overhead). Several firms made it known they would actively resist General Accounting Office probes into expense account records.[15]

The editor of an aerospace trade weekly inveighed against "the most insulting document to both the industry and the military ever to be issued from the Pentagon." He admitted that "booze, blondes, and bashes" have in some cases invaded the relationship between government and contractors, and that the latter frequently hired military officers "purely for their Pentagon influence." Nevertheless, he angrily wrote, ". . . we can assure you, Mr. McNamara . . . that nothing exists any more contemptible than your own attitude toward the industry and the people who work for you." [16] In a later editorial he termed GAO and congressional inquiry into the use of expense accounts by contractors a "sickening affair." They were police-state tactics which led the country "one further step away from the democracy which this industry and the military presumably are pledged to defend." McNamara's actions seemed to be "almost part of a deliberate plot to dismantle what has been a very effective partnership in the past." The editor was equally nauseated by industry's "lack of moral courage": where was the man, he asked, "with the guts to stand up and say that a business luncheon is an accepted practice in this country . . . ? Was there no one to suggest that considerable business of benefit to the government was conducted over luncheon tables?" [17]

One Defense Department contract officer had the courage publicly to come to McNamara's defense,[18] declaring it obvious that gratuities and lunches were not tendered primarily "for friendship's sake." The difficulty of drawing the line between one and one hundred free lunches made it clear, he said, "that McNamara is correct in drawing the line at *no* lunches." The system by which contractors woo procurement officers has reached "the same impasse that giving stamps at supermarkets has reached," the competitors canceling each other's advantage and succeeding only "in raising the cost of doing business." Younger procurement personnel, he testified, "invariably feel odd in accepting lunches and gifts. . . . Their instincts are correct, and it is the old-timers in the business who have conditioned themselves to accept the situation as normal . . . McNamara did the right thing. . . ."

Reforming the contract instrument. Indispensable to progress in cost reduction have been Defense Department efforts to rebuild the in-house technical establishment, to give government laboratories meaningful work and career opportunities, and to attempt to exert control over allowable contractor salaries. On the basis of better technical advice, preparation and evaluation of contract specifications, and better direction of contractor work, adequate in-house would make it possible to switch from cost-plus contracts to open competitive bidding for a larger part of Defense Department procurement. Where negotiated contracts are unavoidable, the department has made a maximum effort to eliminate unwarranted sole-source status and automatic follow-on production contracts. This has been done by inviting multiple-source proposals and converting from fixed to incentive fees, placing pressures on contractors to reduce waste or face the prospect of paying for waste out of fees.

A comprehensive Defense Department program is now in midstream to eliminate contracting-out for technical support of weapons systems; some eleven thousand contractor personnel are in the process of being eliminated or replaced with in-house people.[19] There has been a vigorous effort to upgrade in-house laboratories. A top departmental official (Chalmers W. Sherwin, Deputy Director for Research and Technology) was assigned to direct the work of centralizing control of and building the irreplaceable resources for genuine R&D management. The department sought to recruit its main support from the service laboratories themselves, bringing civil service scientists and engineers into the technical policy process at the highest levels. According to Sherwin, the Defense Department in its future work will emphasize "basic research," with maximum flexibility and initiative for individual laboratories.[20] The purpose is not merely to foster useful and creative work but to keep alive the in-house labs as a management resource and yardstick, and as "windows through which to watch government-contracted university research." The department will seek to define R&D problems before they are turned over to industry, to open areas of unorthodox and pioneering work otherwise by-passed by the contractor bandwagon of applied research and development projects.

The move to maximize fixed-price competition and cost-reduction incentives arose from a reading of the Bell Report, in whose writing McNamara himself was a key participant. Genuine written bids permit the free forces of competition to play their role. Government "is benefited . . ." and ". . . abuses such as favoritism, collusion, nepotism, that attend subjective procurement," are reduced, declared the Joint Economic Committee in 1963. The advantages of competitive bidding, especially when formally advertised and open, "are so overwhelming and conclusive that they give validity to the intent of Congress that such procedure provide the basic rule, and negotiated procurement the exception." [21] The actual trend of procurement, however, had made "the exception" the general rule.

McNamara set fixed goals for each service in the return to more competitive methods. Negotiated procurement in fiscal 1962 was 87 percent of total purchases of $26 billion.[22] Sole-source and cost-plus-fixed-fee awards were 38 percent of the total. The goal was to cut the latter back to 12.3 percent (of new procurements) by 1965, the level that characterized defense contracting prior to the missile race. It was reached one year ahead of time: approximately $5.5 billion was converted from cost-plus-fixed-fee to "more disciplined types, principally fixed-price or incentive types." In 1965 the target was exceeded, reducing the use of non-competitive contracts to 10 percent of new procurement value.

Fixed-price, open competitive bidding rose from 3 percent of contract actions in 1961 to 46.3 percent in 1965. But the fixed-price picture was less rosy in terms of contract value, reaching only 18.6 percent during the first eight months of fiscal 1965 (as compared with 11.9 percent in 1961).[23] Nevertheless, reversal of the old trend was healthy and hopeful. McNamara even put aside the "Buy America" Act in order to bring foreign competition to bear, throwing open to the British, for example, a procurement involving a number of small naval vessels. The act permits this, but the realities of politics have prevented the exercise of such discretion for many years.

In the switch from cost-plus to fixed-price, the Defense Department ran into the problem of contractor prices set on the basis of cost-plus experience, artificially inflated as protection against unfamiliar risks. This fact was reported by the General Accounting Office in 1965 and became the source of a disagreement: the department preferred to maintain the integrity of the fixed-price contract even if it temporarily caused excessive profits. It argued that the cost-base experience thereby acquired for future contracting would ultimatly redound to the government's benefit.[24]

Just as artificially high (and often collusive) prices were used to defeat expansion of fixed-price contracts, so contractors and agencies turned to cost-plus-incentive-fees to satisfy the new regulations where fixed prices were not possible (as in R&D). But the incentive standards were often so soft and permissive as to maintain a high level of tolerance and to transfer most of the risks back to government, thereby defeating McNamara's intent. GAO reports on these practices led in late 1965 to a determined effort by contractors and by some military agencies (especially the Air Force) to exploit the minor differences between the Defense Department and GAO to destroy the latter's independence and objectivity, an effort in which the House Government Operations Committee cooperated and which led to the sudden resignation of the GAO's head.

But the reformers sought ways to control collusive waste and excess profit, moving to place stronger departmental review and control authority over agency contracting, and to spell out strictly and in greater detail the structure of the contract itself. Where fixed prices were not used, several innovations were made: A clear distinction was added to Armed Services Procurement regulations between what the prime contractor makes in his own facilities and what he buys or subcontracts from others. In instances where it was determined that the prime transferred risk to a subcontractor, the profit or fee allowed would be appropriately reduced.[25] Sole-source status would be rigorously re-examined. Automatic R&D follow-on would be resisted by inviting a minimum of two bidders and, where there is a little price history available, inquiring "much more deeply into the independence of the bids." [26] A new regulation required that unsuccessful bidders receive a statement of technical reasons for failing to get a job, forcing the contracting official to find objective and defensible reasons for his decision, and enabling the losing contractors to police the fairness of awards themselves, at least to some extent.[27] In competitive proposals contractors were asked to submit incentive formulas:

Contractors submitting unduly conservative proposals, which involve little or no risk, will endanger their competitive position. Conversely, contractors who are unduly optimistic will be in danger of being awarded the contract at a very low profit or even a loss. Accordingly it can be expected that the arrangements will compel more clarity and integrity in the preparation and submission of proposals for development contracts.[28]

Some examples of the new regime might be cited:

Contracts for building Titan III launch vehicles were converted to cost-plus-incentive-fee (CPIF), imposing penalties on the prime for deviating from target costs, schedule, or performance objectives. The contractor (Glenn Martin Company) could earn from $6 to $35 million profit depending on performance. As a result of the contract form, spending on the program (nearing half a billion) was claimed to be 1 percent below the original cost estimates, the number of engineering changes the lowest of any similar procurement, and the program was on schedule (an event!).[29]

Introduction of a second industry source for the "Bull Pup" air-to-surface missile led to competitive negotiated bidding where none had previously existed, dropping unit prices some 20 percent and saving more than $40 million. The Bull Pup was the first procurement to be successfully second-sourced. This concept was intended to introduce competitive elements in the pre-award stage, to force a maximum amount of the eventual contract to fixed-price and incentive-fee arrangements, while at the same time keeping the final procurement sole-source so as to avoid annual start-up costs and more than one overhead allocation. The Bull Pup case was the first of its kind. The Martin Company, which developed the missile as prime contractor, lost the contract for follow-on production to Maxson Electronics Corporation, a small company which under previous arrangements would have been forced to accept Martin's subcontracts for the work.[30] The same pattern was also applied for procurement of ASTOR (nuclear anti-submarine torpedo), the TALOS (air-defense missile) system, the Sidewinder 1-C, and, most recently, the C-5A, seven-hundred-passenger super-transport and cargo plane.

The Bull Pup award was sharply criticized: "The first time in the industry's history," *Missiles and Rockets* editorialized,

> the prime contractor and developer of a weapon system has watched all the production work go elsewhere . . . if industry firms cannot count on production work to defray the high overhead which results from maintenance of a research and development capability, this inevitably must drive up the cost . . . if the nation pays less for production . . . but more for its development, is there actually a net cost saving? [31]

No major weapons system, the editorial alleged, had been developed without "substantial financial participation by the manufacturer." But the facts confute these allegations, indicating not only that R&D contractors find ways of profiting from this work, but that they also manage to charge to federal contracts their own independent and contract-seeking R&D.

There is ample evidence that the new regime is paying dividends to the public and correcting some outrages of the Contract State. In fiscal 1964 contractors made sizable personnel reductions, giving closer attention to hiring and release of personnel in accordance with real work loads; major reductions were reported in overhead (in one company the ratio of indirect to direct employment dropped from 60 to 40 percent); overtime expenditures have been sharply reduced from an average of 12 to 4 percent; there has been a notable increase in contractor resistance to contract changes which increase costs, thus putting the contractor on the public side in defeating the tendency toward gold-plating; there are instances in which cost estimates are being *underrun,* a revolutionary change.[32]

Project 60. An early move to discipline and unify the far-flung operations of the military was the establishment of the Defense Supply Agency (DSA), which became operational in 1962 and has achieved a remarkable record. At the end of 1965 the DSA had saved the government over $56 million, despite an increase in work load. Since its inception, the united work force assigned to warehousing, inventory, and supply has been cut by more than 25 percent (seven hundred to eight hundred fewer people); facilities have been operated at a higher and more stable level of utilization; and duplication, unwarranted stockpiling, and other forms of waste have been greatly reduced. At the same time, DSA has a much better efficiency record than the older system, maintaining "stock availability" better than 90 percent, and "on-time fill of requisitions" at 85 percent (1965)—both indices having climbed in each successive year.

Continuing study has enabled the Defense Department to turn over to the DSA each year thousands of additional items for centralized management.[33] DSA has permitted top departmental leadership to monitor logistical inventories (now about forty million items) as never before, making visible for the first time to Congress and the public the enormous procurement investment of the military, the dimensions and composition of which prior to the DSA were often unknown to the services themselves.

As part of tightened inventory control, the Defense Department undertook to identify government-owned equipment in the possession of contractors, and began a policy of government ownership of major equipment acquisitions at contractor plants (rather than paying for these as direct costs and permitting the contractor himself the rights of property). In the area of data processing, McNamara determined that the department would purchase some of the computer capability contractors leased for government work, with an annual saving of more than $100 million. (In this area the department did not go quite as far as the GAO wished, but it did seek a fuller utilization of such equipment in areas marked by a high concentration of defense contractors and/or military installations.) The policy of government ownership of major equipment, if effectively implemented, would also enhance the competitive role of small business in seeking government contracts, since large companies would have to give up the government's capital equipment for reassignment to new contractors who previously, for want of access to such equipment, had to include such acquisition cost in their bids.

The success of DSA provided the springboard for "Project 60," an experimental program initiated during 1962 to explore other means of centralizing and consolidating defense procurement, contract administration, and auditing. Its success led to creation of additional unified management services, such as the Defense Communication and Terminal Service, the Contract Administration Service, and the Contract Audit Agency. These mechanisms have begun the process of unifying the logistical activities of the military services, accounting for a substantial portion of recent cost-reduction economies. More important, they have given civilian leadership a source of real decision-making information and control. Project 60 also has forced NASA and other non-defense agencies to imitate the Defense Department's example to some extent, enabling President Johnson to make his "war on waste" something more than a blithe public relations slogan.

In the prosaic area of supply and facility management, many simple and common-sense innovations have become possible which did not exist previously: the compilation of a supply and surplus catalog for the whole federal system; the founding of an in-house surplus store, in which excess supplies of one agency may be discovered and acquired for the needs of another; standardization of on-shelf items and parts among the services, and so on. Consolidation of activities has made possible the identification and elimination of underutilized facilities: by the end of 1964 some 669 locations were declared surplus, 150,000 jobs eliminated, and over a million acres of land released, with a total annual savings in excess of $1 billion, and with the accompanying problems of economic adjustment for the communities affected. (Working with community leaders, the Defense Department's Office of Economic Adjustment has sought to develop taxpaying industries from surplus properties, restoring them to the tax rolls by sale to individuals and small companies, or transferring them to public uses as municipal airports, swimming pools, recreation land, or community development projects.)

In consolidating contract administration, Project 60 began with a regional pilot program centered on Philadelphia (encompassing most of Pennsylvania, the southern counties of New Jersey, and all of Virginia, Maryland, West Virginia, and the District of Columbia). During 1964-1965 it proved to critics that cost reduction, efficiency, convenience, and departmental control were not incompatible with contractor performance. During 1965 the program was extended to the whole nation, which was divided into eleven regional districts. The enthusiasm of the President forced NASA, which deals with the same contractors, to join the system. By the end of fiscal 1966 it was anticipated that 150 existing field offices and over twenty thousand personnel would be re-combined into eleven centers under a single national management.[34]

McNamara was quick to recognize the need for an independent central auditing agency to implement the consolidated contract administration and to replace the self-serving audit resources of the individual services, which were controlled by the same people responsible for procurement actions. Unable to enlist the services of the General Accounting Office, McNamara created (in July 1965) the Defense Contract Audit Agency (DCAA) in order to (in the words of Elmer Staats of the Budget Bureau) "strengthen the central management controls of the Secretary of Defense." [35] The DCAA director reports to the Secretary, and auditing personnel are organizationally independent of procurement and contract administration. The new agency expected to build a staff of approximately 3,600, with 250 operating offices in major cities and in plants of major contractors. In order to break out of the habits of two decades, McNamara was authorized to hire 265 new auditors whose loyalty to professional standards and to the overall interests of the government would be cultivated, preserved, and rewarded.[36]

These new management tools were not to apply to basic research (mostly in universities and in industrial laboratories) which would be protected against over-administration under the more flexible chain of command of the government's refurbished in-house scientific establishment:

> The reason that basic research is not included . . . is that basic research . . . is still primarily a scientist-to-scientist relationship. You need high technical talent to observe the progress under the research contract or grant. You don't have problems of inspection and acceptance of material. You don't have security clearances in the same sense. Basic research contract and grant administration was omitted, and we feel that it should stay outside this, at least until . . . we are further down the road.[37]

Technical management of major weapons programs remained a responsibility of the military services through the department's R&D apparatus, although the post-contract administrative aspects of weapons R&D and procurement were included under the consolidated arrangements.

Another aspect of the unified system was the establishment of a Cost and Economic Information System on a department-wide basis, to make instantly available to contract negotiators material on the past performance and capabilities of contractors. A contractor rating system was to be developed to aid in source selection and in determination of fees. Essentially an information clearinghouse, the system was designed to improve cost estimation, analysis, and progress reporting; enhance the effectiveness of planning, programing, budgeting, contract negotiating, and project management; and provide data for the analysis of the economic impact of federal procurement by geographical area and industry. The Comptroller was given overall responsibility for the program.[38]

Industry reaction to Project 60 innovations was typified by Sylvania Vice President Lawrence J. Straw: the procedures "have made the cost of scrutinizing the dollars . . . as much as the cost of doing the work." Or by Raytheon Vice President Lawrence Levy: "These rigid contractual procedures . . . have tended to stifle new ideas from industry and have caused reduced industry initiative." [39]

The Name of the Game

The campaign to knife through bureaucratic impassivity and reform the folkways of procurement followed a course beset by obstacles. With the innovations of Project 60 and the support of the President—sometimes wavering in political cross-streams but generally sustained—the campaign showed steady progress; but it also showed the resources of devious thrust and parry available to the forces of the Contract State.

Just as NASA from its inception had been penetrated by the Air Force and used to escape control of Eisenhower budget ceilings, and as federal agencies had used their contractors to escape administration surveillance and discipline, so—as McNamara plunged boldly ahead—did contractors and the congressional patrons find means to maneuver against him. To a substantial degree they managed to preserve the old folkways in NASA contracting in spite of formal adoption of the Defense Department's example. Congressional defenders of the status quo remained active, using their traditional power over lawmaking and appropriations to bargain with the President, successfully blocking or diluting the impact of many McNamara policies.

In 1965 the name of the game had been changed, but to a considerable extent it was the same. The very instruments of the McNamara system—competitive procurement, incentive fees, and fixed prices—were inverted. Through agency and contractor collusion, contracts were converted to the new system on the basis of grossly inflated cost figures, often containing unwarranted provisions for implausible contingencies. Incentive fee schedules frequently became cushions against risk, so that at worst the contractor did all right, and at best he made the usual excessive profits. Contractors and agencies were quick to discover that under the law fixed-price contracts protected them against GAO and Defense Department scrutiny, and they turned this discovery to their own purposes, joining the McNamara bandwagon as enthusiastic marchers, relying upon the permissive attitude of contract officers and the lack of governmental yardsticks to assure a continuation of the old game.

NASA's dilemma. Because it deals with the same contractors, NASA felt the pressure of the Defense Department regime and the President's "war on waste." McNamara's program imposed a yardstick against NASA contracting which could not be gainsaid, forcing NASA to join Project 60 and to introduce, at least in form, the same management reforms. Considering virtually all of its procurement as R&D, rather than production, NASA is able to avoid any large-scale use of advertised bidding and to lump into R&D awards even those items (such as construction of facilities and use of off-shelf items) that could be separately contracted by this means.[40] At the end of fiscal 1964 a mere 13 percent of NASA awards were placed under fixed-price contracts. In 1962 NASA did not have a single incentive-type contract; in 1964 it set about remedying this fault, but without much reliance either on competitive methods or on adequate audit. By mid-1965 $1 billion of contract value had been converted to the incentive formula, representing about 10 percent of procurement.[41] It was the declared policy to bring incentive-type contract value to $5.5 billion by the end of fiscal year 1966. But the conversion procedure generally omitted any real multiple-source competitive stage.

The method was illustrated by the Gemini spacecraft contract with McDonnell Aircraft: the contractor himself was invited to submit a proposal for an incentive formula, which was ratified by the Space Agency with only minor changes. In NASA authorization hearings (for fiscal 1966) Senator Clinton Anderson noted the softness of the so-called incentive formula in a case where the contractor had already overrun the negotiated target (by $14.5 million) and the work was seriously behind schedule. Under the incentive formula the minimum fee was set at $1.6 million with a maximum of $3.8 million. In spite of the faulty performance, NASA indicated that the final fee would be established "somewhere between those numbers." Senator Anderson commented: "I thought an incentive fee was given if somebody did the work ahead of schedule, did it at low cost, and turned in a good job. But this has overrun all the way and then you give an incentive fee for doing that."[42] In a bow toward a higher degree of competition, a source selection board was established at NASA headquarters to study competitive proposals for future contracts worth more than $5 million.

NASA did seek to establish something approximating a "step" method of contract awards, like that of the Defense Department. It has declared its intention eventually to eliminate the use of "letter-contracts," under which most NASA contracting has initially been done, and which tends to place the government in the hands of the contractor before any agreement on contract terms. As of mid-1964 the agency required that all letter-contracts, regardless of amount, be authorized by headquarters. On the other hand, the agency reduced headquarter responsibility by raising from $1 million to $2.5 million the dollar level of procurement by the six NASA field centers, awards not subject to approval by Washington.[43]

In the absence of any kind of adequate central and independent auditing, the incentive fee schedule and all aspects of contracting remain somewhat academic. A generous amount of hidden profit and waste can be built into both direct costs and overhead, as we have seen. NASA has never found it necessary to seek independent audits and has enjoyed virtual immunity from the scrutiny of the General Accounting Office, a fact which suggests a number of interesting queries which only a few legislators have enunciated, to no avail. Though tied into the centralized contract management of Project 60, NASA is not subject to the Defense Department's central audit and lacks any sort of independent inspection system of its own. Some congressmen have urged the General Accounting Office to devote more attention to NASA: ". . . this is getting into big money," Senator Holland declared, "and there is so much change, reprograming, and adoption of new programs that have not been specifically submitted to this committee in advance, that it seems to me this committee would be wise to ask the General Accounting Office to give us the benefit of its careful analysis . . ." [44] By mid-1965 only eight minor GAO reports had been made on NASA (compared with 582 for the Defense Department),[45] only four of which dealt with contract matters. The GAO has explained its lack of curiosity on grounds that "the entire operation . . . is a research and development activity . . . there isn't too much in the way of management involved for us to review in that type of contract." [46] In other words, R&D contracts are inauditable.

Supported by a majority of the congressional space committees, Administrator Webb has vigorously opposed all efforts to subject NASA's contracting practices to review, rejecting even a mild proposal by Senator Richard Russell that would have required NASA to maintain a central file on contract communications. Webb testified that "it is not apparent to this agency that more effective machinery is needed . . . than exists today . . . Current practices provide ample opportunity for the committees of the Congress to review and investigate the handling of any particular procurement action . . ." [47]

Bending the new rules. Defense contractors have learned how to play the game by McNamara's rules, setting up their own cost-reduction staffs which were soon discovering "savings" all over the lot—half a billion of which McNamara's auditors could not validate and rejected. Symptomatic was a letter from a cost-reduction official of

General Dynamics, which chastised the editor of *Missiles and Rockets* for the latter's frontal assault on the program.[48] The letter pointed out that the Defense Department's "seven pages of guidelines" merely provided program criteria and might be assimilated by the industry with little inconvenience: "Emphasis throughout is placed on the Defense Department's intent to rely upon, and fully utilize, the contractor's present internal management system for planning, executing, validating, and reporting results." The writer revealed the ambivalence of his own attitude when he termed "somewhat imprudent" the trade journal's attack on the program as an aspersion upon contractor honor.

The cost-plus-incentive-fee contract is not a new instrument, but its use had been very limited prior to McNamara. The padding of such contracts was not unknown in earlier periods, as reported by the Hebert Investigation of 1961: "Rather than the purported 'savings' which we were led to expect, the GAO reports revealed serious failures of government negotiators in costing and pricing for incentive-type and other contracts."[49] Fourteen specific cases involving the Air Force were found in which "prices were negotiated without adequate consideration of available information resulting in . . . about 30 percent profit to the contractor." In 1965 the GAO reported a large number of instances in which the conversion to incentive and fixed-price contracts actually added millions of dollars of cost above what would have been required to complete the work on the basis of existing cost-plus contracts.[50]

GAO blamed contracting officers who were altogether too permissive in judging the reasonableness of costs: "If the decision by the contracting officer—and this is often in the hands of one man—is quite favorable to the contractor, certainly the contractor is not going to appeal, but the government has no means of having the contracting officer's decision reviewed. . . . We think there might be some procedures for having that decision reviewed on the government's behalf."[51]

The 1965 Aerospace Corporation inquiry might be used to judge the penetration of McNamara's reforms. A diligent and frustrated Air Force contract officer summarized the situation: "There has *not* been any improvement, sir."[52] For his candor, Air Force officials removed him from his job and sent him into outer darkness.

A retrograde movement also has become evident in the use of the Bull Pup contracting approach; there exists no feasible method for objective costing except examination of "competitive" multiple-source proposals. The tendency to distribute large awards on a cartelized basis among major contractors and long-standing practice of industry price collaboration make "competitive" bidding something of a formality. The kind of collusion that exists is not exactly a criminal conspiracy, but rather a kind of implicit gentlemanly agreement which keeps cost estimates at the high, risk-cushioning levels that have become "normal," and therefore legitimate, business. There is plenty of work to go around, and temporary imbalances in the contract market are compensated by subcontracting trade-offs, keeping the process of cartelization intact and making the government (and McNamara) more or less helpless to intervene, helpless to make the "free enterprise" contractors accept any financial risks, and helpless to protect the public interest.

If anything, the McNamara reforms are forcing the contractor industry into more explicit collusion than was necessary under the less rigorous system. The Justice Department may eventually find a basis for action here, but the time is not yet and the ultimate impact of reform is still to be determined. The contractors could undo everything in minutes if continuing energy and attention were not directed at these issues from the top of government. The changes of personnel in the Defense Department in 1965 were a mixed bag. Except for Harold Brown, who became civilian head of the Air Force, most of the McNamara team fled to more congenial positions outside government. Most of the replacements were men of questionable allegiance to the new system. McNamara himself has been busy with Santo Domingo, Vietnam, and a wide variety of extra duties thrust upon him by the President. The hope is that the system itself has acquired some built-in momentum which bodes well for the future.

Congress and the GAO. The wave of McNamara reform precipitated a curious reversal of the relations of Congress to the executive branch and to its own creature, the General Accounting Office. Established in 1921 as the result of profiteering and loose contracting practices in World War I, GAO authority as the national "inspector general" has been gradually expanded. Its quest for economy and efficiency has been buttressed by powers "to examine any books, documents, or records" pertaining to negotiated awards, and under certain conditions to order suspension of payments due contractors. Playing the role of "defender of the public purse," Congress has relied upon the GAO for detailed surveillance of the executive.

The McNamara regime has used GAO as a real management tool and has achieved considerable success in mastering the sprawling defense establishment, equaling the GAO in enthusiasm for efficiency and economy, and at long last energetically trying to enforce these values. As a result, a great majority of legislators has manifested growing annoyance and disillusionment with the GAO. McNamara reforms generate mixed emotions. Where Congress previously used GAO charges as a tool of political bargaining with the administration, they now face the prospect of genuine action to eliminate waste. This turnabout unmasks their economizing pretensions, revealing true congressional motives to be not economy and efficiency in the use of taxpayer funds, but a maintenance of bargaining power over the ways in which the public purse should be both used and abused.

This situation resulted in a 1965 congressional investigation of the GAO in response to contractor complaints against McNamara rigor. Asking Congress to clip GAO's wings, the spokesmen of the contractor community made it clear that their real concern was with the combined impact of the McNamara reforms, aided and abetted by GAO audits. Boeing's Howard W. Neffner declared:

> As a result of its expanding role within government, GAO has become intimately involved in the daily activities of many government agencies. In the exercise of its more recently acquired right to audit contractors, GAO is attempting a similarly deep penetration of industry. The increasing number of audit reports, the scope and uncompromising nature of GAO criticism, and the efforts made to enforce its recommendations are already dominant factors in procurement policy and practice.[53]

H. M. Horner, chairman of United Aircraft, declared: "Increased congressional guidance of GAO is needed to make sure that that agency does not, through its wide powers, seek to impose unwise policies on the executive agencies, which may lead, among other things, to increased defense procurement costs and to undermining the integrity of government contracts." [54]

In McNamara's reforms, based upon a long history of GAO reports and recommendations, contractors saw a philosophy fundamentally subversive to "free enterprise": in the words of Lockheed's president, "a contrary philosophy that contractors whose business is primarily with the government are, in reality, agents of the government rather than private businesses . . ." [55] A trade association officer declared that, in the Comptroller General's view, defense industry was "analogous to a regulated industry and hence profits on negotiated defense contracts are to be limited in somewhat the same fashion as utility company profits are limited by regulatory commissions." [56]

The resignation of Comptroller General Joseph Campbell in the middle of the 1965 GAO hearings "for reasons of health" embarrassed the House Military Operations Subcommittee (Committee on Government Operations) and Chairman Chet Holifield: "Let me say emphatically, to remove any misunderstandings on this score," Holifield declared, "that our subcommittee recognizes the need and importance of the General Accounting Office as an arm of the Congress." [57] But the "misunderstandings" persisted, fostered by the unusually sycophantic treatment accorded to industry spokesmen, viz:

MR. CALLAWAY. I wonder if I might be granted the special privilege of welcoming my close friend, Dan Haughton. He is a former Georgian. He was head of Georgia Lockheed, the largest single employer in Georgia. He was a great Georgian in every way and we hated to lose him to California, when he became president of Lockheed, and it is a pleasure to have him here today.

MR. HOLIFIELD. I am certainly glad to have that comment and we are glad to have you in California, in the parent company, I might say.

MR. HORTON. We wish you would come to New York.

MR. HOLIFIELD. Any comment from you, Mr. Latta?

MR. LATTA. I would be glad to have them in Ohio.

MR. HOLIFIELD. Mr. Moorhead?

MR. MOORHEAD. Pennsylvania's door is open.

MR. HOLIFIELD. You may proceed. You find yourself welcomed with open arms.

MR. HAUGHTON. It is more than I deserve.[58]

On the other hand, committee members behaved like prosecuting attorneys in dealing with GAO officials. At this writing the committee report and recommendations have not appeared, but there are indications that some revisions of GAO authority are contemplated, including a proposal to set a statute of limitations on its post-audit authority and to limit its investigatory powers. Congressional attempts to weaken the government's "inspector general" will face the determined opposition of a small but dedicated group of legislators who value and respect the GAO's work, including the leadership of the Joint Economic Committee and the House Armed Services Committee. Even without new legislation, however, the investigation had already achieved some effect, alerting the GAO to congressional apprehensions, and in effect warning that its forthright role might involve it in further difficulties.

The Defense Department's new Central Audit Agency can do an important part of the job which the GAO has performed heretofore alone, giving the civilian leadership of the department greater access and control over the $24 billion annual negotiated procurements, certain to grow still larger. But the new agency is neither government-wide nor independent of the department. After McNamara its role may be drastically altered and the old system, still very much alive, may once again emerge, the new unified Defense Department powers distributed again among the parochial and self-serving services. The only government-wide agency is the Bureau of the Budget, which has only incidental post-audit responsibilities and of course lacks independence from the on-going interests of the administration. In spite of congressional interest in GAO as a weapon of political bargaining (for whatever special interests influential congressmen may serve), the preservation of its role is indispensable to providing Congress and the public a source of information on government expenditures independent of the executive and of the agencies of the vast federal establishment.

The House Select Committee on Government Research in 1964 found it "extraordinary that the Congress has spent so much time articulating rules for business deductions" in the tax code, but "practically no time in defining costs under government contracts, including costs under research and development contracts." [59] It is only "extraordinary" in terms of the myth by which Congress interprets its own motives and actions. While vigorously and scathingly inquiring into federal grants and contracts to universities and, in the instance of Aerospace Corporation, into non-profit organizations, the major government contractors are always treated with deference, sympathy, and an elaborate ritual of manners. There has been a deliberate muting of information about impropriety, questionable legality, waste, and profiteering. Instead of Lockheed, GAO is investigated; the brass knuckles of congressional power—though gloved in velvet—are raised against hapless scapegoats by the protectors of the public purse.

Less than a week before President Kennedy's murder, James Reston observed that seldom has the Congress "appeared more futile . . . and morally indifferent than it does today." He noted a number of instances involving "not minor or eccentric members of the Congress but prominent and influential men," to wit the Bobby Baker case and the accepted practice of doing favors for corporations and taking cash rewards. The prevailing "easy, anything-goes approach . . . makes influence-peddling possible." The more the government "has got into the economy of the nation with huge defense and space contracts," the "more casual" has become the congressional attitude toward "what anyone thinks about their conduct."

With the growing crisis of escalating warfare in Vietnam, Congress in the summer of 1965 opened a direct attack on McNamara, charging that military unpreparedness and shortages of equipment were the products of shortsighted, cost-saving campaigns; it called for a return to the methods of emergency mobilization (with the attendant disregard for profits and waste). Administration requests for supplemental appropriations to support a build-up of American forces triggered the attack. Congressional leaders sought to prevail upon the President to declare a national emergency and to loosen the purse strings on all manner of defense contract pump-priming. Reston commented that the administration was

> now discovering what President Eisenhower meant in his farewell warning to the nation about the power of the industrial-military complex in this country. . . . All kinds of powerful forces are involved in this campaign. Congressmen who have been complaining about military bases being closed . . . now want them reopened in the name of preparedness. . . . The leaders of communities whose military programs have been cut back . . . are quietly lobbying to get the old programs restored or new orders assigned to their idle machines. . . . Top officers in the Pentagon who have been nursing their wrath for years over McNamara's reduction of old programs or rejection of new and expensive programs now feel that Viet Nam provides new arguments for their ideas and new excuses for lobbying with willing senators for their pet schemes.[60]

The legislators conveniently forgot their own complicity in starving the nation's conventional warfare capability for two decades while plumping for infinite redundancy of expensive and useless strategic systems, sacrificing the Army and Navy to the ambitions of the Air Force and to the norms of the Contract State. Conveniently forgotten was the vigorous effort of McNamara and Kennedy to give the nation adequate means for conducting limited war and counter-insurgency.

McNamara denied the "equipment gap" and asserted that the Vietnam problem was a limited one and limited measures of preparedness were appropriate. It was something new to have a Secretary of Defense argue for the need to preserve the political and social purposes of the nation. After a long August weekend of decision-making, President Johnson made McNamara's views official policy, gently turning aside the impetuous urgings of his industrial and congressional friends, persevering in his desire to cull and winnow space and defense spending in order to increase the national investment in improving the quality of American life.

The congressional attack on McNamara was blunted but continued to simmer near the surface, ready for the next opportunity. It revealed its presence later in the month when Congress passed the Military Construction Act of 1965 with a congressional veto over military base shutdowns. President Johnson was forced to apply the executive veto. Congress returned the bill to the White House without the objectionable provision, but a new ingredient had been added to legislative-executive relations. LBJ previously had not found it necessary to exercise the veto power, managing through delicate political maneuver to receive legislation from Congress which already embodied the administration position.

In a sense the 1965 veto symbolized the beginning of a breakdown of the formerly smooth-operating channels of Johnson's leadership. In avoiding a further clash with Congress, the President was forced into informal understandings concerning future base closings, understandings which while not legally damaging to the executive's military flexibility would tend to impose a higher degree of consideration for special interests in future Defense Department actions. The methods of quiet political maneuver had failed; Congress had carried the issue to an open test, winning some of its objectives. The new phase meant that the Johnson honeymoon was beginning to wear out and that Congress, relishing its success, would be less reluctant to use the threat of an open clash once again to enhance its influence and to salvage the Contract State. In mid-1966 a number of top military men resigned in protest against what they called the "over-centralization" of the Defense Department, and rumors flourished of the imminent departure of Robert McNamara.

In the throes of new crises the President himself began to waver. Internal unrest, represented by the Negro Revolution, student protests, crime and riots in the streets, began to take on a character as serious as international troubles. Both were increasingly coupled in the nation's reflexes, triggering dangerous impulses both domestically and internationally. A nervous premonition seemed to stir the nation. Vague and ominous shadows darkened—shadows cast by the Contract State, the chronic sickness of an economy flushed with artificial stimulants, renewed inflation, social division, and memories of a murder in Dallas.

CHAPTER

XIX

WHO WILL CHOOSE?

The nature of U.S. involvement in world affairs and the breathless pace of technological change do not present us with so unprecedented a situation as some assert. To a considerable degree the pace and direction of technological change are matters of choice and can be controlled. However rich, no nation enjoys unlimited resources, nor can it escape the necessity for choice in their allocation. It is through such choices that a nation makes itself and gives meaning to its existence. To imitate our enemies is to ignore those things worth doing for ourselves. Further, we run the risk of forcing our rivals to do the same, thereby merely "increasing the cost of doing business" and setting in motion what may prove to be a vicious escalation of unrewarding and depleting exertions in a world of smoke and shadows.

Blind faith in technological innovation ("progress") for the sake of "science," or on behalf of its "inevitability," obscures and evades the problems of political choice. The syllogism blankets the hard and concrete issues concerning the relative social utility and cost of specific investments in technological innovation. "Technological progress is not only natural, but the secret of American affluence," so why is it necessary to quibble whether research and development goes into this or that field, is conducted by universities or private contractors, is concentrated in this or that portion of the country, involves private or public funds? The words "science" and "technology" have become a staple part of solemn declarations by every special interest seeking consideration for its own welfare. They have joined the lexicon of "God" and "motherhood" as self-evident eternal verities.

Popular faith in the mystique of innovation, almost an end in itself, has provided a cover for the emergence of an industrial R&D and systems-engineering management cult with unparalleled private economic and public decision-making power. The interaction between government and its contractors has brought a kind of backhanded national planning which tends to confuse the definition of legitimate defense needs, the requirements of economic health, and the manpower and educational needs of the nation with demands for preservation of a subsidized and sheltered process of industrial and political empire-building. The public consensus for defense, space, and science is distorted to serve the interests of the private contractors who penetrate government at all levels and inevitably interpret narrow special interests as those of the nation. Business and industry have traditionally been close to the centers of political power, but never before have so few enjoyed so broad an acceptance of their role as a virtually independent branch of government.

Cost competition, the traditional safeguard against waste and misallocation of resources, has been suspended. A special immunity against the laws of economic realism is granted to a favored few, and the economic marketplace is subverted by the political cloakroom. For almost three decades the nation's resources have been commanded by military needs, and political and economic power have been consolidated behind defense priorities. What was initially sustained by emergency has become normalized through a cabal of vested interests. In the system which evolved, R&D contracts became the initial installment paid by the nation to ever-fewer corporate empires able to use public decision-making power and hidden profits to extend their reach into the civilian economy. The result is a new wave of concentration, intensifying the gap between developed and underdeveloped sectors of the economy, between active and stagnant geographical regions, between have and have-not Americans.

High-level civilian stagnation, prosperity for some and growing desperation for others, the poultices of new welfare programs and new forms of educational and scientific pork barrel are characteristic. They deny the nation what may be the ultimate basis of diplomatic strength and the only means to maintain the impetus of a mature economy, namely the fullest enjoyment by all our people of the immense bounty of equity and well-being almost within our grasp.

The surviving myths of private enterprise insulate the industrial giants from social control, distorting the national reading of realities at home and abroad, concealing the galloping pace of corporate mergers and economic concentration, protecting the quasi-public status of narrow private interests. In contractor management of public programs are all of the most pernicious attributes of incestuous bureaucracy.

In addition to claims of security, national prestige, and prosperity, the sacred name of science is hailed as a surrogate consensus, an alibi to soften, defer, and deflect the growing divisions of American society. Sputnik I and the Soviet lead in space shook the nation because they provided a clarion call to all groups who sensed or suffered the social standstill of the 1950's and wanted change. Every energetic pressure group sought to capitalize on the alleged mortification, offering a variety of formulas certain to repair the national image. In the ripples of movement and change, the missile race, science, and technology took precedence over most other values. A race with the Soviet Union in these areas became the vehicle which fortified the power of the military-industrial elite.

Broader demands for revitalizing American life and undertaking bold new programs to meet social and environmental needs were held in abeyance until the middle of the present decade when first steps were taken. But here too research was emphasized as a substitute for expensive and controversial action programs. The politics of consensus appeared to dictate shallow ameliorative efforts while basic conditions went largely untouched. The program generated not real social advance but a treadmill of palliatives in the form of tax cuts and increased subsidies to welfare and education. In confronting urgent domestic problems, tokenism rather than real reform became the characteristic response. The science-technology race provided an avenue of substitute pump-priming which maintained personal income without increasing civilian goods, further aggravating inequities in the structure of purchasing power which commands and organizes national resources. These measures and the emphasis on further research did not inconvenience the economically powerful and of course will not touch the sources of unease. Even such first steps were aborted in 1966 by the escalating crisis of Vietnam.

The so-called military-industrial complex is not a conspiracy but rather a culmination of historical trends. It is a fact of contemporary public life that is eating the heart out of our society, reducing potential for real economic and social growth and eroding the foundation of democratic pluralism. In the words of President Eisenhower's Farewell Address, the "conjunction of an immense military establishment and a large arms industry . . ." exerts an influence "felt in every city, every statehouse, every office of the federal government. . . . In the councils of government, we must guard against the acquisition of unwarranted influence. . . . The potential for the disastrous rise of misplaced power exists and will persist."

Before this century, America, safeguarded by two great oceans and the British Navy, endured as a minor nation for more than a hundred years, enjoying considerably less international prestige and power than the great European nations. In this century we have risen to manhood and have witnessed how two catastrophic wars collapsed the power of Europe; we have seen the emergence of Russian power to challenge our own; we have shuddered at the evaporation of our insularity under the shadow of nuclear-tipped intercontinental missiles. The confrontation of Russian and American power has already completed its classic phase: both sides have probed and consolidated the limits of their power, discovering at last their mutual interest in the reduction of fear. For both, the stabilization of the arms race was accompanied by sharpening domestic confrontations and declining power to dominate other nations except at escalating costs imposed by simple people armed only with desperate hope.

Of the two great powers, the United States is the unmistakable leader in wealth, industrial capacity, and political resiliency. As such, we are endowed with a special responsibility. We possess the means to control the pace of the science-technology race throughout the world. But our greatness holds temptations as well as opportunities. There is the temptation to resume the search for infinite security by forcing the arms race onward and upward in the hope that rivals will break before we do. And we are tempted to overcome the inherent limits upon what wealth and power can do by flexing them at a higher level, a process rendered profitless by Russian exertions but also by the inherent inefficacy of these instruments to create long-term influence and world stability.

If we manage to avoid temptation we may stabilize the balance of fear at a reasonable level of cost and concentrate upon deepening domestic crises and the ambiguous tasks of creative diplomacy. In the long run we must be judged by the success of our free society in liberating, humanizing, and conserving values.

In the words of Jerome Wiesner, security means more than arms. "It means good relations with others, a strong economy, and a healthy people, and science contributes in a major way to all these objectives." There is an honest and honorable place for science, but it is a limited place, rationed in accordance with a responsible ordering of national interests. Innovation is, after all, a human activity, and its purpose must lie in its affirmative effect on the total human situation. There is no escape from the allocation of values and the immemorial agony of choice.

The growth of national government and its omnipresence as an organizing force in all activities is inevitable. It is concomitant with the growth of massive private organizations in the economy and increasing interdependence within and between nations. This growth raises the spectre of impersonal and arbitrary power and the corruption of democratic pluralism, but these tendencies are not inevitable. It is within our hands to humanize authority and preserve the multiple forums of a free society so that many voices will still be heard. This is the most profound challenge implicit in the issues raised by public science and private power.

The nation cannot much longer evade the necessity for choice and for positive democratic planning. The familiar controversy over the abstract issue of planning conceals the fact that planning is already here, embodied in the structure of special-interest subsidies, taxes, regulation, fiscal policy, government contracting, and so on. The real heart of the controversy concerns planning for what and for whom. Those who are most vociferous in attacking planning per se are the beneficiaries of the Contract State, the present system of inchoate and disguised planning. Our society must shake itself out of this scratchy groove in order to dispute the real issues: toward what objectives should the national energies and resources be directed? and how can democratic values be preserved?

There is already general agreement about certain broad national priorities—full employment, improvement in the quality of American life, victory over poverty. These priorities are mutually supporting and reinforcing. There is much in the Great Society program to engender hope in spite of conflicts, compromises, and imperfections. The politics of presidential success impose a variety of conflicting values, including the requirements of international diplomacy, domestic commitments, political limits upon the power of government, and the legitimate concern of the President to preserve a high degree of consensus in order to maximize his freedom of action.

Both Presidents Kennedy and Johnson understood and feared the recurring cycle of economic recessions which had under Eisenhower largely been permitted to run their course. Both were convinced of the validity of Keynesian economics and of their responsibility to stabilize and advance prosperity, not only through monetary policy, but also by active sponsorship of new large-scale government programs. The Council of Economic Advisers has recognized the need to improve the structure of income and effective demand by augmented programs to improve the quality of American life. Unfortunately, most of the groups who might benefit by such programs are socially and politically isolated, apathetic, and inarticulate. Practical executive action is shaped more directly by those active, vocal, and influential interest groups already pressing the President's arm and seeking to advance or preserve their stake in the Contract State. To bring these groups to support even modest changes in the structure of income will be difficult; Kennedy's baffling three-year frustrations with Congress established that fact.

Johnson is the beneficiary of JFK's three years of educating the country to more positive values. The shock of the assassination softened many divisive influences and LBJ made himself the instrument of the great wave of conciliation that swept the nation. Now a hopeful beginning has been made in concerting all the instruments of federal power to achieve a balanced collection of social values. The distortions of political and economic power will continue to hag-ride the future, but the real questions must become visible so that the larger implications of the contemporary predicament may be forced into open and forthright dialogue, enabling political leadership to build consensus behind measures to strengthen the whole nation, first and most importantly for its own sake.

U.S. emergence as a super-power with global responsibilities at a time of rapidly changing technology has not destroyed the usefulness of traditional standards of comparative value in allocating resources and energies. In fact, such standards are necessary to redress the quality of American life and the success of American lives, as well as to preserve and enhance the institutions of a free society.

We do not face easy choices, and we shall always make mistakes. But we must understand what is happening to us and what we can do about it. Government cannot pass bills ordaining happiness, but there are many areas in which the needs are apparent and the means of action at hand. As never before we have the power to stand up to nature, to soften and tame the environment to fit our needs, and to humanize our institutions. If we fail to choose, we shall abdicate the decision to the Contract State.

NOTES

CHAPTER I

The Science-Technology Race

1. David E. Lilienthal, *Change, Hope, and the Bomb* (Princeton: Princeton University Press, 1963), p. 29.
2. U.S. Congress, Senate, Committee on Government Operations, Subcommittee on National Policy Machinery, *Hearings on Organizing for National Security,* February 23, 1960–August 24, 1961 (Washington: Government Printing Office, 1961), Part 5, p. 777.
3. Congressman Chet Holifield, Chairman of the Joint Committee on Atomic Energy, quoted in U.S. Congress, House, Committee on Science and Astronautics, *Hearings, Ways and Means of Effecting Economies in the National Space Program*, 87th Cong., 2nd Sess. (Washington: Government Printing Office, 1962), p. 77.
4. Jess Gorkin, *Parade Magazine,* inserted into the *Congressional Record* by Sen. Robert C. Byrd, quoted in U.S., NASA *Astronautics and Aeronautics, 1963* (Washington: NASA Scientific and Technical Information Division, 1964), p. 462.
5. U.S. Congress, House, Committee on Appropriations, *Independent Offices Appropriations for 1963, Part III,* 87th Cong., 1st Sess. (Washington: Government Printing Office, 1962), p. 424.
6. U.S. Congress, Senate, Committee on Astronautics and Space, *Hearings, NASA Authorization for Fiscal Year 1966, Part I, Scientific and Technical Programs and Program Management,* S. 927, 89th Cong., 1st Sess. (Washington: Government Printing Office, 1965), p. 28.
7. U.S. Congress, House, Committee on Science and Astronautics, *Hearings, 1964 NASA Authorization, Part I, March 4 and 5, 1963*, H.R. 5466, 88th Cong., 1st Sess. (Washington: Government Printing Office, 1963), p. 5.
8. *Ibid.,* p. 24.
9. U.S. Congress, House, Committee on Science and Astronautics, Staff Study, *The Practical Values of Space Exploration,* 87th Cong. 1st Sess. (Washington: Government Printing Office, 1961), p. 22.
10. Sen. Stuart Symington, quoted in U.S. Congress, Senate, Committee on Aeronautical and Space Sciences, *Hearings, NASA Authorization for Fiscal Year 1966, Part II, Program Budget Detail,* 89th Cong., 1st Sess. (Washington: Government Printing Office, 1965), p. 971.
11. Dr. Wernher von Braun, quoted in U.S. Congress, House, Select Committee on Government Research, *Hearings, Federal Research and Development*

Programs, Part I, 88th Cong., 1st Sess. (Washington: Government Printing Office, 1964), p. 528.

12. U.S. Congress, Senate, Committee on Aeronautical and Space Sciences, *Hearings, NASA Authorization for Fiscal Year 1964, Part II, Program Detail,* 88th Cong., 1st Sess. (Washington: Government Printing Office, 1963), p. 971.
13. Letter to editor, *Missiles and Rockets,* May 24, 1965, p. 6.
14. U.S. Congress, Senate, Committee on Aeronautical and Space Sciences, *Hearings, Scientists' Testimony on Space Goals,* 88th Cong., 1st Sess. (Washington: Government Printing Office, 1963), pp. 28-31.
15. *New York Times,* June 3, 1965, p. 21.
16. *New York Times,* June 29, 1965, p. 8.
17. U.S. Congress, Senate, Committee on Aeronautical and Space Sciences, *Hearings, NASA Authorization for Fiscal Year 1965, Part II, Program Detail,* S. 2446, 88th Cong., 2nd Sess. (Washington: Government Printing Office, 1964), p. 331.
18. *Federal Research and Development Programs, Part I,* p. 462.
19. *Ways and Means of Effecting Economies in the National Space Program,* p. 59.
20. *Scientists' Testimony on Space Goals,* p. 119.
21. *Ibid.,* p. 52.
22. L. V. Berkner, *The Scientific Age* (New Haven: Yale University Press, 1964), p. 20.
23. U.S. Congress, House, Committee on Science and Astronautics, Report of the National Aeronautics and Space Administration, *Astronautical and Aeronautical Events of 1962,* Committee Print, 88th Cong., 1st Sess. (Washington: Government Printing Office, 1963), pp. 115-161.
24. Dr. Charles L. Dunham, Director, Division of Biology and Medicine, quoted in U.S. Congress, House, Committee on Government Operations, Subcommittee, *Hearings, Civil Defense—1961,* 87th Cong., 1st Sess. (Washington: Government Printing Office, 1961), p. 134.
25. *Science,* September 25, 1964, pp. 1413-1415.
26. U.S. Congress, Senate, Committee on Aeronautical and Space Sciences, *Hearings, NASA Authorization for Fiscal Year 1965, Part I, Scientific and Technical Programs,* S. 2446, 88th Cong., 2nd Sess. (Washington: Government Printing Office, 1964), p. 5.
27. *Astronautical and Aeronautical Events of 1962*, p. 67.
28. *Ibid.*, p. 68.
29. Vernon Van Dyke, *Pride and Power: The Rationale of the Space Program* (Urbana: University of Illinois Press, 1964), p. 85.
30. U.S. Congress, Joint Committee on Atomic Energy, *Hearings, AEC Authorizing Legislation Fiscal Year 1965, Part II, Reactor Development,* 88th Cong., 2nd Sess. (Washington: Government Printing Office, 1964), p. 574.
31. Gen. Bernard A. Schriever, quoted in U.S. Congress, House, Committee on Government Operations, Subcommittee, *Hearings, Systems Development and Management, Part III,* 87th Cong., 2nd Sess. (Washington: Government Printing Office, 1962), p. 816.
32. U.S. Congress, House, Committee on Science and Astronautics, Subcommittee on Manned Space Flight, *Hearings, 1964 NASA Authorization, Part II (b),* 88th Cong., 1st Sess. (Washington: Government Printing Office, 1963), p. 833.
33. *Astronautical and Aeronautical Events of 1962,* p. 117.
34. Walter F. Dornberger, "Military Necessity: The Overriding Reason in

Space and National Security—Symposium," *Space Digest,* November 1961, p. 76.
35. *Air Force–Space Digest,* July 1962, quoted in Van Dyke, p. 55.
36. *Astronautics and Aeronautics, 1963,* p. 131.
37. *Astronautical and Aeronautical Events of 1962*, p. 42.
38. Adm. William F. Raborn, quoted in *Astronautical and Aeronautical Events of 1962,* p. 283.
39. U.S. Congress, House, Committee on Science and Astronautics, *Report, National Meteorological Satellite Program,* House Report No. 1281, Union Calendar No. 549, 87th Cong., 1st Sess. (Washington: Government Printing Office, 1961), p. 5.
40. Dr. Simon Ramo, quoted in *Astronautics and Aeronautics, 1963,* p. 138.
41. U.S. Congress, Senate, Committee on Government Operations, Subcommittee on National Security and International Operations, *Hearings, Conduct of National Security Policy, Part II,* 89th Cong., 1st Sess. (Washington: Government Printing Office, 1965), p. 87.
42. William J. Coughlin, editorial, *Missiles and Rockets,* April 20, 1964, p. 46.
43. Sen. Henry M. Jackson, quoted in *Conduct of National Security Policy, Part II,* p. 85.
44. Jerome B. Wiesner and Herbert F. York, *Scientific American,* October 1964, quoted in *New York Times,* September 25, 1964, p. 16.
45. U.S. Congress, Senate, Committee on Foreign Relations, *Hearings, To Amend the Arms Control and Disarmament Act,* 89th Cong., 1st Sess. (Washington: Government Printing Office, 1965), p. 2.
46. U.S. Congress, House, Committee on Science and Astronautics, *Panel on Science and Technology, Fifth Meeting, No. 1,* 88th Cong., 1st Sess. (Washington: Government Printing Office, 1963), p. 24.
47. U.S. Congress, Senate, Committee on Aeronautical and Space Sciences, *Hearings, Educational Programs of NASA; Facilities, Training, and Research Grants Programs of the National Aeronautics and Space Administration,* 88th Cong., 1st Sess. (Washington: Government Printing Office, 1964), p. 39.
48. *Federal Research and Development Programs, Part I,* p. 177.
49. Dr. Walter Brattain, quoted in *Conduct of National Security Policy, Part II,* p. 106.
50. Dr. Harold Brown, quoted in *Systems Development and Management, Part II,* p. 465.
51. U.S. Congress, House, Committee on Science and Astronautics, *Hearings, Panel on Science and Technology, Fourth Meeting, No. 3,* 87th Cong., 2nd Sess. (Washington: Government Printing Office, 1962), p. 78.

CHAPTER II
Celestial Roulette

1. *Missiles and Rockets,* April 26, 1965, p. 16.
2. Dr. U. S. Troitsky of the Radiophysic Institute in Gorky, and Dr. Frank Drake of Cornell University, reported in *New York Times,* September 15, 1965, p. 25.
3. U.S. Congress, Senate, Committee on Aeronautical and Space Sciences, *Hearings, NASA Authorization for Fiscal Year 1965, Part II, Program Detail,* March 17, 1965, p. 601.

4. Dr. George E. Mueller, *ibid.*, p. 489.
5. William J. Coughlin, reported in *Missiles and Rockets,* March 1, 1965, p. 46.
6. Dr. John A. Hornbeck, head of Bellcomm, quoted in U.S. Congress, House, Committee on Science and Astronautics, Subcommittee on Space Sciences and Advanced Research and Technology, *Hearings, 1964 NASA Authorization, Part III,* 88th Cong., 1st Sess. (Washington: Government Printing Office, 1963), p. 384.
7. Dr. Simon Ramo, quoted in *Scientists' Testimony on Space Goals,* p. 38.
8. Rex Pay, "Apollo Astronauts Painstakingly Protected from Solar Flare Protons," *Missiles and Rockets,* July 20, 1964, pp. 22-26.
9. *Panel on Science and Technology, Fourth Meeting, No. 3,* p. 94.
10. *Astronautics and Aeronautics, 1963,* p. 471.
11. *Panel on Science and Technology, Fourth Meeting,* p. 95.
12. See discussion of the difficulties of knowing accurately the location of tracking stations on earth and the precise nature of the earth's gravitational field in U.S. Congress, House, Committee on Science and Astronautics, Subcommittee on Space Sciences, *Hearings, Project Anna–Geodetic Satellite System,* 87th Cong., 2nd Sess. (Washington: Government Printing Office, 1962), pp. 12-30.
13. Dr. Brainerd Holmes, quoted in U.S. Congress, House, Committee on Science and Astronautics, *Hearings, Space Posture,* 88th Cong., 1st Sess. (Washington: Government Printing Office, 1963), p. 163.
14. John W. Finney, *New York Times,* October 14, 1962, p. 14.
15. Statement of George Low, Deputy Director, Office of Manned Space Flight, *NASA Authorization for Fiscal Year 1964, Part II, Program Detail,* p. 849.
16. Dr. Robert C. Seaman, Jr., Associate Administrator, NASA *NASA Authorization for Fiscal Year 1964, Part I, Scientific and Technical Programs,* pp. 596-597.
17. Quoted in *Astronautics and Aeronautics, 1963,* p. 50.
18. See his statement in *1964 NASA Authorization, Part III,* pp. 27, 30. He suggests that we would have to "start over to develop spacecrafts different from the Apollo."
19. Proposed by John N. Cord and Leonard N. Seale in a paper presented to the Institute of Aerospace Sciences in Los Angeles, June 1962, reported in *Astronautical and Aeronautical Events of 1962,* p. 112.
20. Quoted in Arthur Herzog, *The War-Peace Establishment* (New York: Harper and Row, 1963), p. 148.
21. James Webb, quoted in *NASA Authorization for Fiscal Year 1965, Part II, Program Detail,* p. 275.
22. U.S. Congress, Senate, Committee on Aeronautical and Space Sciences, *Hearings, NASA Authorization for Fiscal Year 1966, Part I, Scientific and Technical Programs and Program Management,* March 1963, p. 27.
23. Quoted in *Astronautics and Aeronautics, 1963,* p. 274.
24. *Ibid.*, p. 303.
25. *Ibid.*, p. 305.
26. *Ibid.*, p. 276.
27. Quoted in *Science,* October 25, 1963, p. 470.
28. *New York Times,* October 27, 1963, p. 1.
29. *Astronautics and Aeronautics, 1963,* p. 401.
30. *Ibid.*, p. 409.
31. *New York Times,* August 24, 1965, p. 1.

32. Interview with Dr. Wernher von Braun, *U.S. News and World Report,* June 1, 1964, p. 55.
33. U.S., NASA, *Astronautics and Aeronautics, 1964, Chronology on Science, Technology, and Policy,* prepared by the NASA Historical Staff, Office of Policy Planning, NASA SP-4005 (Washington: NASA, Scientific and Technical Information Division, 1965), p. 205.
34. *Ibid.,* p. 356.
35. "The Challenge of International Competition," Appendix A, in U.S. Congress, Senate Doc. 56, *Report,* August 1965, p. 447.
36. *Ibid.,* pp. 450-451.
37. Quoted in *Washington Post,* October 19, 1963, p. A4.
38. Public Law 89-128 (79 Stat. 534), approved August 16, 1965.
39. U.S. Congress, Senate, Committee on Aeronautic and Space Sciences, *Hearings, National Space Goals for Post-Apollo Period,* August 1965, p. 116.
40. *Ibid.,* p. 115.
41. *Ibid.,* p. 323.
42. *Astronautics and Aeronautics, 1964,* p. 76.

CHAPTER III
The Politics of Space

1. R. Cargill Hall, "Early U.S. Satellite Proposals," in Eugene M. Emme, ed., *The History of Rocket Technology: Essays on Research, Development, and Utility,* published in cooperation with the Society for the History of Technology (Detroit: Wayne State University Press, 1964), p. 74.
2. See testimony of George B. Kistiakowsky, *Panel on Science and Technology, Fourth Meeting,* p. 61.
3. *Ibid.,* p. 62.
4. U.S. Congress, House, Committee on Science and Astronautics, *Report, A Chronology of Missile and Astronautic Events,* House Report No. 67, Union Calendar No. 34, March 8, 1961, 87th Cong., 1st Sess. (Washington: Government Printing Office, 1961), p. 17.
5. Wernher von Braun, "The Redstone, Jupiter, and Juno," in Emme, *The History of Rocket Technology,* p. 111.
6. *A Chronology of Missile and Astronautic Events,* p. 23.
7. *Ibid.,* p. 23; also U.S. Congress, House Committee on Governmental Operations, *Eleventh Report, Organization and Management of Missile Programs,* House Report No. 1121, Union Calendar No. 483, 86th Cong., 1st Sess. (Washington: Government Printing Office, 1959), pp. 129-130.
8. John P. Hagen, "The Viking and the Vanguard," in Emme, *The History of Rocket Technology,* p. 128.
9. This material and a full report of the 1954 through 1959 struggle is contained in *Eleventh Report, Organization and Management of Missile Programs,* September 1959.
10. *Ibid.,* Section VII.
11. The Redstone Arsenal had for two years already conducted configuration and design studies of this "super" booster capable of generating 1.5 million pounds of thrust and lifting multi-ton payloads into earth orbit or smaller payloads to deep space missions. See von Braun, *op. cit.,* pp. 119-120.
12. U.S. Congress, Senate, Special Committee on Space and Astronautics,

National Astronautics and Space Act, *Hearings on Sen. 3609,* 1958, especially Part I.

13. In mid-1960 he resigned to become senior vice president of the Northrop Corporation.
14. Maj. Gen. John D. Medaris, *Countdown for Decision* (New York: Putnam, 1960), p. 267.
15. This was done by the paper transfer of Gen. Ostrander to NASA where a new division of launch vehicles was set up under his control to superintend the work of the von Braun team. This information is reported in *A Chronology of Missile and Astronautic Events,* p. 97.
16. *Ibid.,* p. 965.
17. U.S. Congress, Senate, Committee on Aeronautical and Space Sciences, *Hearings, Investigation of Governmental Organization for Space Activities,* 86th Cong., 1st Sess. (Washington: Government Printing Office, 1959), p. 413.
18. *Ibid.,* p. 428.
19. *A Chronology of Missile and Astronautic Events,* p. 87.
20. The letter was leaked to the *New York Times* about a month later and is cited in *Aeronautical and Astronautical Events of 1961,* p. 136.
21. Letter in response to an inquiry by the House Science and Astronautics Committee in *ibid.,* p. 11.
22. U.S. Congress, House, Committee on Science and Astronautics, *Military Astronautics* (Preliminary Report) (Washington: Government Printing Office, 1961), p. 36.
23. See statement of Dr. Hugh L. Dryden, Deputy Administrator of NASA, in *1962 NASA Authorization, Part III,* p. 104.
24. Quoted in *Aeronautical and Astronautical Events of 1961,* p. 54.
25. Speech made on floor of House, September 6, 1962, quoted in *Astronautical and Aeronautical Events of 1962,* p. 176.
26. Quoted at the time of his retirement in *New York Times,* July 4, 1965, p. 16.
27. U.S. Congress, House, Committee on Science and Astronautics, Subcommittee on Science, Research and Development, *Hearings, Government and Science,* 88th Cong., 1st Sess. (Washington: Government Printing Office, 1964), p. 187.
28. See the stories offered by *Time,* June 21, 1963, p. 21; William E. Baggs, "Man on the Moon," *New Republic,* August 3, 1963, pp. 6-7; and Richard Austin Smith, "Agonizing Reappraisal of the Moon Race," *Fortune,* November 1963, p. 269.
29. *Astronautics and Aeronautics, 1963,* p. 21.
30. Marvin Miles, *Washington Post,* April 8, 1963, p. 1.
31. *1963 NASA Authorization, Part II (a),* pp. 629-700.
32. U.S. Congress, House, Select Committee on Government Research, *Hearings, Federal Research and Development Programs, Part I,* H. Res. 504, 88th Cong., 1st Sess. (Washington: Government Printing Office, 1964), p. 81.
33. See U.S. Congress, House, Committee on Science and Astronautics, Subcommittee, *Development of Large Solid Propellant Boosters* (Washington: Government Printing Office, 1962).
34. U.S. Congress, House, Committee on Science and Astronautics, *Report, Centaur Launch Vehicle Development Program,* House Report No. 1959, Union Calendar No. 812, 87th Cong., 2nd Sess. (Washington: Government Printing Office, 1962), p. 2.

35. Testimony of Homer E. Newell, Director, Office of Space Sciences, NASA, *Centaur Program,* p. 6.
36. *Ibid.,* p. 44.
37. *Ibid.,* pp. 31-32.
38. See statement of Finn J. Larsen, Assistant Secretary of the Army for Research and Development, U.S. Congress, House, Committee on Science and Astronautics, *Hearings, Project Advent—Military Communications Satellite Program,* 87th Cong., 2nd Sess. (Washington: Government Printing Office, 1962), p. 3.
39. *Ibid.,* p. 20.
40. U.S. Congress, House, Committee on Science and Astronautics, *Report, Project Advent: Military Communications Satellite Program,* House Report No. 2558, Union Calendar No. 1068, Pursuant to House Resolution 55, 87th Cong., 1st Sess. (Washington: Government Printing Office, 1962), p. 6.
41. *NASA Authorization for Fiscal Year 1964, Part I, Scientific and Technical Programs,* p. 612.
42. Statement of Harold Brown, Director of Defense Research and Engineering, *NASA Authorization for Fiscal Year 1965, Part II, Program Detail,* S. 2446, p. 461.
43. Quoted in *Missiles and Rockets,* March 30, 1964, p. 71.
44. Statement of Secretary of the Army Cyrus Vance, U.S. Congress, Senate, Committee on Aeronautics and Space Sciences, *Hearings, NASA Authorization for Fiscal Year 1966, Part I, Scientific and Technical Programs and Program Management,* S. 927, 89th Cong., 1st Sess. (Washington: Government Printing Office, 1965), p. 560.
45. *National Space Goals for Post-Apollo Period,* p. 303.
46. Statement of Secretary of the Army Cyrus Vance, *NASA Authorization for Fiscal Year 1966, Part I, Scientific and Technical Programs and Program Management,* p. 560.
47. James J. Haggerty, Jr., quoted in *Astronautics and Aeronautics, 1964,* p. 10.
48. *National Space Goals for Post-Apollo Period,* p. 29.
49. *Ibid.,* pp. 108-109.

CHAPTER IV
Innovation and Economic Growth

1. Kenneth Boulding: "Imperial adventure or political coercion is simply an investment with a much lower rate of return than investment in applied science and technological progress at home." This can be seen, for instance, "in the case of Portugal which now has probably the largest per capita empire and the lowest per capita income in Europe. By contrast the Scandinavian countries and Switzerland, which have refrained from imperial adventures, have probably done better economically than their imperial counterparts." The dismantling of empires during this century springs not only from the changing balances of world power but also from the "recognition that in terms of the values of a modern society, empires simply do not pay." *The Meaning of the Twentieth Century: The Great Transition,* World Perspectives, Vol. 34, Ruth Nanda Anshen, ed. (New York: Harper and Row, 1964), p. 12.
2. William McChesney Martin, Jr., Chairman, Board of Governors of the

Federal Reserve System, *Conduct of National Security Policy, Part IV,* p. 197.

3. U.S. Congress, House and Senate, Joint Economic Committee, *1965 Economic Report,* p. 19.
4. John H. Rubel, Assistant Secretary of Defense (Deputy Director, Defense Research and Engineering), *Trends and Challenge in Research and Development* (mimeographed speech, 1964), p. 47.
5. Reported by Kathleen McLaughlin, *New York Times,* July 14, 1965, p. 57.
6. U.S., President's Council of Economic Advisers, *Report of the Committee on the Economic Impact of Defense and Disarmament* (Washington: Government Printing Office, 1965), p. 8.
7. Based on the analysis of 1956 through 1964, Council of Economic Advisers and Bureau of Labor Statistics, submitted for the record by Walter Reuther. U.S. Congress, Joint Economic Committee, *Hearings, January 1965 Economic Report of the President, Part IV, Invited Comments,* 89th Cong., 1st Sess. (Washington: Government Printing Office, 1965), p. 3.
8. U.S. Congress, Joint Economic Committee, *Invention and the Patent System,* Materials Relating to Continuing Studies of Technology, Economic Growth, and the Variability of Private Investment, Joint Committee Print, 88th Cong., 2nd Sess. (Washington: Government Printing Office, December 1964), p. 20.
9. *Invention and the Patent System,* p. 33.
10. Excerpts from his address printed in *Science,* November 29, 1963, p. 1129.
11. Robert Solo, cited in Amitai Etzioni, *The Moon-Doggle: Domestic and International Implications of the Space Race* (Garden City: Doubleday, 1964), p. 73.
12. *Ibid.,* p. 68.
13. Quoted by Lawrence Galton, "Will Space Research Pay Off on Earth?" *New York Times Magazine,* May 26, 1963, p. 29.
14. Dr. James A. Van Allen, testifying in *Panel on Science and Technology, Fifth Meeting,* p. 23.
15. Rubel, *op. cit.*, p. 1.
16. According to former Secretary of Commerce Luther Hodges, quoted in *Astronautics and Aeronautics, 1963*, p. 127.
17. See Etzioni, p. 74.
18. Speech to the Public Affairs Conference of the U.S. Chamber of Commerce, quoted in *Astronautics and Aeronautics, 1963*, p. 24.
19. *Address to the National Security Industrial Symposium, March 1963* (mimeograph press release).
20. U.S. Congress, Joint Committee on Atomic Energy, *Hearing, Stanford Accelerator Power Supply,* 88th Cong., 2nd Sess. (Washington: Government Printing Office, 1964), pp. 24-25.
21. Dr. J. Herbert Hollomon, Assistant Secretary of Commerce for Science and Technology, U.S. Congress, House, Committee on Science and Astronautics, Subcommittee on Science, Research, and Development, *Hearings, Government and Science; Distribution of Federal Research Funds; Indirect Costs re Federal Grants,* 88th Cong., 2nd Sess. (Washington: Government Printing Office, 1964), p. 221.
22. Arthur Kramish, quoted in *Astronautics and Aeronautics, 1964,* p. 88.
23. Minority Views, U.S. Congress, Joint Economic Committee, *Report, January 1965 Economic Report of the President with Minority and Additional Views* (Washington: Government Printing Office, 1965), p. 101.

24. Statement by Dr. Thomas L. K. Smull, NASA Director of Grants and Research Contracts, quoted in *ibid.*, p. 226.
25. U.S. Congress, House, Select Committee on Government Research, *Report, Impact of Federal Research and Development Programs, Study No. VI,* House Report No. 1938, Union Calendar No. 831, H. Res. 504 as amended by H. Res. 810, 88th Cong., 2nd Sess. (Washington: Government Printing Office, 1964), p. 73.
26. According to Luther Hodges in an address at Virginia Polytechnic Institute, cited in *Astronautics and Aeronautics, 1963,* p. 127.
27. *January 1965 Economic Report of the President with Minority and Additional Views,* p. 101.
28. Seymour Melman, *Our Depleted Society* (New York: Holt, Rinehart and Winston, 1965), p. 50.
29. Cited in *Astronautics and Aeronautics, 1963,* p. 36.
30. *Impact of Federal Research and Development Programs, Study No. VI,* p. 129.
31. Piel, *op. cit.,* p. 108.
32. Lilienthal, p. 73.
33. Quoted in "News and Comment," *Science,* June 28, 1963, p. 1380.
34. Robert Colburn, *International Science and Technology,* June 1962, p. 11.

CHAPTER V
Entropy and Pump-Priming

1. Figures by Lester R. Brown, staff economist, U.S. Department of Agriculture, quoted in *New York Times,* May 30, 1965, p. 8.
2. Quoted in *New York Times,* June 10, 1965, p. 23.
3. Quoted in *New Republic,* September 18, 1965, p. 6.
4. Committee on Population, National Academy of Sciences–National Research Council, "Report, The Growth of U.S. Population," in *Science,* May 28, 1965, p. 1205.
5. Quoted in *New York Times,* June 20, 1965, p. 15.
6. Reported in *Milwaukee Journal,* August 8, 1965, p. 1.
7. Interview in *U.S. News and World Report,* February 3, 1964, p. 76.
8. Source of data, Survey of Current Business Economic Indicators, cited in *January 1965 Economic Report of the President, Part IV, Invited Comments*, p. 4.
9. Data from the Survey of Current Business, 1962, cited in *ibid.*, p. 5.
10. U.S. Congress, Joint Economic Committee, Subcommittee on Fiscal Policy, *Hearings, Fiscal Policy Issues of the Coming Decade,* 89th Cong., 1st Sess. (Washington: Government Printing Office, 1965), p. 3.
11. *Ibid.,* p. 12.
12. U.S., President's Council of Economic Advisers, *Report of the Committee on the Economic Impact of Defense and Disarmament,* July 1965, p. 20.
13. *Ibid.,* pp. 25-26.
14. *Fiscal Policy Issues of the Coming Decade,* p. 27.
15. Dr. Philip H. Abelson, Director of Carnegie Institute's Geophysics Laboratory and editor of *Science,* quoted in *Astronautics and Aeronautics, 1963,* p. 164.
16. Rubel, *op. cit.,* p. 3.

17. *January 1965 Economic Report of the President, Part II,* pp. 13-14.
18. *New York Times,* June 14, 1965, p. 36.
19. Cited in *Astronautics and Aeronautics, 1963,* p. 167.

CHAPTER VI
Science: Process and Ideology

1. Gerald Holton, "Modern Science and the Intellectual Tradition," in *The New Scientist: Essays on the Methods and Values of Modern Science,* Paul C. Obler and Herman A. Estrin, ed. (Garden City: Doubleday, 1962), p. 24.
2. C. F. von Weizsacker, *The Relevance of Science: Creation and Cosmogony, Gifford Lectures 1959-1960* (New York: Harper and Row, 1964), p. 17.
3. Benjamin Lee Whorf, "Science and Linguistics," reprinted from *Language, Thought and Reality,* John B. Carroll, ed. (Boston: The Technology Press of Massachusetts Institute of Technology, 1957), p. 216.
4. Warren Weaver, "Science and People," in Obler and Estrin, *The New Scientist,* pp. 107-108.
5. Quoted by James R. Newman, *Science and Sensibility* (New York: Simon and Schuster, 1961), II, 20.
6. Quoted in *International Science and Technology,* May 1962, p. 75.
7. Cyril Smith goes one step further and contends: "There is a definite interplay between his knowledge of the properties of matter and the kinds of things he thought worth doing." "Materials and the Development of Civilization and Science," *Science,* May 14, 1965, p. 908.
8. Jerome B. Wiesner, "Technology and Society," *Science as a Cultural Force,* The Shell Companies Foundation Lectures, Harry Woolf, ed. (Baltimore: The Johns Hopkins Press, 1964), p. 46.
9. See Walter R. Dornberger, "The German V-2," in Emme, *The History of Rocket Technology*, p. 30.
10. Interview in *U.S. News and World Report,* June 1, 1964, p. 40.
11. "Time and the Space Traveler," *Atlantic Monthly,* October 1957, pp. 153-158.
12. Richard G. Hewlett and Oscar E. Anderson, Jr., *The New World, 1939-1946: A History of the United States Atomic Energy Commission* (University Park: Pennsylvania State University Press, 1962), p. 142.
13. *Ibid.,* pp. 174-180.
14. Interview with Robert R. Wilson, *International Science and Technology,* May 1965, p. 43.
15. Quoted in Newman, I, 291.
16. Max Born is in the process of preparing for publication his voluminous correspondence with Professor Einstein. *New York Times,* July 1, 1965, p. 4.
17. Vannevar Bush, *Science,* December 27, 1963, p. 1623.
18. Quoted in *New York Times,* January 8, 1963, p. 37.
19. *New York Times,* November 20, 1964, p. 30.

CHAPTER VII

The Scientists: High Priests or Vestal Virgins?

1. Cited by Don K. Price, "Organization of Science Here and Abroad," *Science,* March 20, 1959, p. 759.
2. As described in D. C. Beardslee and D. D. Dowd, "The College Student Image of the Scientist," *Science,* No. 3457, pp. 997-1001; quoted by A. Cornelius Benjamin, *Science, Technology and Human Values* (Columbia: University of Missouri Press, 1965), p. 159.
3. Francis Bello, "The Young Scientists," in Obler and Estrin, *The New Scientist,* pp. 65-68.
4. Judge Alexander Haltzoff made the comment in sentencing Bernhard Deutch to ninety days in jail. *New York Times,* December 14, 1956, p. 15.
5. U.S. Congress, Senate, Committee on the Judiciary, Subcommittee to Investigate the Administration of the Internal Security Act and Other Internal Security Laws, Staff Analysis, *The Pugwash Conferences,* Committee Print, 87th Cong., 1st Sess. (Washington: Government Printing Office, 1961), pp. 72-73.
6. *Ibid.,* p. 49.
7. *Ibid.,* p. 21.
8. Boulding, *op. cit.,* p. 45.
9. C. P. Snow, *The Two Cultures and the Scientific Revolution* (New York: Cambridge University Press, 1959), p. 11.
10. Warner R. Schilling, "Scientists, Foreign Policy, and Politics," in *Scientists and National Policy Making,* Robert Gilpin and Christopher Wright, ed. (New York: Columbia University Press, 1964), p. 170.
11. Robert C. Wood, "Scientists and Politics: The Rise of an Apolitical Elite," in *ibid.,* p. 69.
12. Article in *Christian Century,* quoted in *Time,* April 7, 1958, p. 57.
13. Quoted in *Science,* March 30, 1962, p. 1114.
14. See Robert M. Hutchins, *Science, Scientists, and Politics: An Occasional Paper* (Santa Barbara: The Fund for the Republic, Center for the Study of Democratic Institutions, 1963), pp. 1-2.
15. See Wood, *op. cit.,* p. 48.
16. Hewlett and Anderson, pp. 533-535.
17. Letter to D. Landucci, quoted by Arthur Koestler, *The Watershed: A Biography of Johannes Kepler* (Garden City: Anchor Books, 1959), p. 185.
18. *Scientists' Testimony on Space Goals,* June 10 and 11, 1963, p. 154.
19. NASA contract award of $80,000 announced on September 16, 1962.
20. Interview with Wernher von Braun, *U.S. News and World Report,* June 1, 1964, p. 56.
21. The title of a recent book by Ralph E. Lapp, *The New Priesthood* (New York: Harper and Row, 1965).
22. Cited by Jerry E. Bishop, "Brains for Barter," *Wall Street Journal,* December 30, 1963, p. 20.
23. Quoted by Daniel S. Greenberg, "News and Comment," *Science,* September 25, 1964, p. 1416.
24. Quoted by Daniel S. Greenberg, "News and Comment, *Science,* October 30, 1964, p. 2144.

25. Source: 1962 Federal Reserve Board estimate, projected to 1965 in *U.S. News and World Report,* October 11, 1965, p. 119.
26. Quoted in U.S. Congress, Senate, Committee on Government Operations, Staff Analysis and Summary, *Science and Technology Act of 1958,* Doc. No. 90 on S. 3126, 85th Cong., 2nd Sess. (Washington: Government Printing Office, 1958), p. 110.
27. Cited by Dr. Frederick Seitz, President, National Academy of Sciences, U.S. Congress, House, Committee on Science and Astronautics, Subcommittee on Science, Research, and Development, *Hearings, Government and Science,* p. 5.
28. See Daniels letter to Edison in Josephus Daniels, *The Wilson Era: Years of Peace, 1910-1917* (Chapel Hill: University of North Carolina Press, 1944), p. 491.
29. Cited in *Aeronautics and Astronautics, 1915-1960,* p. 48.
30. *Science,* September 18, 1964, p. 1282.
31. Newman, II, p. 105.
32. Quoted by Jerome B. Wiesner, *Where Science and Politics Meet* (New York: McGraw-Hill, 1961), p. 11.

CHAPTER VIII
The Politics of Artful Brutality: Oppenheimer and Strauss

1. For full description of these events see Hewlett and Anderson, pp. 198-203.
2. Reported in *ibid.,* p. 355; also see James F. Byrnes, *All in One Lifetime,* p. 284, cited by Lewis L. Strauss, *Men and Decisions* (Garden City: Doubleday, 1962), p. 191.
3. Cited by J. Robert Oppenheimer in U.S. Atomic Energy Commission, *In the Matter of J. Robert Oppenheimer,* transcript of hearing before Personnel Security Board (Washington: Government Printing Office, 1954), p. 45.
4. *Ibid.,* p. 69.
5. *Ibid.,* pp. 66-69.
6. *Ibid.,* p. 75.
7. *Ibid.,* pp. 76-77.
8. See testimony of Isidor Rabi in *ibid.,* p. 453.
9. *Ibid.,* pp. 79-80.
10. *Ibid.,* p. 710.
11. *Ibid.,* p. 718.
12. Testimony of J. Robert Oppenheimer in *ibid.,* p. 83.
13. *Eleventh Report, Organization and Management of Missile Programs,* p. 117.
14. *In the Matter of J. Robert Oppenheimer,* p. 763.
15. *Ibid.,* p. 764.
16. *Ibid.,* p. 754.
17. *Ibid.,* p. 94.
18. *Ibid.,* p. 95.
19. White House News Conference, transcript, June 30, 1954.
20. Letter of April 13, 1954, quoted by Dwight D. Eisenhower, *Mandate for Change,* Part 13, *Washington Post,* October 25, 1963, p. A6.
21. Cited by Marquis Childs, *Eisenhower, Captive Hero* (New York: Harcourt, Brace, 1958), p. 186.
22. See Piel, *op. cit.,* pp. 126-127.

23. *In the Matter of J. Robert Oppenheimer*, p. 710.
24. Quoted by Giorgio de Santillana, "Galileo and J. Robert Oppenheimer," *Reporter*, December 26, 1957, p. 12.
25. See Harry Kalven, Jr., "Case of J. Robert Oppenheimer Before the AEC," *Bulletin of the Atomic Scientists*, September 1954, p. 262.
26. Piel, *op. cit.*, p. 129.
27. De Santillana, *op. cit.*, pp. 10-18.
28. *Story of a Friendship* (New York: George Braziller, 1965).
29. Strauss, *op. cit.*
30. *New York Times*, December 27, 1954, p. 5.
31. Letter to editor, *New York Times*, November 29, 1960, p. 5.
32. See U.S. Congress, Senate, Committee on Interstate and Foreign Commerce, *Hearings, The Nomination of Lewis L. Strauss to be Secretary of Commerce*, 86th Cong., 1st Sess. (Washington: Government Printing Office, 1959).
33. *Ibid.*, p. 373.
34. Strauss, pp. 164-169.
35. *The Nomination of Lewis L. Strauss to be Secretary of Commerce*, p. 373.
36. *New York Herald Tribune*, June 18, 1954, p. 16.
37. U.S. Congress, Senate, Floor Debate, June 4, 1959, *105 Congressional Record* 9809.
38. Quoted by Dr. David L. Hill, *The Nomination of Lewis L. Strauss to be Secretary of Commerce*, p. 726.
39. Quoted in *The Pugwash Conferences*, pp. 33-35.
40. C. P. Snow, *Science and Government*.
41. Quoted in *New York Times*, January 29, 1955, p. 7.

CHAPTER IX
The Emergence of Pluralism

1. John W. Finney, *New York Times*, December 1, 1957, Section IV, p. 4.
2. Interview with George Kistiakowsky, *International Science and Technology*, October 1964, pp. 49-50.
3. U.S. Congress, Senate, Committee on Government Operations, Subcommittee on National Policy Machinery, *A Study, Organizing for National Security*, Pursuant to S.R. 20, Committee Print, 87th Cong., 1st Sess. (Washington: Government Printing Office, 1961), pp. 3-4.
4. *Conduct of National Security Policy, Part II*, p. 105.
5. See the transcript of Hearings, U.S. Congress, Joint Committee on Atomic Energy, *Technical Aspects of Detection and Inspection Controls of a Nuclear Test Ban*, 86th Cong., 2nd Sess. (Washington: Government Printing Office, 1960).
6. See report by Warren Roger, Jr., *Washington Post*, April 24, 1960, p. 1.
7. See A. M. Rosenthal, *New York Times*, May 9, 1960, p. 1.
8. Reported in *New York Times*, May 12, 1960, p. 1.
9. See *Bulletin of the Atomic Scientists*, September 1960, p. 303; and *New York Times*, July 19, 1960, p. 3.
10. Statement of Dr. Paul Doty, *The Pugwash Conferences*, p. 90.
11. See U.S. Congress, Senate, Committee on the Judiciary, Subcommittee to Investigate the Administration of the Internal Security Act and Other

Internal Security Laws, *Hearings, Testimony of Dr. Linus Pauling,* 86th Cong., 2nd Sess. (Washington: Government Printing Office, 1960).

12. *The Pugwash Conferences,* 1961.
13. Meg Greenfield, writing in *Reporter,* quoted by Etzioni, p. xiii.
14. *Astronautical and Aeronautical Events of 1962,* p. 3.
15. Statement of Dr. George E. Mueller, Associate Administrator of NASA, *NASA Authorization for Fiscal Year 1965, Part I, Scientific and Technical Programs,* p. 67.
16. Statement of Robert C. Truax, U.S. Congress, Senate, Committee on Aeronautical and Space Sciences, *Economies in the Space Program* (Washington: Government Printing Office, 1962), p. 140.
17. Speech delivered in Houston, Texas, reported in *Astronautics and Aeronautics, 1963*, p. 440.
18. Cited by John W. Finney, *New York Times,* October 14, 1962, p. 46.
19. Wiesner, p. 53.
20. "Communication," *Science,* January 31, 1964, p. 429.
21. Quoted in *Science,* January 22, 1965, p. 357.
22. Quoted in *Science,* October 9, 1964, p. 232.
23. "Willy Higinbotham," Letter from the Chairman, *FAS Newsletter,* June 1965, p. 4.
24. The board of directors includes Ruth Adams, non-scientist managing editor of the *Bulletin of the Atomic Scientists;* Maurice Fox, MIT associate professor of biology; Jerome D. Frank, professor of psychiatry at Johns Hopkins; Matthew Meselson, professor of biology at Harvard; James Patton, head of the National Farmers Union; and Charles Pratt, Jr., a New York photographer.
25. Quoted in *New York Times,* June 18, 1962, p. 12.
26. See testimony, Lt. Gen. James G. Ferguson, Deputy Chief of Staff, Research and Development, Air Force, *Space Posture,* p. 231.
27. Quoted in Lapp, p. 30.
28. Dr. Philip Abelson, *Scientists' Testimony on Space Goals,* p. 22.
29. See testimony of Dr. Schwarzchild, *ibid.,* pp. 160-161; also that of Abelson, *ibid.,* p. 11.
30. *Panel on Science and Technology, Fifth Meeting,* p. 32.
31. Quoted in *Astronautics and Aeronautics, 1963,* p. 162.
32. Lapp, p. 170.
33. *Scientists' Testimony on Space Goals,* p. 147.
34. Lapp, p. 179.
35. Words of Sen. E. L. Bartlett, *Astronautics and Aeronautics, 1963,* p. 289.
36. Interview with Clinton P. Anderson, *International Science and Technology,* April 1964, p. 62.
37. *Ibid.,* p. 56.
38. *Ibid.,* pp. 58-59.
39. Newman, II, 105.

CHAPTER X
The Contract State

1. Robert Colborn, editorial, *International Science and Technology,* April 1965, p. 21.
2. *Systems Development and Management, Part 1,* pp. 51-52.

3. James McCamy, *Science and Public Administration* (Birmingham: University of Alabama Press, 1960), pp. 58-59.
4. *Industrial Research,* December 1959, pp. 8-9; cited in U.S. Congress, House, Committee on Science and Astronautics, Staff Study, *The Practical Values of Space Exploration.* 87th Cong., 1st Sess. (Washington: Government Printing Office, 1961), p. 74.
5. *Astronautics and Aeronautics, 1963,* p. 400.
6. Bell Report, printed in Appendix I, *Systems Development and Management, Part I,* p. 205.
7. U.S. Congress, House, Select Committee on Government Research, *Report, Impact of Federal Research and Development Programs, Study No. VI,* 1964, p. 55.
8. See *Aerospace Facts and Figures, 1962* (Washington: Aerospace Industries Association of America, 1963), p. 20; *U.S. Aeronautics and Space Activities, 1962,* Bureau of the Budget, cited in *Missiles and Rockets,* January 21, 1963, p. 12; see *Aviation Week,* January 21, 1963, p. 30; also *NASA Authorization for Fiscal Year 1965,* p. 56.
9. *Eleventh Report, Organization and Management of Missile Programs,* p. 49.
10. See Carl E. Barnes, "Industrial Research, Is It Outmoded?" *Business Horizons,* Summer 1964.
11. U.S. Congress, House, Select Committee on Government Research, *Report, Contract Policies and Procedures for Research and Development, Study VII,* House Report No. 1942, Union Calendar No. 835, H. Res. 504 as amended by H. Res. 810, 88th Cong., 2nd Sess. (Washington: Government Printing Office, 1964), p. 58.
12. U.S. Congress, House, Committee on Armed Services, Subcommittee for Special Investigations, *Hearings, Sole-Source Procurement, Part I,* H. Res. 78, 87th Cong., 1st Sess. (Washington: Government Printing Office, 1961), pp. 17-18.
13. Quoted in *Missiles and Rockets,* July 20, 1964, p. 15.
14. *Harper's,* January 1960, p. 43.
15. Testimony of David E. Bell, Director, Bureau of the Budget, *Systems Development and Management, Part I,* p. 4.
16. *Panel on Science and Technology, Sixth Meeting,* p. 65.
17. Ralph J. Cordiner, "Competitive Private Enterprise in Space," in *Peacetime Uses of Outer Space,* Simon Ramo, ed. (New York: McGraw-Hill, 1961), p. 222.
18. Statement by William H. Ryan, then lobbyist for the AFL-CIO, now New York congressman, testifying, *Systems Development and Management, Part I,* p. 127.
19. Don K. Price, "The Scientific Establishment," in *Scientists and National Policy Making,* p. 39.
20. William J. Coughlin, *Missiles and Rockets,* August 2, 1965, p. 46. Governor Brown's statements are cited here as well.
21. James M. Gavin, Address to the International Bankers Association, December 1958, quoted in *The Practical Values of Space Exploration,* p. 25.

CHAPTER XI
The Ramo-Wooldridge Story

1. Hewlett and Anderson, pp. 186-187.
2. *Eleventh Report, Organization and Management of Missile Programs,* p. 70.
3. Term used in *ibid.,* p. 70.
4. Robert L. Perry, "The Atlas, Thor, Titan, and Minuteman," in Emme, *The History of Rocket Technology,* p. 144.
5. *Eleventh Report, Organization and Management of Missile Programs,* p. 72.
6. *Ibid.*
7. *Systems Development and Management, Part III,* p. 951.
8. *Ibid.,* pp. 163-164.
9. *Eleventh Report, Organization and Management of Missile Programs,* p. 72.
10. *Ibid.*
11. *Ibid.,* pp. 77-78.
12. *Ibid.,* p. 74.
13. Statement of Herbert Roback, staff director, *Systems Development and Management, Part II,* p. 489.
14. *Eleventh Report, Organization and Management of Missile Programs,* p. 81.
15. *Ibid.,* pp. 81-82.
16. Quoted in *ibid.,* p. 82.
17. *Ibid.,* p. 86.
18. At the time of the congressional investigation, a new eighteen-month contract was signed and "the fee to be paid has in no way been reduced." *Ibid.,* p. 87.
19. Quoted in *ibid.,* p. 89.
20. *Ibid.*
21. *Ibid.,* p. 90.
22. *Ibid.,* p. 91.
23. *Ibid.*
24. *Ibid.,* p. 92.
25. *Ibid.,* p. 91.
26. *Aeronautical and Astronautical Events of 1961,* p. 66.
27. *Scientists' Testimony on Space Goals,* p. 27.
28. *New York Times,* October 16, 1961, p. 1.
29. Reported in *Missiles and Rockets,* December 21, 1964, p. 35.
30. *New York Times,* April 20, 1965, p. 61.
31. Statement of Herbert Roback, *Systems Development and Management, Part III,* p. 979.
32. *Sole-Source Procurement, Parts I and II,* May and June 1961.
33. *Ibid.,* I, 443.
34. *Ibid.,* p. 452.
35. *Ibid.,* pp. 53-54.
36. Reported in *Missiles and Rockets,* November 30, 1964, p. 29.
37. *Eleventh Report, Organization and Management of Missile Programs,* p. 93.
38. *Ibid.,* p. 94.
39. *Ibid.,* p. 95.
40. *Ibid.,* p. 96.
41. *Ibid.,* p. 97.

42. *Ibid.*, p. 98.
43. *Systems Development and Management,* Part I, p. 67.
44. Rule announced September 17, 1962, reported in *Astronautical and Aeronautical Events of 1962,* p. 191.
45. Navy official quoted in *Missiles and Rockets,* November 9, 1964, p. 16.

CHAPTER XII
Throwing Away the Yardstick

1. 10 *U.S. Code* 4532.
2. 10 *U.S. Code* 9532.
3. *Conduct of National Security Policy, Part II,* p. 116.
4. U.S. Congress, Senate, Committee on Aeronautical and Space Sciences, *Hearings, Investigation of Governmental Organization for Space Activities,* March-May 1959 (Washington: Government Printing Office, 1959), p. 425.
5. Eisenhower policies expressed, for example, by *Bureau of Budget Bulletin 60-2,* dated September 21, 1959, which states that maximum military and space R&D shall be performed by private contractors.
6. U.S., President, Science Advisory Committee, *Report, Strengthening American Science* (Washington: Government Printing Office, 1958), pp. 16-17.
7. U.S. Congress, House, Committee on Science and Astronautics, *Hearings, Miscellaneous Business* (Washington: Government Printing Office, 1961), p. 37.
8. U.S. Congress, House, Committee on Science and Astronautics, Subcommittee on Space Sciences and Advanced Research and Technology, *Hearings, NASA Authorization, Part III (b),* 88th Cong., 1st Sess. (Washington: Government Printing Office, 1963), p. 3.
9. Statement of David E. Bell, *Systems Development and Management, Part I,* p. 6.
10. *Ibid.*, p. 127.
11. U.S. Congress, House, Select Committee on Government Research, *Manpower for Research and Development, Study Number II,* H. Res. 504, 88th Cong., 2nd Sess. (Washington: 1964), p. 34.
12. *Government and Science,* p. 290.
13. Quoted from statement of chairman, *Panel on Science and Technology, Fifth Meeting,* No. 1, p. 88.
14. *Scientists' Testimony on Space Goals,* p. 158.
15. Reported by Michael Getler, *Missiles and Rockets,* August 24, 1964, p. 24.
16. William J. Coughlin, editorial, *Missiles and Rockets,* June 28, 1965, p. 50.
17. *New York Times,* July 4. 1965, p. 16.
18. Quoted in *Systems Development and Management, Part I,* p. 112.
19. *Scientists' Testimony on Space Goals,* p. 16.
20. *Ibid.*, pp. 25-26.
21. John H. Rubel, *op. cit.*, p. 7.
22. Statement of Dr. James A. Shannon, Director, National Institutes of Health, U.S. Congress, House, Committee on Government Operations, Subcommittee, *Hearings, The Administration of Grants by the National Institutes of Health,* 87th Cong., 2nd Sess. (Washington: Government Printing Office, 1962), p. 75.
23. See statement of Dr. Jerome B. Wiesner, *Government and Science,* p. 80.
24. *Strengthening American Science,* p. 23.

25. See statement of Dr. Robert C. Seamans, Jr., *NASA Authorization for Fiscal Year 1964, Part I,* p. 137.
26. *Space Posture,* p. 7.
27. *Systems Development and Management. Part I,* pp. 164-165.
28. U.S. Congress, House, Committee on Science and Astronautics, *Hearings, Missile Development and Space Sciences* (Washington: Government Printing Office, 1959), pp. 40-41.
29. *NASA Authorization for Fiscal Year 1964, Part II,* p. 853.
30. *Centaur Launch Vehicle Development Program,* p. 12.
31. Statement by Krafft Ehricks, Director, Advanced Studies, General Dynamics/Astronautics, *ibid.*
32. *Astronautics and Aeronautics, 1964,* p. 73.
33. *Ibid.,* p. 324.
34. *Eleventh Report, Organization and Management of Missile Programs,* p. 51.
35. Editorial, *Science,* September 10, 1965, p. 1179.
36. U.S. Congress, House, Committee on Science and Astronautics, Subcommittee on NASA Oversight, *Report, Project Ranger,* House Report No. 1487, Union Calendar No. 637, 88th Cong., 2nd Sess. (Washington: Government Printing Office, 1964), p. 18.
37. *1964 NASA Authorization, Part II (a),* p. 453.
38. *Ibid.,* p. 6.
39. *New York Times,* April 16, 1963, p. 3.
40. *New York Times,* May 5, 1964, p. 36.
41. Reported in *Science,* May 15, 1964, p. 825.
42. Statements of Newell and Nicks reported in *Missiles and Rockets,* May 4, 1964, p. 14.
43. *Missiles and Rockets,* May 11, 1964, p. 15.
44. *Ibid.*
45. *Project Ranger,* pp. 30-31.
46. *Ibid.,* p. 27.
47. *Ibid.,* p. 29.
48. *Ibid.,* p. 24.
49. *Ibid.,* p. 31.
50. Quoted in *New York Times,* May 5, 1964, p. 36.
51. Richard C. Lewis, "At Sea on the Moon," *Bulletin of the Atomic Scientists,* November 1964, p. 361.
52. *Missiles and Rockets,* July 6, 1964, p. 8.
53. *Missiles and Rockets,* June 1, 1964, p. 15.
54. *NASA Authorization for Fiscal Year 1966, Part II,* p. 728.
55. William J. Coughlin, *Missiles and Rockets,* May 11, 1964, p. 46.
56. *1964 NASA Authorization, Part I,* p. 6.
57. William H. Pickering, "Public Fears and the Race into Space," *Armed Forces Management,* April 1959, p. 12.
58. *Federal Research and Development Programs, Part I,* p. 298.

CHAPTER XIII
The New Braintrusters

1. James Killian, Jr., "Toward a Research-Reliant Society: Some Observations on Government and Science," in Harry Woolf, ed., *Science as a Cultural Force* (Baltimore: Johns Hopkins Press, 1964), p. 32.

2. Statement of F. R. Collbohm, RAND President, *Systems Development and Management, Part III,* p. 922.
3. *Ibid.,* pp. 920-921.
4. See statement of Dr. S. J. Lawwill, President, Analytical Services, Inc., *ibid.,* pp. 1074-1075.
5. See testimony of M. O. Koppler, President, Systems Development Corporation, *ibid.,* p. 991; *New York Times,* December 30, 1961, p. 1; *Science,* January 12, 1962, p. 88.
6. See statement of Ivan A. Getting, President, Aerospace Corporation, *Systems Development and Management, Part III,* p. 961.
7. U.S. Congress, House, Committee on Armed Services, Subcommittee for Special Investigations, *Report, The Aerospace Corporation: A Study of Fiscal and Management Policy and Control,* 89th Cong., 1st Sess. (Washington: Government Printing Office, 1965), p. 7.
8. *Ibid.,* p. 56.
9. *Ibid.,* p. 55.
10. *Ibid.,* p. 1.
11. *Ibid.,* p. 3.
12. Quoted in *ibid.,* p. 23.
13. *Ibid.,* p. 3.
14. U.S. Congress, House, Appropriations Committee, *Hearings, Ballistic Missile Program* (Washington: Government Printing Office, 1957), p. 7.
15. *Eleventh Report, Organization and Management of Missile Programs,* pp. 133-135.
16. *Ibid.,* p. 154.
17. *Ibid.,* pp. 136-137.
18. *Systems Development and Management, Part II,* p. 459.
19. *Eleventh Report, Organization and Management of Missile Programs,* p. 479.
20. *Ibid.,* p. 517.
21. *Ibid.,* p. 140.
22. *Ibid.,* p. 141.
23. Lapp, p. 35.
24. Reported in *Missiles and Rockets,* April 5, 1965, p. 9.
25. *Report, The Aerospace Corporation,* p. 48.
26. *Ibid.,* p. 31.
27. *Ibid.,* p. 3.
28. *Ibid.,* p. 21.
29. Lapp, p. 35.
30. *Report, The Aerospace Corporation,* p. 14.
31. Cited by Heather M. David, *Missiles and Rockets,* May 10, 1965, p. 16.
32. *Hearings, The Aerospace Corporation: A Study of Fiscal and Management Policy and Control,* p. 2.
33. *Ibid.,* p. 18.
34. Cited in *New York Times,* May 27, 1965, p. 2.
35. See testimony of Max Golden, Air Force counsel, *Systems Development and Management, Part III,* pp. 889-900.
36. According to John W. Finney, *New York Times,* May 20, 1965, p. 46.
37. See statement of John A. Hornbeck, President, Bellcomm, *Systems Development and Management, Part V,* pp. 1752-1753.
38. *1964 NASA Authorization, Part II (a),* p. 380.
39. *Report, NASA Authorization for Fiscal Year 1964,* p. 50.
40. Quoted by Geoffrey Gould, *Washington Post,* March 31, 1962, p. 6.

41. *1964 NASA Authorization, Part II (a),* p. 386.
42. *Ibid.,* p. 591.
43. *1964 NASA Authorization, Part II (b),* p. 1133.
44. *1964 NASA Authorization, Part II (a),* p. 383.
45. *Ibid.,* p. 381.
46. *Ibid.,* p. 1118.
47. *NASA Authorization for Fiscal Year 1966, Part I, Scientific and Technical Programs and Program Management,* p. 224.
48. *1964 NASA Authorization, Part II (a),* pp. 376-377.
49. Dr. Joseph F. Shea, Deputy Director (Systems), Office of Manned Space Flight, NASA, *1964 NASA Authorization, Part II (b),* p. 1101.
50. Dr. Robert C. Seamans, Jr., *Systems Development and Management,* p. 1705.
51. See statement of John A. Hornbeck, President, Bellcomm, *ibid.,* p. 1754; see also NASA statement summarizing Bellcomm's operating instruction, *1964 NASA Authorization, Part II (b),* p. 1128.
52. *Ibid.,* p. 1093.
53. Reported by John W. Finney, *New York Times,* May 20, 1963, p. 46.
54. *1964 NASA Authorization, Part II (b),* p. 1106.
55. Dr. Joseph F. Shea, *ibid.,* p. 1108.
56. Provisions described by Hornbeck, Bellcomm president, *ibid.,* p. 1754.
57. *Ibid.,* p. 1755.
58. *Systems Development and Management, Part V,* p. 1764.
59. *Ibid.,* p. 1756.
60. *Ibid.,* p. 1760.
61. Statement of Roback, *ibid.,* p. 1766.
62. See statement of David E. Bell, *Systems Development and Management, Part I,* p. 69.
63. *1964 NASA Authorization, Part II (a),* p. 380.
64. See statement of John A. Hornbeck, *ibid.,* p. 1125.
65. Congressman James G. Fulton, *ibid.,* p. 1126.
66. U.S. Congress, Joint Economic Committee, Subcommittee on Federal Procurement and Regulation, *Report, Economic Impact and Federal Procurement,* Joint Committee Print, 89th Cong., 1st Sess. (Washington: Government Printing Office, 1965), p. 8.
67. *1964 NASA Authorization, Part II (b),* p. 1109.
68. *Hearings, The Aerospace Corporation,* p. 59.
69. George L. Heller, Vice President, General Electric, *Systems Development and Management, Part V,* p. 1727.
70. *1964 NASA Authorization, Part II (b),* p. 1119.
71. *Report, The Aerospace Corporation,* p. 25.
72. Statement of George L. Heller, *Systems Development and Management, Part V,* p. 1728.

CHAPTER XIV
"The Sporty Course": Waste and Profit

1. *Hearings, The Aerospace Corporation,* p. 128.
2. Phrase used by R. D. Stephens, a long-time Washington representative of a government contract company; letter to the editor, *Missiles and Rockets,* May 4, 1964, p. 5.

3. Here will be cited only some few typical illustrations. If the reader is interested in the full catalog, he may begin with the following: U.S. General Accounting Office, U.S. Accounting and Auditing Division, "Summary of Audit Reports issued to the Congress for the Period May 1, 1963, through May 1, 1964," printed in *Congressional Record* (daily edition), June 12, 1965, pp. 13147-13153; U.S. General Accounting Office, U.S. Accounting and Audit Division, *Index of Reports on Defense Activities,* Appendices 3 and 4, printed in U.S. Congress, Joint Economic Committee, Subcommittee on Federal Procurement and Regulation, *Background Material on Economic Impact of Federal Procurement—1965,* Joint Committee Print, 89th Cong., 1st Sess. (Washington: Government Printing Office, 1965), pp. 57-212; Department of Defense Comments on General Accounting Office Audit Reports, Appendices 2a through 2e, U.S. Congress, House Committee on Government Operations, Subcommittee, *Hearings, Comptroller General Reports to Congress on Audits of Defense Contracts,* 89th Cong., 1st Sess. (Washington: Government Printing Office, 1965), pp. 735-835.
4. See statement of Chairman Hebert, U.S. Congress, House, Armed Services Committee, Subcommittee for Special Investigations, *Hearings, Overpricing of Government Contracts,* H. Res. 78, 87th Cong., 1st Sess. (Washington: Government Printing Office, 1961), p. 16.
5. *Comptroller General Reports to Congress on Audits of Defense Contracts,* p. 46.
6. Joseph Campbell, Comptroller General, *Overpricing of Government Contracts,* p. 14.
7. Cited by Sen. George McGovern, *Congressional Record* (daily edition), June 12, 1965, p. 13146.
8. Statement of Joseph Campbell, *Comptroller General Reports to Congress on Audits of Defense Contracts,* p. 40.
9. GAO Report cited in *Missiles and Rockets,* January 4, 1965, p. 10.
10. Interview, *U.S. News and World Report,* August 16, 1965, p. 60.
11. See Edward B. Roberts, "How the U.S. Buys Research," *International Science and Technology,* September 1964, p. 72.
12. For an extended list to demonstrate this point, see U.S. Congress, House, Committee on Armed Services, Subcommittee for Special Investigations, *Hearings, Sole-Source Procurement, Parts I and II.* Also see U.S. Congress, Senate, Committee on the Judiciary, Subcommittee on Antitrust and Monopoly, *Hearings, Administered Prices, 1960 and 1961,* Parts I through XXVI, 86th Cong., 2nd Sess. (Washington: Government Printing Office, 1961).
13. Airman 1st class John A. Eiby and Percy Branscom, reported in *Milwaukee Journal,* October 5, 1965, p. 1.
14. *Missiles and Rockets,* July 6, 1964, p. 311.
15. See U.S. Congress, House, Committee on Armed Services, Subcommittee for Special Investigations, *Hearings, Negotiation and Award of Research Contracts by the Air Force, A Case Study,* H. Res. 84, 88th Cong., 1st Sess. (Washington: Government Printing Office, 1964).
16. Statement of Dr. Norman W. Rosenberg, *ibid.,* p. 65.
17. Reported by Jules Duscha, "Costly Mysteries of Defense Spending," *Harper's,* April 1964, p. 63.
18. U.S. Congress, House, Committee on Armed Services, *Report, Employment of Retired Military Officers by Ten Leading Defense Contractors, 1959,*

86th Cong., 1st Sess. (Washington: Government Printing Office, 1960), p. 11.

19. Testimony to *Systems Development and Management, Part III,* p. 821.
20. Testimony of Dr. Frederic H. Smith, Deputy Director, GAO, *Government and Science, Distribution of Federal Research Funds; Indirect Costs re Federal Grants,* p. 272.
21. *Report, The Aerospace Corporation,* p. 1.
22. U.S. Congress, House, Subcommittee of the Committee on Appropriations, *Hearings, Department of Defense Appropriations for 1966, Testimony of Vice Adm. Hyman G. Rickover, U.S. Navy,* 89th Cong., 1st Sess. (Washington: Government Printing Office, 1965), p. 29.
23. Tony Berger, Burbank, California, in letter to the editor, *Missiles and Rockets,* April 27, 1964, p. 7.
24. Alfred Dreyfus, Los Angeles, California, letter to the editor, *Missiles and Rockets,* May 31, 1965, p. 6.
25. *Comptroller General Reports to Congress on Audits of Defense Contracts,* p. 647.
26. *Centaur Program,* pp. 31-35.
27. *Ibid.,* p. 10.
28. *Ibid.*
29. *NASA Authorization for Fiscal Year 1964, Part I,* p. 157.
30. *1964 NASA Authorization, Part II (a),* p. 581.
31. *Ibid.,* p. 499.
32. Quoted in *Missiles and Rockets,* January 4, 1965, p. 10.
33. Above instances cited by GAO, *Congressional Record* (daily edition), June 12, 1965, pp. 13147-13153.
34. Cited in *Comptroller General Reports to Congress on Audits of Defense Contracts,* p. 401.
35. As explained by a GAO lawyer: "It is our custom when we encounter cases where we feel there might be a suspicion of fraud to refer such cases to Justice." *Overpricing of Government Contracts,* p. 11.
36. Summary of cases reported to the Department of Justice for period July 1, 1958–December 31, 1964, U.S. General Accounting Office, printed in *Comptroller General Reports to Congress on Defense Contracts,* pp. 121-127.
37. *Department of Defense Appropriations for 1966, Testimony of Vice Adm. Hyman G. Rickover,* p. 21.
38. Reported by John G. Norris, *Washington Post,* December 31, 1963, p. 84.
39. Reported in *New York Times,* April 19, 1962, p. 12.
40. Reported in *Wall Street Journal,* April 1, 1962, p. 11.
41. Quoted in *Comptroller General Reports to Congress on Audits of Defense Contracts,* p. 398.
42. Summary of Audit Report Issued to Congress for the period May 1, 1963, through May 1, 1964, *Congressional Record* (daily edition), June 12, 1965, p. 13150.
43. *Overpricing of Government Contracts,* p. 23.
44. Statement of Navy contracting official in *Sole-Source Procurement, Part I,* p. 292.
45. Reported in *Washington Post,* December 31, 1963, p. A4.
46. *Government and Science,* p. 79.
47. *Department of Defense Appropriations for 1966, Testimony of Vice Adm. Hyman G. Rickover,* p. 17.
48. *Ibid.,* p. 18.

49. "Summary of Audit Report Issued to Congress for the Period May 1, 1963, through May 1, 1964," *Congressional Record* (daily edition), June 12, 1965, p. 13150.
50. See U.S. Congress, House, Committee on Armed Services, Subcommittee for Special Investigation, *Hearings, Cost of Morale and Recreation Benefits for Defense Contractor Employees,* H. Res. 84, 88th Cong., 2nd Sess. (Washington: Government Printing Office, 1964).
51. *Ibid.,* p. 7.
52. *Ibid.,* p. 8.
53. *Ibid.,* p. 9.
54. Statement of Joseph Campbell, *ibid.,* p. 11.
55. *Report, The Aerospace Corporation,* p. 35.
56. *Ibid.,* p. 42.
57. *Comptroller General Reports to Congress on Audits of Defense Contracts,* p. 654.
58. Declared Jerome B. Wiesner, testifying, *Systems Development and Management, Part I,* p. 161.
59. *Manpower for Research and Development, Study Number II,* pp. 53-54.
60. Statement of Dr. Frederic H. Smith, Deputy Director, GOA, testifying, *Government and Science, Distribution of Federal Research Funds, Indirect Costs re Federal Grants,* p. 273.
61. *Comptroller General Reports to Congress on Audits of Defense Contracts,* p. 378.
62. Rep. A. Paul Kitchin, *Sole-Source Procurement, Part I,* p. 452.
63. Reported in *Missiles and Rockets,* February 8, 1965, p. 9.
64. *Department of Defense Appropriations for 1966, Testimony of Vice Adm. Hyman G. Rickover,* pp. 8-9.
65. See discussion of large-diameter, solid-fueled launch vehicle programs and nuclear space devices, *Space Launch Vehicles,* pp. 96-97.
66. Congresswoman Martha W. Griffiths, U.S. Congress, Joint Economic Committee, Subcommittee on Fiscal Policy, *Hearings, Fiscal Policy Issues of the Coming Decade,* July 1965, p. 78.

CHAPTER XV
Patents and Power

1. *Invention and the Patent System,* p. 130.
2. *Ibid.,* p. 50.
3. *Ibid.,* p. 34.
4. *Ibid.,* p. 44.
5. See *ibid.,* pp. 7-8.
6. U.S. Congress, House, Select Committee on Government Research, *Report, Contract Policies and Procedures for Research and Development, Study No. VII* (Washington: Government Printing Office, 1964), p. 60.
7. Summarized in *ibid.,* pp. 60-61; quoted in detail in NASA Statement, *NASA Authorization for Fiscal Year 1966, Part II,* pp. 664-665.
8. See Stacey V. Jones, *New York Times,* September 28, 1963, p. 33.
9. Quoted in *Science,* February 9, 1962, p. 416.
10. See statement of Philip H. Trechse, Acting Assistant Secretary of State, *Washington Post,* March 27, 1962, p. 21.

11. Quoted in *Missiles and Rockets,* January 4, 1965, p. 14.
12. Quoted in *New York Times,* November 12, 1961, p. 66.
13. *Report, Contract Policies and Procedures for Research and Development, Study, No. VII,* p. 58.
14. Richard J. Barber, *Economic and Legal Problems on Government Patent Policy,* report prepared for the Select Committee on Small Business, June 1963.
15. U.S. Congress, Senate, Committee on Foreign Relations, *Hearings, United States Foreign Policy: Possible Non-military Scientific Developments and their Potential Impact on Foreign Policy Problems of the United States; Worldwide and Domestic Economic Problems and their Impact on the Foreign Policy of the United States; and United States Foreign Policy—Africa,* 86th Cong., 2nd Sess. (Washington: Government Printing Office, 1960), pp. 30-31.
16. *Missiles and Rockets,* July 12, 1965, p. 15.
17. *Report, Impact of Federal Research and Development Programs, Study No. VI,* p. 137.
18. Contractor comment summary in Denver Research Institute–University of Denver, *The Commercial Application of Missile/Space Technology,* Parts I and II (Denver: University of Denver, 1963), p. 204.
19. Quoted in *Science,* April 2, 1965, pp. 55-56.
20. *Invention and the Patent System,* p. 137.
21. *Ibid.,* p. 58.
22. *Ibid.,* p. 87.
23. *Ibid.,* p. 88.
24. According to *ibid.,* p. 135.
25. See statement of Admiral Knickerbocker, *Administered Prices, Part 24, Administered Prices in the Drug Industry (Antibiotics),* pp. 13790-13791.
26. Reported in *New York Times,* July 7, 1965, p. 58.
27. See hearings on government solicitation of bids on medical drugs from Italy, *Administered Prices, Part 24, Administered Prices in the Drug Industry (Antibiotics),* pp. 13812-13814.
28. Testifying in U.S. Congress, Senate, Committee on Small Business, Hearings, Subcommittee on Monopoly, reported in *Astronautics and Aeronautics, 1963,* p. 93.
29. *Government and Science, Distribution of Federal Research Funds, Indirect Costs re Federal Grants,* p. 272.
30. Reported in *Missiles and Rockets,* September 7, 1964, p. 11.
31. *1964 NASA Authorization, Part II (b),* p. 1106.
32. See *ibid.,* pp. 292-293.
33. Cited by GAO, Summary of Audit Report, Issued to Congress for the Period May 1, 1963, through May 1, 1964, reprinted in *Congressional Record* (daily edition), June 12, 1965, p. 13150.
34. *Report, Economic Impact of Federal Procurement,* p. 12.
35. *Report, Contract Policies and Procedures for Research and Development, Study No. VII,* p. 65.
36. Statement of Joseph Campbell, *Comptroller General Reports to Congress on Audits of Defense Contracts,* pp. 168-169.
37. *Report, Contract Policies and Procedures for Research and Development, Study No. VII,* p. 55.
38. S. C. Gilfellan, *Invention and the Patent System,* p. 90.

CHAPTER XVI

The Economics of Ambiguity: Comsat and SST

1. The Bell Report, reprinted in *Systems Development and Management, Part I,* pp. 208-209.
2. U.S. Congress, House, Committee on Science and Astronautics, *Report, Commercial Applications of Space Communications Systems,* H.R. No. 1279, Union Calendar No. 547, 87th Cong., 1st Sess. (Washington: Government Printing Office, 1961), p. 1.
3. *Quoted in 1964 NASA Authorization, Part IV*, pp. 3295-3296.
4. Statement of Rep. William F. Ryan, *ibid.,* p. 3296.
5. Statement of Sen. Clinton P. Anderson, U.S. Congress, Senate, Committee on Aeronautical and Space Sciences, *Hearings, Nomination of Incorporators,* 88th Cong., 1st Sess. (Washington: Government Pritning Office, 1963), p. 73.
6. See *Report, Commercial Applications of Space Communications Systems,* pp. 4-8; also U.S. Congress, Senate, Committee on Aeronautical and Space Sciences, *Communication Satellites: Technical, Economic, and International Developments,* Staff Report, 87th Cong., 2nd Sess. (Washington: Government Printing Office, 1962), p. 47.
7. *Report, Commercial Applications of Space Communications Systems,* pp. 24-29.
8. Represented on the committee were the American Cable and Radio Corporation, AT&T, the Hawaiian Telephone Company, Press Wireless, Inc., Radio Corporation of Puerto Rico, RCA, the South Puerto Rico Sugar Company, the Tropical Radio and Telegraph Company, the U.S. Liberia Radio Corporation, and the Western Union Telegraph Company; *ibid.* p. 27.
9. *Communications Satellite Act of 1962,* p. 6.
10. *Ibid.,* pp. 4-6.
11. *Nomination of Incorporators,* p. 87.
12. *Astronautics and Aeronautics, 1963,* p. 480.
13. See testimony of Cyrus Vance, *NASA Authorization for Fiscal Year 1966, Part I,* p. 554.
14. Reported in *Missiles and Rockets,* October 12, 1964, p. 10.
15. *NASA Authorization for Fiscal Year 1966, Part I,* p. 554.
16. *New York Times,* June 12, 1965, p. 13.
17. *New Republic,* October 23, 1965, p. 8.
18. *Communication Satellites: Technical, Economic, and International Developments,* p. 130.
19. Reported in *Astronautics and Aeronautics, 1963,* p. 112.
20. Reported in *Science,* May 10, 1963, p. 738.
21. Reported in *New York Times,* February 16, 1964, p. 28.
22. Frank G. McGuire, "Europeans Shying at Possibility of U.S.-Dominated COMSAT Corporation," *Missiles and Rockets,* May 4, 1964, p. 28.
23. Statement of Ivan Cheprov, *ibid.,* September 28, 1964, p. 34.
24. *New York Times,* October 18, 1965, p. 14.
25. *Ibid.,* June 29, 1965, p. 12.
26. Statement of L. B. Maytag, Jr., in speech to National Aerospace Education Council, Miami Beach, Florida, July 10, 1963, quoted in *ibid.,* p. 720.

27. N. E. Halaby, Administrator, Federal Aviation Agency, testifying, *Panel on Science and Technology, Sixth Meeting*, p. 45.
28. *Ibid.*, pp. 56-58.
29. N. E. Halaby, interviewed, *U.S. News and World Report,* December 23, 1963, p. 57.
30. *Astronautics and Aeronautics, 1963,* p. 230.
31. *Ibid.*, pp. 407-408.
32. *Ibid.*, p. 487.
33. *Panel on Science and Technology, Sixth Meeting,* p. 42.
34. *Wall Street Journal,* March 2, 1964, p. 2.
35. Press conference, July 24, quoted in *Missiles and Rockets,* August 3, 1964, p. 11.
36. Robert Hotz, *Aviation Week,* August 3, 1964, p. 11.
37. Reported in *New York Times,* April 15, 1965, p. 15.

CHAPTER XVII
The Bell Report: Rationality and Reaction

1. Declaration was published by the Bar Association as "Conflict of Interest in Federal Service," cited in *International Science and Technology,* April 1962, p. 55.
2. Quoted in *New York Times,* October 20, 1961, p. 3.
3. Report to the President on government R&D contracting, April 1962, printed in Appendix I, *Systems Development and Management, Part I,* p. 209.
4. *Ibid.*, p. 192.
5. *Ibid.*, pp. 43-44.
6. *Ibid.*, p. 192.
7. *Ibid.*, p. 195.
8. *Ibid.*, p. 173.
9. *Ibid.*, pp. 228-229.
10. *Ibid.*, pp. 467-468.
11. See statement of Bell, *ibid.*, p. 19.
12. *Ibid.*, pp. 99-100.
13. *Ibid.*, p. 97.
14. *Ibid.*, pp. 111-112.
15. *Ibid.*, p. 476.
16. According to Congressman Chet Holifield, *ibid.*, p. 16.
17. *Report, The Aerospace Corporation,* p. 34.
18. *Ibid.*, p. 39.
19. See statement of Dr. Harold Brown, Director, Defense Research and Engineering, Department of Defense, *Systems Development and Management, Part II,* pp. 482-483.
20. *Ibid.*, p. 473.
21. Statement of Bell, *ibid.*, I, p. 54.
22. *Ibid.*, p. 162.
23. See *International Science and Technology,* February 1965, p. 35.
24. *Systems Development and Management, Part II,* p. 439.
25. See statement of Harold Brown, *ibid.*, p. 438.
26. See statement of Finn J. Larsen, Assistant Secretary of the Army, *ibid.*, pp. 1374-1375.

27. Reported in *Astronautical and Aeronautical Events of 1962,* p. 108.
28. As characterized in *Missiles and Rockets,* March 29, 1965, p. 42.
29. *Missiles and Rockets,* May 31, 1965, p. 26.
30. Judson A. Lovingood, letter to the editor, *Missiles and Rockets,* June 15, 1965, p. 5.
31. Reported in *Missiles and Rockets,* May 31, 1965, p. 24.
32. *NASA Authorization for Fiscal Year 1966, Part II,* p. 763.
33. *NASA Authorization for Fiscal Year 1966, Part I,* p. 35.
34. Reported by John W. Finney, *New York Times,* September 6, 1964, p. 37.
35. Bell, *Systems Development and Management, Part I,* p. 8.

CHAPTER XVIII
The Battles of McNamara

1. Johnson and Schultze's press conference, reported in *New York Times,* August 26, 1965, p. 17.
2. William J. Coughlin, editorial, *Missiles and Rockets,* April 6, 1964, p. 54.
3. *Report, Economic Impact of Federal Procurement,* p. 3.
4. Statement of Paul R. Ignatius, Assistant Secretary of Defense, *Comptroller General Reports to Congress on Audits of Defense Contracts,* pp. 5, 141.
5. See U.S., President, Bureau of the Budget War on Waste, *Cost Reduction Through Better Management* (Washington: Government Printing Office, 1964).
6. Reported in U.S. Congress, Joint Economic Committee, Subcommittee on Federal Procurement and Regulation, *Background Materials on Economic Impact of Federal Procurement—1965,* 89th Cong., 1st Sess. (Washington: Government Printing Office, 1965), p. 1.
7. William J. Coughlin, editorial, "Alice in McNamara Land," *Missiles and Rockets,* August 3, 1964, p. 46.
8. Reported by Evert Clark, *New York Times,* July 15, 1965, p. 3.
9. Statement of Charles J. Hitch, Assistant Secretary of Defense, *Systems Development and Management, Part II,* p. 514.
10. Secretary of Defense McNamara, quoted by Jack Raymond, *New York Times,* March 28, 1962, p. 16.
11. Statement of Bell, *Systems Development and Management, Part I,* p. 70.
12. "Report to the President on Government R&D Contracting," April 1962, printed in Appendix I, *ibid.,* pp. 193-194.
13. Reported in *Missiles and Rockets,* October 5, 1964, p. 16.
14. Both quoted in *Missiles and Rockets,* October 26, 1964, p. 10.
15. See *Missiles and Rockets,* November 23, 1964, pp. 9, 20.
16. William J. Coughlin, editorial, *Missiles and Rockets,* October 12, 1964, p. 50.
17. *Missiles and Rockets,* November 23, 1964, p. 46.
18. S. R. Mayerhofer, stationed at Pacific Missile Range in Hawaii, in letter to the editor, *Missiles and Rockets,* November 30, 1964, p. 9.
19. Reported in *Missiles and Rockets,* August 16, 1965, p. 9.
20. Quoted in *Missiles and Rockets,* March 30, 1964, p. 108.
21. U.S. Congress, Joint Economic Committee, Report of the Subcommittee on Defense Procurement, *Impact of Military Supply and Service Activities on the Economy,* Joint Committee Print, 88th Cong., 1st Sess. (Washington: Government Printing Office, July 1963), pp. 2-3; the law giving preference

to competitive bidding is Public Law 413, Armed Services Procurement Act of 1947, see sections 2 and 3.

22. *Ibid.*, p. 3.
23. *Report, Economic Impact of Federal Procurement,* p. 11.
24. See statement of Graeme C. Bannerman, Assistant Secretary of the Navy (Installations and Logistics), *Comptroller General Reports to Congress on Audits of Defense Contracts,* p. 24.
25. *Ibid.*, p. 13.
26. *Ibid.*, p. 26.
27. See *Negotiations and Award of Research Contracts by the Air Force; A Case Study,* p. 193.
28. Statement of Thomas D. Morris, Assistant Secretary of Defense, *Systems Development and Management, Part II,* p. 554.
29. Cited by Secretary of Defense McNamara, news conference reported in *Missiles and Rockets,* July 13, 1964, p. 11.
30. See Michael Getler, *Missiles and Rockets,* August 3, 1964, pp. 18-22.
31. William J. Coughlin, editorial, *ibid.*, p. 46.
32. See statement of Graeme C. Bannerman, *Cost of Morale and Recreation Benefits for Defense Contractor Employees,* p. 146.
33. *Report, Economic Impact of Federal Procurement,* p. 5.
34. See statement of Lt. Gen. William J. Ely, Army Deputy Director for Administration and Management, testifying, *Government and Science; Distribution of Federal Research Funds; Indirect Costs re Federal Grants,* pp. 24-25.
35. *Comptroller General Reports to Congress on Audits of Defense Contracts,* p. 182.
36. See Department of Defense statement printed in *ibid.*, pp. 14-16; also *Report, Economic Impact of Federal Procurement,* p. 11.
37. Statement of Gen. Ely, *Government and Science; Distribution of Federal Research Funds; Indirect Costs re Federal Grants,* pp. 24-25.
38. See discussion in *Comptroller General Reports to Congress on Audits of Defense Contracts,* pp. 71-82.
39. Speeches made to the Third Annual Management Conference on Marketing in the Defense Industries, held at Boston, reported in *Missiles and Rockets,* June 1, 1964, p. 12.
40. See statement of Congressman James G. Fulton, *Space Posture,* p. 165.
41. See statement of Dr. Robert C. Seamans, Jr., *NASA Authorization for Fiscal Year 1966, Part I,* p. 80; and NASA, *Eleventh Semiannual Report to Congress, January 1–June 30, 1964,* p. 187.
42. *NASA Authorization for Fiscal Year 1966, Part II,* pp. 766-767.
43. NASA, *Eleventh Semiannual Report to Congress, January 1–June 30, 1964,* p. 186.
44. *NASA Authorization for Fiscal Year 1964, Part II,* p. 1087.
45. Letter from Comptroller General to Congressman Holifield, March 11, 1965, printed in *Comptroller General Reports to Congress on Audits of Defense Contracts,* p. 678.
46. Statement of Adolph T. Samuelson, GAO staff member, *Government and Science; Distribution of Federal Research Funds; Indirect Costs re Federal Grants,* p. 301.
47. Quoted in *Astronautics and Aeronautics, 1963,* p. 493.
48. W. J. Ridge, letter to the editor, *Missiles and Rockets,* April 13, 1964, p. 7.
49. *Overpricing of Government Contracts,* p. 1.

50. See statement of Comptroller General, *Comptroller General Reports to Congress on Audits of Defense Contracts,* p. 170.
51. Statement of Keller of the GAO in *ibid.,* p. 640.
52. Statement of Colonel Hamby, *Hearings, The Aerospace Corporation,* p. 149.
53. *Comptroller General Reports to Congress on Audits of Defense Contracts,* p. 452.
54. *Ibid.,* p. 453.
55. Statement of Daniel A. Haughton, President, Lockheed, *ibid.,* p. 450.
56. Statement of Charles I. Derr, Sr., Vice President, Machinery and Allied Products Institute, *ibid.,* p. 494.
57. Statement in *ibid.,* p. 574.
58. *Ibid.,* p. 446.
59. *Report, Contract Policies and Procedures for Research and Development, Study No. VII,* p. 74.
60. *New York Times,* November 17, 1963, p. E8.

INDEX

A NOTE ON THE AUTHOR

H. L. Nieburg was born in Philadelphia, attended the public schools there, and worked part time for the *Philadelphia Inquirer*. He received his bachelor's degree from the University of Chicago, then worked briefly for the old *Chicago Sun*. After a period devoted to writing fiction, he completed his master's degree at the University of Chicago, spent two years in the Air Force, and returned to Chicago to take his Ph.D. with Hans Morgenthau. He has since taught at Illinois State University and Case Institute of Technology, and was assistant director of the Center for Programs in Government Administration at the University of Chicago. He first became interested in science and public policy in the late fifties; his first book, *Nuclear Secrecy and Foreign Policy,* was published in 1964 to wide acclaim. He has also written many articles for magazines and journals, including the *Bulletin of the Atomic Scientists, Science,* the *American Political Science Review,* and *Technology and Culture*. Mr. Nieburg is now Associate Professor of Political Science at the University of Wisconsin, Milwaukee. He is married and has four children.